D1483132

PRAISE FOR *HOLISTIC GOAT CARE*

"In *Holistic Goat Care*, Gianaclis Caldwell demonstrates that modern, practical, holistic, small-farm methods are the true state of the art. Her unique understanding of goat care—connecting science, real-world farming, and the healing arts—comes alive through her clear and inspiring writing. This much-needed book gifts goat owners with a ready, reliable reference for whatever nature sends their way."

—**Fred Walters**,
editor, *Acres U.S.A.* magazine

"*Holistic Goat Care* is a rare and refreshing synergy of commonsense goat lore and sound holistic principles—worthy of a place in every goat owner's library. I am impressed with the book's layout, as well as the content."

—**Richard J. Holliday**, DVM, holistic veterinarian;
coauthor of *A Holistic Vet's Prescription for a Healthy Herd*

"*Holistic Goat Care* is practical, well written, and comprehensive. Giancalis Caldwell covers everything from browse to barns, breeding to birthing, vitamins to vaccination, and parasites to pizzle rot, in an easy-to-read format. The book reflects both her hands-on experience with goats and her extensive knowledge of their physiological needs and their sometimes idiosyncratic behavior. The examples of goat farms in various climates and countries make the information broadly applicable to many regions. With a light touch of humor and a hearty helping of practical advice, Gianaclis shares her depth of knowledge and her appreciation of the role and value of goats in agriculture both currently and historically. Read this book cover to cover, or keep it handy as a reference for all aspects of goat care."

—**Sarah Flack**, author of
The Art and Science of Grazing

"*Holistic Goat Care* is far and away the most complete guide to goat keeping I've ever seen. Gianaclis Caldwell handles basic care and advanced subjects, such as on-farm necropsies, cud transplants, and scur removal, with equal aplomb. Whether you have two goats or two hundred, you need this book!"

—**Sue Weaver**,
author of *The Backyard Goat*

"The next best thing to learning about goat keeping through years of experience is to read *Holistic Goat Care*. Gianaclis Caldwell not only provides all the essential details, she frankly describes unhappy events along with successes, to save readers the anguish of making similar mistakes and to demonstrate that sometimes bad things happen even to the most conscientious goat keeper. Rather than dictating a single rigid approach to holistic goat management, Caldwell offers numerous natural and alternative options to help you develop practices that best suit your goals and your herd's specific needs."

—**Gail Damerow**, author of *The Backyard
Homestead Guide to Raising Farm Animals*

"I wish this book had been around when I started raising goats. Gianaclis Caldwell has a friendly, clear style of writing that makes a huge subject much less daunting. I highly recommend that beginners read *Holistic Goat Care* before starting out on their own goat adventure. Experienced goat owners will definitely find this book a useful reference as well. As a firm believer in providing holistic and humane care for all my animals, I am pleased to finally find a book that covers this slant for goats."

—**Molly Nolte**, founder,
Molly's Herbals and FiasCoFarm.com

"*Holistic Goat Care* is an excellent resource for raising healthy goats. Gianaclis Caldwell stresses the importance of preventative medicine, which is a critical aspect of raising goats. The information in this well-written book will be very beneficial for beginners, agriculture-oriented students, veterinary students, and veterinarians. The author stresses the importance of early recognition of conditions and diseases. She also discusses treatment through conventional medicine and alternative or integrative medicine, as well as the use of complementary therapies such as herbal and homeopathic if needed. All sixteen chapters are very informative, and the text is referenced to excellent resources in goat medicine. I believe this book will be an excellent source of information in raising goats."

—**Lionel J. Dawson**, BVSc, MS, DACT,
professor, Center of Veterinary Health Sciences,
Oklahoma State University

Also by Gianaclis Caldwell

Mastering Artisan Cheesemaking

Mastering Basic Cheesemaking

The Small-Scale Dairy

The Small-Scale Cheese Business

Holistic Goat Care

A Comprehensive Guide to Raising Healthy
Animals, Preventing Common Ailments,
and Troubleshooting Problems

GIANACLIS CALDWELL

Chelsea Green Publishing
White River Junction, Vermont

Copyright © 2017 by Gianaclis Caldwell.
All rights reserved.

Unless otherwise noted, all photographs and
illustrations copyright © 2017 by Gianaclis Caldwell.

No part of this book may be transmitted or reproduced
in any form by any means without permission in writing
from the publisher.

Project Manager: Alexander Bullett
Editor: Fern Marshall Bradley
Copy Editor: Laura Jorstad
Indexer: Linda Hallinger
Designer: Melissa Jacobson
Page Layout: Abrah Griggs

Printed in the United States of America.
First printing May 2017.
10 9 8 7 6 5 4 3 2   19 20 21 22

Our Commitment to Green Publishing

Chelsea Green sees publishing as a tool for cultural change and ecological stewardship. We strive to align our book manufacturing practices with our editorial mission and to reduce the impact of our business enterprise in the environment. We print our books and catalogs on chlorine-free recycled paper, using vegetable-based inks whenever possible. This book may cost slightly more because it was printed on paper from responsibly managed forests, and we hope you'll agree that it's worth it. *Holistic Goat Care* was printed on paper supplied by Versa Press that is certified by the Forest Stewardship Council.®

Library of Congress Cataloging-in-Publication Data
Names: Caldwell, Gianaclis, 1961- author.
Title: Holistic goat care : a comprehensive guide to raising healthy animals, preventing common
   ailments, and troubleshooting problems / Gianaclis Caldwell.
Description: White River Junction, Vermont : Chelsea Green Publishing, [2017]
   | Includes bibliographical references and index.
Identifiers: LCCN 2016059354| ISBN 9781603586306 (hardcover) | ISBN 9781603586313 (ebook)
Subjects: LCSH: Goat farming. | Goats. | Holistic veterinary medicine.
Classification: LCC SF383 .C25 2017 | DDC 636.3/9—dc23
LC record available at https://lccn.loc.gov/2016059354

Chelsea Green Publishing
85 North Main Street, Suite 120
White River Junction, VT 05001
(802) 295-6300
www.chelseagreen.com

FSC
www.fsc.org
MIX
Paper from
responsible sources
FSC® C005010

For goats and their people

CONTENTS

My Path to Goats and the Holistic Approach

I've had livestock all my life, which began in 1961, but it's only in the last decade that I've delved deeply into studying and practicing the care and nurturing of a single species. As with every topic that I've ever attempted to master, including nursing, printmaking, and cheesemaking, the most important thing I've learned is that learning never ends. So it is with the greatest humility that I share with you the information in this book. As long as I have and care about goats they will continue to teach me more. As long as I manage livestock or work with animals there will be more to experience, perfect, adapt, and revise.

Goats deserve our respect and admiration. In much of mainstream Western culture, however, goats have been relegated to the role of second-class farm citizens. I, like many, didn't pay much attention to my first goats and their possible needs. I kept those Pygmy or Pygmy-cross goats simply as cute pets and companions for my horses. Fortunately for me, my cavalier treatment had no ill effects on them, and Sundance, Amigo, and Goatee-goat lived full lives. It wasn't until 2003, when my youngest daughter and I talked my husband into the idea of buying a small group of registered Nigerian Dwarf dairy goats, that my caprine learning curve began in earnest.

A Foundation in Health and Healing

I was raised to view human health as a continuum based primarily on a person's nutritional and physical fitness with a dash of genetics thrown in. Both of my parents were chiropractors, and during their training they were exposed to what were then (it was the 1950s) radical and extreme ideas about health—including the danger of things such as chemical residues (this was the age of DDT, when science said it was safe, even to drink!), trans fats, and the reusing of fat in deep fryers—that now have solid scientific backing. My parents, sister, and I lived on 220-plus acres of woodland in rural southern Oregon that had been homesteaded in the late 1800s and early 1900s. One of the original pioneers' log cabins was our home. We grew a large organic garden, raised our own meat, and rarely visited a doctor (when we did, we consulted a naturopath). My parents' goal (one never realized) was to construct a holistic healing center on the land where people would come to discover wellness. It would be staffed by alternative practitioners, but also by medical doctors, nutritionists, and physical therapists—a true holistic approach.

Figure I.1. The modern goat's prime ancestor, the Bezoar ibex. *Photo courtesy of Alexander Bisseling*

My own dream of working at that healing center inspired me to become a licensed practical nurse. That was in the early 1980s, and it was a shock to observe the poor state of health in which many people lived their lives and the aggressive use of pharmaceuticals by doctors attempting to help these people. Of course it's far easier to get a person to take a pill than to change lifestyles—much as it's simpler to spray herbicides and pesticides on crops than it is to pull weeds and accept the blemishes that pests inflict. But the simple thing is rarely the right thing.

Another important influence on my views of medicine—for people and animals—was our farm veterinarian. Dr. Archibald was the archetype vet of that era and place—a kind, wise, educated person who knew how to heal and care for animals while at the same time treating farmers with respect and patience. I adored him. He made me want to be a vet. In fact, as a teenager I spent a couple of spring breaks riding around the countryside with him, observing procedures, helping when I could, and listening to his wonderful tales—stories akin to those told in one of my favorite books, *All Creatures Great and Small*, by country veterinarian James Herriot. Observing the sheer physicality of Dr. Archibald's job, one in which brute strength might mean the difference

between life and death—such as trying to restrain a terrified cow in order to pull a dead calf she could not deliver on her own—was inspiring, but also sobering. I came to realize that becoming a large-animal veterinarian would not be right for me—or, more correctly, I couldn't become the type of vet I would want to be. Nonetheless, Dr. Archibald's way of practicing medicine and caring for livestock left a deep mark on my development.

In the early years of Pholia Farm, the management principles I applied to our goat herd were far more conventional than those I promote in this book. They were the practices I had learned from my first goat mentors, practices that aligned with the advice of most publications of that time. I began to change my approach after I stumbled across the website of Molly Nolte (who, along with her company Molly's Herbals, is profiled in this book). Rather than proposing that herbs can singularly treat every ailment, Molly espouses a holistic approach—sharing the wisdom that any disruption in health must not simply be treated, whether with herbs or prescribed drugs; instead its underlying cause must be considered. Her website promoted her herbal remedies but also had an extensive list of conventional drugs along with the details regarding dosage, withdrawal times, and more. I loved the balance! I printed out almost every page of the wisdom she shared online and plunked the whole stack in a three-ring binder, which I still use as a reference today.

Over time my list of influences came to include several other books, among them *Natural Goat Care*, by Pat Coleby; *A Holistic Vet's Prescription for a Healthy Herd*, by Richard J. Holliday and Jim Helfter; *Guide to Regional Ruminant Anatomy Based on the Dissection of the Goat*, by Gheorghe M. Constantinescu; and *Treating Dairy Cows Naturally*, by Hubert J. Karreman. The more I read, the further I pushed the management principles of our herd in the holistic direction. I also had the opportunity to listen to many experts speak at conventions and conferences. Sometimes their words inspired me, but more often I felt frustrated by their dismissal of the anecdotal wisdom that goat producers with decades of experience possess—as if a PhD or other degree is the conclusive proof of consummate wisdom. At the other end of the spectrum, I sometimes heard alternative healers and purveyors of remedies purporting their views and goods as the only solution to goat health problems. It's my belief that no single expert or remedy can provide all the answers. There needs to be balance among academic wisdom, producer wisdom, and the treatment spectrum. *Holistic Goat Care* is meant to offer that balance.

About This Book

I've divided this book into five parts, all intended to guide you to a big-picture view of holistic goat care that you can apply to daily management. The very nature of the holistic approach means that it will be constantly evolving, changing, and adapting. My holistic approach to caring for my goats is likely to look different from the one you have or will develop. I'll be sharing what I do here at our farm, but all the while encouraging you to apply foundational principles to the needs of your own herd.

Part 1 focuses on the developmental history of goats—how the very nature and needs of this species must be understood based within this context in order to manage goats in the most natural fashion possible. I explain how to evaluate your land and resources, how to pick the right goat for the job, and other topics that will help set the stage for holistic success. I also delve into practices such as organics, extensive management, and even a bit about biodynamics.

Part 2 includes the topics of housing, containing, and feeding goats in a way that supports their natural needs. I dig deeply into the subject of how the goat's rumen and digestive system function and the extra focus that mineral requirements must receive in order to provide a foundation of health and vitality. I also cover the many options for feeding goats as naturally as possible with browsed and grazed crops and the role of supplementation.

Part 3 focuses on skills and knowledge that will help you troubleshoot and treat health issues. This includes how to do a skilled assessment of the animal's health status, how to use basic remedies and nutritional support when the goat's health falls below ideal, and other skills, including hoof management and dealing with parasite issues. I also describe some advanced techniques such as fecal egg counts, drawing blood, euthanasia, and even how to perform a basic necropsy.

Part 4 delves into the many considerations surrounding reproduction, including breeding, birthing, and options for raising kids that allow for flexibility and promote efficiency while still meeting the natural needs of the animals. These include important management skills such as disbudding, tattooing, and options for weaning. I give special attention to how the needs of the young animal differ from those of the adult goat.

Part 5 is really a book within a book—an in-depth guide to the most common disorders that affect goats. Each chapter in Part 5 focuses on a particular body system or systems. A summary table in each chapter will help you correlate signs and symptoms with disorders. I also discuss the cause of each disorder described and the treatment and prevention options. This section of the book will help you understand these disorders and work with your veterinarian to properly manage your herd's health.

Holistic Goat Care in Practice

I like to think of Pholia Farm as my own laboratory, where I have the luxury of applying my theories and testing different approaches. Seeking to improve things, whether that's efficiency of operations or the overall health of the animals, is my greatest pleasure. I've always loved a challenge, and working with goats has certainly provided that! I'm by no means done learning, and I hope you will take the tales of personal experiences and practices that I've interspersed throughout the book as the sharing of a philosophy you can adapt to best suit your own unique herd. Although no prestigious-degree letters follow my name, I am a great respecter of the scientific process. I turn to authoritative research and publications for all the information provided in this book that isn't noted as simply a personal preference or theory. One of the most profound things I've learned from these sources is that even the authorities and data have yet to answer all the questions surrounding goat health.

If you are a goat farmer, be patient with the experts, understanding that their point of view comes from much studying and experience but is also affected by the mechanisms that drive research and the financial pressures they face, including malpractice insurance and possible litigation. If you are a veterinarian or academic expert, be patient with the producers, respecting the value of their anecdotal evidence. Appreciate their attempts to educate themselves—even if that's just through a Google search—and remind yourself that much of our knowledge of goats is still incomplete.

The holistic approach to caring for goats includes the whole animals. We seek to understand their nature, their ideal environment, their physiology, their production cycles, their nutritional needs, and their disorders. I hope to gird you with this wide base of knowledge and then coax your observational abilities and management skills to a higher level. Let's give our goats the care that our respect and admiration of them demand.

PART I
Starting Out Right

CHAPTER ONE

Understanding Goats

Understanding an animal's complete needs—including mental and physical requirements—is the basis for giving proper care. Although we humans often eagerly apply this holistic, whole-animal approach to our pets, we tend to treat livestock—hardworking farm animals—more one-dimensionally. It doesn't help that goats, or caprines, have a reputation for being trouble-free in regard to feeding and housing, but troublesome in their behavior. (Indeed, the word *capricious*, which means "unpredictable and impulsive," has at its root the word *capra*, meaning "goat.") It's a reputation that is sometimes upheld by reality and sometimes not. This stereotype, unfortunately, often leads people to put goats in situations where they don't thrive and that might even lead to animal suffering, which also results in unhappy and unsuccessful goat owners.

Getting inside the psyche of goats is the best way to understand not only why they behave the way they do, but why you must be considerate of that when working with them. To help you understand the nature of goats, it's revealing to learn about the ancestors of the modern goat—wild goats, which are the forebears of all domestic goats, and the first types of goats used as livestock. This includes understanding the historical habitat of the goat and the unusual ability of all goats, even today, to successfully return to a wild, feral state. It's amazing that the diversity of today's specialized caprine categories of meat, milk,

and fiber goats all arose from a common ancestor. Goat psychology is important, too—being a small prey animal defines a goat's behavior and reactions to situations including confinement, handling by humans, and herd dynamics. And each category of working goats faces specific health challenges.

The World's First Farm Animal

Goats hold the title of "first livestock": the first animals successfully raised for food, materials, and labor. The partnership between goats and humans began at least 100 centuries ago during Neolithic times, when humans also first began to grow crops. Humans tamed dogs even before that, but although many types of dogs are farm dogs, they aren't considered livestock. Before working with goats, humans also tried to domesticate the elegant gazelle. Both gazelles and goats could provide meat and hides, but goats proved much easier to breed successfully in captivity. Sheep became a part of the Neolithic farm about 500 years after goats.

Historians agree that the manner in which humans first managed goats was similar to the way goats existed naturally before their domestication. The animals roamed and browsed over large areas of land in a rather arid climate where a diverse selection of shrubs and leafy plants abounded. Today when goats are allowed to forage freely over extensive acreage (this type of management is called *extensive*), they are usually at their best, because this style of management more closely resembles how nature and time tweaked their physiological functions for optimal health.

There are several species of wild goats, or ibex, still capriciously climbing and leaping in wild and remote areas of the world. These all belong to the genus *Capra*, the same as our modern domestic goat. However, the so-called Rocky Mountain goat of North America is not actually a goat; it's part of the same family as cattle and antelope. Current research indicates that a single species from the genus *Capra* provided the genetics for our farm goats of today—the Bezoar ibex, or goat, *Capra aegagrus*.[1] The domestic goat is considered a subspecies of the Bezoar; its scientific name is *Capra aegagrus hircus*. The magnificent Bezoar—whose males have horns that are proportionally (compared with body size) the longest of any living creature—can still be found roaming its native range throughout several countries in the Middle East, Asia, and Eastern Europe. If you're a fan of ancient medicine or Harry Potter, you might know that a bezoar is a hard mass, or stone, formed in the stomach of goats and some other animals, that was believed to have the power to neutralize poisons. In the fictional world created by J. K. Rowling, unlike medieval medicine, bezoars actually worked.

The farming of plant crops combined with the ability to grow successive crops of animals allowed for the switch from what anthropologists refer to as a dead animal economy, in which animals are hunted or captured only for slaughter, to a live animal economy, in which animals are valued as mothers with the ability to create new generations and therefore the possibility of wealth, trade, and sustenance. Goats continue to provide these amazing assets for peoples throughout the world today.

Figure 1.1. This 12-year-old Guernsey buck, named Swind Plymouth, shows an impressive set of horns rarely seen on domesticated goats. The average life span for most bucks is 10 years. *Photo courtesy of Mary Wilson*

The Development of Diversity

The Oklahoma State University Department of Animal Science website lists and describes an amazing 81 breeds of domestic goats. (The site is well worth visiting and spending some time there enjoying the interesting variety of caprines throughout the world; see appendix D for the link.) Many of these breeds are in danger of disappearing because of increased pressures on farmers to improve productivity; others are newly developed or revived.

How did the transformation from one uniformly standard wild goat, the Bezoar, to the incredible variety of goats today come about? The wild goat has erect ears; many domestic goats around the world have long, floppy ears. The Bezoar has distinctive horns that arch backward with a sharp edge; many domestic goats have spiraling horns.

The Angora goat, a very ancient breed, has silky fleece. The Valais Blackneck goat of the Alps has a long skirted coat. Most domestic goats have amber-colored eyes, but some Nigerian Dwarf goats have blue eyes. The list of variations goes on. The impressive diversity seen in today's goats is the result of two distinct factors: the forces of domestication and selective breeding by humans and the isolation of populations from the influence of humans and outside genetics.

Today close to 100 goat breeds are claimed across the globe.[2] Most have developed relatively slowly thanks to the previously mentioned forces. Humans continue to develop new breeds by crossing goats of different breeds and then selecting offspring for the desired traits. In some cases breeders trademark the name of the newly developed breed, as is the case with the meat breeds Genemaster, TexMaster, and

Figure 1.2. The Bagot goat, an endangered breed in England, developed living semi-feral on the large estate of Sir John Bagot, where they were first brought in the late 1300s. The isolation of the population from outside genetic influences has allowed it to develop a beautiful and distinctive look. *Photo courtesy of Tom Blunt*

The New Frontier in the Goat Family Tree

About the time I was wrapping up work on this book, some folks approached me at a homesteading event and showed me photographs of their hybrid goats—a cross between the Alpine ibex and the domestic goat. I was amazed. I had no idea that these hybrids existed. I did some computer research and found ibex of all types for sale at trophy animal game parks. Male ibex, with their impressive horns, find a ready market among those who like to hunt for trophy kills. As a surplus by-product of these operations, ibex offspring are finding their way into breeding situations with domestic goats, primarily meat crosses.

All members of the *Capra* genus can mate and produce fertile offspring, so I don't know why it came as a surprise to me that this was occurring. It's an appealing prospect—hybrid vigor for health and hardiness combined with the wild goat's survival instincts and ability to thrive on marginal lands—really, it sounds like the perfect meat goat. Whether these crosses—and I found them for almost all of the seven or eight different wild goat species—take off or not for commercial breeders remains to be seen. From the standpoint of genetics, it's a good thing that DNA testing to trace the origins of our modern domestic goat has already been done, because with these crosses we're throwing a whole new set of genetics—an ancient set—into the mix.

Figure 1.3. These feral Australian goats, also known as rangeland goats, have been captured from the extensive open rangelands on which they thrive. They have adapted so successfully, in fact, that they are now a major source of goat meat throughout the world. *Photo © iStock.com/Martin Auldist*

Tennessee Meat Goat. Some breeders also continue to improve the stock they have through selection and crossing with other breeds, while others maintain a strictly purebred approach, which keeps the genes of other breeds off the family tree.

The available gene pool of breeds in any country is somewhat restricted by import and export laws designed to limit the spread of animal-borne illnesses. This can be an obstacle for those desiring to improve an existing breed or to outcross to another that has desirable characteristics, because the genetics may not be available to them. For example, a Nigerian Dwarf goat breeder in Australia contacted my farm about bringing our proven genetics into their herd. However, the import of live goats or frozen semen to Australia is not allowed. Only embryos that are harvested at a controlled facility in the United States can be brought into Australia, and then only after the donor doe has been euthanized and conclusively tested free of disease. Basically the donor doe must be sacrificed for the potential of a new generation. It wasn't a scheme I was willing to commit our animals to, although I do feel sympathy for farmers in these predicaments. It's also understandable that nations are concerned about animal and human health.

Isolated populations of feral or semi-feral goats will grow to have uniform distinguishing characteristics and so over time create their own look, or phenotype, and genetics, or genotype. Examples include the San Clemente goat from the island of the same name off the coast of Southern California and the Arapawa Island goat from the island of Arapawa off the coast of New Zealand. Both breeds developed from domesticated goats left on the islands by sea explorers to provide a meat supply upon the sailors' return.

No matter how much apparent diversity goats have developed, as a whole they have retained their original preferences for arid land and climate—there's no such thing as "swamp goats." This long-enduring characteristic is a source of potential problems for those trying to raise goats in any climate that differs too much from their ancestral habitats.

Goats Gone Wild

Among domesticated livestock, goats win the prize for being the most capable of adapting to the wild if they escape from their farm's confines. The San Clemente and Arapawa goats as well as goats on the Hawaiian Islands were all too successful utilizing their island environment, to the detriment of native species, and ultimately had to be controlled

Figure 1.4. If there's a way for a goat to get into a predicament, it will probably find it. This Nigerian Dwarf got her head through the handle of a bucket, then—instead of backing out—tried to climb out over the bucket. In fact, bolt cutters should be one of the tools that every goat owner keeps on hand!

(albeit through extermination programs that were distasteful in the extreme to many people).

In Australia and New Zealand the feral goat population is made up of goats released by sailors in the 1700s, goats that have escaped from farms, and Angora goats purposefully released in the 1920s when the fiber market became unprofitable. In New Zealand feral goats are estimated to number several hundred thousand; in Australia, over three million. Although these animals, many of which sport the Bezoar's coat pattern, are considered an invasive species and a pest, they are also the basis of a huge goat meat industry. The United States alone imports just under 20,000 metric tons (that's about 22,000 US tons) of goat meat per year, 97 percent of which comes from Australia.[3]

There are news reports of feral goats wreaking destruction and chaos on gardens and yards in New Zealand. In parts of Ireland, though, residents regard feral goats fondly; the goats have learned to live in an amicable balance with humans and native animals.

> At its core, the goat is more likely to trust its primal instincts than to trust humans.

However you feel about feral goats running amok, it's smart to respect the innate adaptability of the species and try to use that to your advantage in managing your goat farm. At its core, the goat is more likely to trust its primal instincts than to trust humans. As a goat farmer you must anticipate these instincts. For example, you'll need to find a way to protect tasty ornamental or poisonous plants on your property from becoming a goat's dinner or to keep a hormonally needy doe from wriggling her way into the buck pen. You may decide to hand-raise kids that you want to remain tame.

The Goat Psyche

I have a saying that sums up much of what you need to know about working with goats: "If you love being around human toddlers, then you'll love goats." It's no exaggeration to say that most goats have about the same level of intelligence, penchant for mischief, and ability to delight as does a two-year-old human. And just as with humans, goat health isn't only about the physical; it's also mental. Throughout this book I point out ways you can help avoid situations that expose your goats to too much emotional stress—whether from herdmates, humans, or other causes.

Three main dynamics drive goat behavior: They are meant to be hunted by predators; they are meant to live in large groups (herds); and they evolved in fairly arid climates. Let's look at how each of these dynamics affects their psyche.

Flight or Fight

In nature, mammals are either at their core predators (for instance, a lion) or prey (say, a goat). You can observe one of the most amazing differences between these two categories just a few hours after a baby goat is born—the kid is standing up, nursing, and even ready to run. Baby predators are born helpless, by contrast, and have a long infancy period. Goats face an interesting dilemma: They are prey animals being cared for by predators—humans.

Prey animals vary in the way they deal with their fear of being attacked and eaten. It's the classic set of responses called flight-or-fight behavior. The way animals react to a threat, flight or fight, depends on the situation, their size, and also on the type of environment that their ancestors evolved in and the abilities that they developed in response to their environment. For example, horses are large animals, and they developed their instincts on wide-open grassy plains where running from a predator was usually the best bet for survival. Donkeys or burros, on the other hand, are smaller and developed in terrain that is more treacherous. Instead of flight, donkeys are likely to react aggressively toward a threat. From this fighting instinct comes their reputation as accurate kickers. (This is also why they can make good livestock guardian animals—more on that in chapter 3.) Cattle, which are large but slower moving and evolved on open terrain, find their strength in both flight and fight, usually through acting together as a herd. Goats also deal with threats through a combination of flight and fight. They are most likely to flee if the threat is dire, but if it isn't, they may stand their ground and use their horns. Their small size seems to make them quite aware of dangers, and they avoid placing themselves in situations where predators might be more likely to appear. For example, even when an entire herd has access to a forested area, goats are not likely to venture far into it unless in the company of their livestock guardian animal or their human. Similarly, when something unaccustomed occurs, such as a low-flying airplane passing overhead, they are likely to run for shelter.

This awareness and anticipation of threats sometimes causes prey animals to act in ways that don't make sense to us humans. For example, a goat being pulled up a ramp to the milk stand for the first time struggles not because she wants to annoy you, but because she is being dragged by the neck, which her instincts tell her is not a good thing! Finding ways to work with your goats' prey-animal instincts instead of against them helps prevent stress, and that's fundamentally important in a holistic care model.

If goats sense aggression—and that might be a human's frustration at the goats' behavior, or rushing through chores to avoid being late for work—they

Figure 1.5. Understanding goat psychology can help when you're making choices about practical matters such as housing and fencing. This herd of Alpine dairy goats at Pat Morford's Rivers Edge Farm roams a pasture that's unfenced at the forest side. The goats' natural instincts keep them from venturing into the forest.

respond in sometimes not-so-subtle ways that can be totally counterproductive to their owners' goals. It becomes a real exercise in patience to modify your behavior in a way that soothes the goats.

Keep in mind that each goat has an individual flight zone—the personal space that when entered by a human or other predator animal causes the animal to feel threatened and move away. Here are some tips for avoiding predator-like behavior when you're working with goats:

- When you want to approach wary goats, don't move straight toward them in a stealthy, hunched-over fashion. Instead, move slowly but purposefully in a zigzag pattern, speaking quietly to them as you move. This way your odds of catching them will improve. Stopping frequently if they are especially nervous is also helpful.
- When trying to move a goat forward, don't push on the animal's back or haunches—that tends to make a goat want to drop down. Instead, either pull up the collar high on the neck, to apply pressure behind the head (rather than under the jaw), and/or push from behind by placing your hand along the thigh to lift and move the animal forward. If a goat doesn't have a collar but is tame enough to wear one, you can use your hand with your thumb and index finger spread apart and placed across the back of the goat's head, just behind their ears, to guide them forward.
- Even when you are frustrated, try to keep your voice calm and your mannerisms nonthreatening.
- Whatever your goal, you're unlikely to succeed if you start chasing a goat. If a goat slips out of your grasp, instead of giving chase look around for ways to either attract them to you, such as food or a treat, or funnel them into a smaller area where they can be captured. You will have to outthink them, rather than outrun them.

> It becomes a real exercise in patience to modify your behavior in a way that soothes the goats.

Goats that have been handled since birth have a decreased flight response and smaller flight zone. (This makes handling such goats much easier than their dam-raised counterparts. Goats that regard their human handler as the head of the herd will not even need herding but will readily follow that person—their goatherd. You can use this behavior to great advantage when you're taking goats out for pasture and browse or trying to return them to a pen.

Safety in Numbers

Given goats' natural fear of potentially becoming a tasty meal for a predator, it shouldn't come as much of a surprise that goats, along with the majority of prey animals, including horses, cattle, and sheep, are meant to live in herds. In nature these large groups would develop over time and have a hierarchy—one that is frequently challenged but is also an important part of the herd's day-to-day life. The hierarchy, with herd leaders, peer groups, and mother/offspring groups, helps guide the herd to new pastures and teach new generations how to behave. A goat kept alone will suffer anxiety and discontent. Even a small herd of two or three goats may have behaviors that are different from what you'd see in a herd that develops organically over time and generations. Of course, very few goat farmers will ever see a truly natural hierarchy in their herds. It's not realistic to simply allow the generations to evolve without your interference. Farmers must sell animals, bring in new genetics, separate mothers from their young, and make many other choices that prevent this natural evolution. When these management decisions do happen, however, it's important to keep the hierarchy in mind when trying to understand not only the goat's behavior but possibly decreased milk production, poor mothering skills, or even a failure to thrive. If you are watching for these and anticipating them with countermeasures, you are likely to be successful at keeping the animals healthy and as content as possible.

Out in the Rain? No Way!

This last bit of goat psychology is directly related to the goat's innate evolutionary needs—and it's an inside joke for goat farmers! When it comes to water, goats act like the Wicked Witch of the West—afraid that even a bit will make them melt. Centuries of adaptation to dry, arid climates created this mind-set, and goat physiology—the way a goat's body functions—has adapted to match that reality. For example, lush, wet pasture and land hold many hazards for goats, including parasites such as liver flukes and stomach worm larvae. No, they won't melt, but there's a good reason that goats avoid wet areas. If grazing goats on pasture is part of your management plan, you'll need to pay close attention to moisture—the wetness or dryness of the grasses and herbs when you turn the goats out to graze, and the availability of shelter, even if that is adequate tree canopy or rocky overhangs, in case the weather turns less goat-friendly. Goats born on full extensive management learn to accept the reality of being a bit wet, but they will still need some sort of natural shelter under which to tuck their kids.

The Job Demands of Working Goats

The main career path of most specialized working goats is to produce meat, milk, or fiber. But other job options exist, including carrying the packs of hikers, pulling carts, clearing brush, and living as a beloved pet. It's important to remember that no matter what a breed has been developed for, whether that's to make as much milk as possible or to pack on bulky muscles for future goat chops, every goat is still a very versatile animal. Any dairy goat might end up as a meat goat. Every female meat goat has the potential to give milk. Every pet goat could also clear brush. All goats have the same fundamental needs, and each is at risk of a common set of health problems, but specialization has brought with it increased risks and problems that are specific to the animal's job.

Dairy Goats

Most goat breeds used primarily for milking have been bred for many generations and even centuries to perform that task well. These breeds have the ability

Goat Statistics

It's both interesting and helpful to spend a moment looking at the trends in goat populations over the years. These numbers from the United States show the skyrocketing popularity of breeding meat goats. Dairy goats have increased in numbers as well, thanks to goat's milk and goat's-milk products becoming more popular.

DAIRY GOATS

1997—190,588 (8% of total goat population)
2002—290,789 (11.5%)
2015—375,000 (14%)

ANGORA GOATS

1997—829,263 (37% of total)
2002—300,756 (11.9%)
2015—150,000 (6%)

MEAT GOATS

1997—1,231,762 (55% of total)
2002—1,938,924 (76.6%)
2015—2,100,000 (80%)

POUNDS OF MOHAIR SOLD

1997—5,287,312
2002—2,416,376
2015—765,000

Source: USDA's National Agricultural Statistics Service, goat and sheep survey census.

to put a great portion of the energy they consume into the production of milk, just as modern dairy cows do. Milk goats might even appear skinny to those accustomed to beef cattle or meat goats—instead of developing large muscles they, as dairy goat people say, "put it in the pail."

Like any hardworking goat, rather than a backyard pet, a milk goat must be built well so that she can stay strong through many years of grazing, browsing, being bred, having babies, and walking to and from the milking parlor. A good work ethic and temperament are valuable qualities for milking goats because they are handled frequently by the farmer. The strain of producing large volumes of milk demands a different diet from that of a meat or fiber goat. For example, the milking doe might need more energy feeds, often in the form of grain, and feeds higher in the minerals her body is using to make milk (calcium, phosphorus, magnesium, and more), often in the form of a hay such as alfalfa (also known as lucerne). The milk goat's day-to-day care might include more frequent hoof trimming, and her offspring are more likely to be hand-raised by the farmer than left with the doe (dam-raised).

Worldwide, within the group of goats considered milking goats, many breeds are still used as multipurpose animals. In Western culture, however, attention to genetics and focused breeding programs have pushed dairy goats to increasingly higher production levels, and consequently increasing physical stress. For example, you can't put a herd of heavy milking Saanen goats that are producing 2 gallons (8 L) of milk per animal per day on 200 acres (81 ha) of a low-energy feed such as sagebrush and other scrub and expect them to thrive. In fact, their strong "will to milk," as it's called, might even starve their bodies to continue producing a large quantity of milk, despite the poor-quality feed. Managing such a herd holistically will demand feeding them in a way that keeps their body condition prime, including perhaps more high-energy feed such as grain, while at the same time retaining and breeding animals that are vigorous and more able to maintain their body condition on less grain.

Figure 1.6. No one knows how or why "wattles"—the fleshy appendages that dangle from the head or necks of some goats—developed. They serve no function other than adornment and are removed at birth by some goat farmers who consider them unsightly. I think they are adorable and, if nothing else, a fun conversation piece!

Most of the major dairy breeds of today were developed in Europe and the United States from European goats. The Swiss breeds—Saanen, Alpine, Toggenburg, and Oberhasli—were originally developed in the mountainous regions of Europe. The long-eared Nubian descends from Indian and Egyptian goats that were crossed with milking goats in England (with Continental origins), and the short-eared LaMancha goat was developed in the United States from naturally short-eared goats with Spanish ancestry. The diminutive Nigerian Dwarf dairy goat was also created in the United States from the selections made from original stock of small goats brought from West Africa to zoos and animal parks.

The different origins of some of these breeds play a role in their management and health considerations. For example, Nubian kids born in a very cold climate are at risk for losing their long ears to frostbite. But in the hot equatorial regions where Nubian

Fainting Goats—Why Are We Laughing?

If you spend much time browsing the Internet, no doubt you've seen the videos, often with several million views, of fainting or Myotonic goats. These goats have an inherited disorder called myotonia congenita, which causes their muscles to remain contracted for several seconds when the animals are suddenly startled or stimulated. They don't actually faint, but remain fully aware during the episode. Consider the discussion earlier in this chapter about a goat's prey-animal instincts, and then place yourself in a Myotonic goat's hooves and imagine being unable to move when your fear instinct is telling you to run. It would not be a pleasant experience.

These goats are considered a landrace breed from the United States that developed in Tennessee and later in Texas. The first documentation of the goats, which go by several other names, including Tennessee Fainting and Texas Wooden Leg, was in 1880. According to The Livestock Conservancy's entry on the breed, "an itinerant farm laborer named John Tinsley came to central Tennessee, reputedly from Nova Scotia. Tinsley had with him four unusual, stiff goats." From there the animals passed on their condition and the breed grew in popularity, often simply as a source of amusement.

Humans can also inherit this disorder. In fact, Myotonic goats have been used to study the condition, which causes stiffness and a range of pain from constant to intermittent. One must assume that the goats also experience similar discomfort. The breed has been valued for its heavy muscling and used as a meat goat and for creating meat goat crosses. But these animals are also more susceptible to predators.[4]

Although there are many people who want to maintain the breed as pure and value it for all its qualities, this rationale doesn't sit well with me. How is perpetuating a disorder that causes discomfort, and quite likely anxiety, a humane choice?

ancestors thrived, those basset-hound-style ears helped release heat and keep the animals cool. These natural differences between breeds are important for new goat owners to learn about before deciding what breed to raise. It is possible to raise almost any dairy goat breed anywhere, except perhaps the South Pole, but you might have to make modifications in your management to keep certain breeds comfortable and healthy in your local climate.

Meat Goats

Goats have been a source of meat and hides since before they were domesticated. Even today in most parts of the world goat meat is the most popular animal protein source, surpassing the other top contenders: lamb, pork, beef, and chicken. The Western world, however, is a relatively new convert to the appreciation of goat meat, also called cabrito and chevon. But the Western appetite is rapidly expanding to appreciate this lean, sustainable meat source. For example, the United States now imports close to $100 million worth of goat meat per year according to the USDA. And even though the population of goats in the US has grown significantly, US producers still cannot keep pace with the demand for goat meat.

Until fairly recently in caprine history, there were no specialized breeds of meat goats. Any breed of goat might be raised for meat. Meat goats as a specialized breed category began in the early 20th century with the development in South Africa of the Boer goat. The Kiko breed was developed later in the 1980s in New Zealand, and other breeds are still being thought up today. (I've even had some fun with that myself, dubbing our Nigerian Dwarf/LaMancha crosses "Lagerians.") In most cases the emphasis is on selecting for genetics that produce sturdy, heavily muscled animals that are resistant to parasites, have good mothering instincts, and can thrive on marginal lands and native feed. The Spanish goat is a landrace breed (meaning it developed naturally over time but from domesticated stock that was left to fend for itself) from the US Southwest, especially Texas. These hardy animals descended from original Spanish goat stock brought to the United States in the

Figure 1.7. This meat goat farmer in North Carolina, Curtis Burnside, deals with the issues of parasitism that a highly humid, moist environment can bring.

1500s, as well as other feral goats that joined them over time. In the case of the Spanish goat, as well as many feral goat populations, natural selection, rather than farmer selection, produced an animal that is quite hardy and with high parasite resistance.

These inherent qualities of the Spanish goat are good news, because animals that are naturally hardy and resistant to parasites are well suited for holistic, extensive management.

When Boer goats were first imported to the United States in 1993, they were so popular and in such high demand that incredibly high prices were paid for breeding stock—in the range of $25,000 to $35,000, and running as high as $80,000.[5] There was a sad consequence of this frenzied willingness to sell and buy at these prices. Because these goats were so expensive and valuable, breeders understandably lost sight of the importance of culling for vigor, hardiness, disease and pest resistance, and good mothering skills. In fact, Boer goat breeders in the US are still dealing with the fallout. The goats might have arrived in North America with many desirable qualities, but after a few generations much of this was lost.

In almost a reverse conundrum, there is concern today that the hardy Spanish goat might lose its vigor and genetic strength by being outcrossed with the Boer and Boer crosses in the attempt to increase muscling and yield. Such crosses also increase the hardiness and parasite resistance of the Boer. Although these are desirable changes, the fear is reduction of an already dwindling population of purebred Spanish goats and the work of centuries of natural selection.

Successfully breeding meat goats today requires an eye to improving both yield and growth rates while also selecting for goats that will have decreased management and input costs. It demands choosing not simply the animals that weigh the most at butcher time, but also those that get to that point with the least amount of trouble and expense—including feed, health management, housing, and so on. This goal is easy to state, but it's not easy to attain. Record keeping is critical in being able to objectively analyze these topics, and then the producer must be willing to select future breeding stock based on the data.

Fiber Goats

Some of the world's most luxurious garments are made from the hair of goats. Not just any goats, though, but those selected and bred to produce long, soft hair that can be spun into supple strands. Most goats have a coat primarily made up of coarse hair called *guard hair*. In cold months goats, like most dogs, also grow a downy undercoat. The undercoat is called *cashmere*. Breeds that produce abundant and long cashmere are called cashmere goats. Their fiber, harvested by pulling, combing, or shearing, is the most valuable, but also produced in the lowest quantities. Cashmere-type goats are often multipurpose meat and fiber goats.

Angora goats produce long silky fiber called *mohair* that is quite a bit thicker than cashmere but still a wonderful fiber for fabrics. Mohair is produced in large quantities compared with cashmere and is harvested by shearing.

Angora goats are one of the oldest types of goats noted in historical documents and images. The breed's roots in the mild, dry climate of Turkey and

Figure 1.8. Angora goats do well in dry climates with warmer temperatures—especially after shearing in the spring. *Photo by Ignite Lab/Bigstock.com*

its dependence on its thick coat for protection make it more vulnerable to harsh weather than many other breeds, especially when young or after shearing. Cashmere-producing goats are, as noted, either shorn, combed, or pulled to remove the cashmere. When shorn, they, too, experience extra physical stress, because shearing is deliberately timed to be done before a goat begins to shed its coat, in order to maximize the yield of the valuable fiber. And unfortunately for the goats, that means shearing takes place at a time of year when it's still cold at night. If you are raising fiber goats, take this stress into consideration by providing proper shelter from the elements and making sure the goats are in good body condition to have the best chance of keeping your goats healthy and productive.

The Angora breed is as a rule less prolific than dairy and meat breeds, having more single births on average than other breeds. Much plays into this, including nutrition. The fiber goat breeder faces a bit of a paradox—better-fed goats grow fiber that isn't quite as fine, but they also might produce more offspring whose fiber is even more valuable thanks to having softer, finer texture than that of adult goats. (See chapter 8 for more on this.)

The Pack and Cart Goat

Any goat can, in theory, be trained to pull a cart or carry a pack saddle, but for goats that are going to be "professionals," size, structure, and demeanor are important. Pack and cart goats are often large-breed dairy goat wethers (males that have been castrated). Castrated males of all livestock species grow taller than their intact counterparts, providing the procedure is done before they are mature. The energy that would have been used developing secondary sex characteristics—such as a thick neck and luxurious beard—is instead diverted to increased height.

All castrated goats are considered at increased risk for the development of often lethal urinary stones—mineral stones that develop in the bladder and then block the drainage of urine, causing extreme pain to the goat. To limit the likelihood of these stones, or *calculi*, it's vital to monitor the diet of the wether for mineral balance. (See chapter 5 for more about mineral balancing.) The professional pack goat wether is most likely to subsist primarily on browse found along the trails he travels. And luckily, this is the perfect diet for the overall health of an adult wether. Those that only pack or pull part-time are at greater risk.

The Brush Goat

Using goats for organic brush and invasive plant remediation has become quite popular. Private landowners and governmental agencies at all levels are learning to appreciate the efficient abilities of goats to thin, clear, and control the growth of plants that are otherwise almost impossible to manage—even with the use of the latest and greatest herbicides—and to do so with little adverse impact on the ecosystem. Goats are especially useful in reducing or limiting invasive plants, including these pesky plants in the United States: multiflora rose (*Rosa multiflora*), kudzu (*Pueraria montana* var. *lobata*), Scotch broom (*Cytisus scoparius*), and Himalayan blackberry (*Rubus armeniacus*).

Figure 1.9. These hearty and healthy Alpine wethers enjoy a good life working on Willow-Witt Ranch in the mountains of southern Oregon with their owners and trainers Lanita Witt [*left*] and Suzanne Willow [*right*].
Photo courtesy of Dave Baldwin

Any kind of goat can do the job of brush control, but large, hardy breeds are the most effective. They can reach higher into the canopy to browse, have larger mouths to eat heavier stems and larger thorns, and are at less risk from predators. Brush goats might be adult

Figure 1.10. Goats are great at clearing unwanted shrubs and brush; here a LaMancha enjoys a meal of poison oak.

Figure 1.11. Pygmy goats were originally viewed as multipurpose animals, providing milk, meat, and companionship. Today the breed, developed in the United States from animals of West African descent, has a reputation for kidding difficulties thanks to having been selectively bred for a stocky, short build. *Photo by Swimpuppy/Bigstock.com*

wethers (past their prime for the early, high-value cabrito market), meat or fiber goat brood does, culled dairy does that are healthy but past their peak of milk production, or young meat goats that work as brush goats until heading to their final destination in the food chain. Keep in mind, however, that older animals that have been managed more intensively are likely to have difficulty changing careers to one that requires such a different diet and lifestyle. As you might expect, the health needs of brush goats vary depending on whether they are wethers, does, or young goats.

When on the job, brush goats need some oversight by the owner for their safety. The goats might need shelter from harsh weather, protection from predators, and even protection from mischievous humans. (For more information and guidelines for managing brush goats see appendix D.)

The Pet Goat

You might not think of pet goats as working animals. Indeed, their lives rarely feature physical stress, but they still have a job to do. One of the beauties of goats, unless you plan on eating them, is that they might all be classified as pets—working pets.

Pet goats are found on many suburban and rural small lots as well as on large farms and in big cities. Some city pet goats are even milk goats, providing both companionship and tasty milk. Most pet goats are small breeds such as Pygmy and Nigerian Dwarfs. Some are retired does from working farms, but the vast majority are castrated males that otherwise would have more likely ended up on the barbecue.

The biggest health problem facing pet goats is the same as that facing pampered dogs and cats: too much love. In other words, too much food and too many treats, resulting in obesity and overall poor nutrition. Pet goats usually enjoy a leisurely life and receive lots of attention. Without much physical work to do, milk to make, or babies to rear, the adult pet goat has minimal calorie requirements. Pet wethers also face the same risk of urinary stones as other adult wethers, and likely more risk because we all love to feed our pets, and our pets love treats . . . but treats usually don't provide the proper balance of minerals.

CHAPTER TWO

Building a Strong Herd

Whether you raise dairy, meat, fiber, or other types of goats, you face a common set of considerations and choices that are important for developing a strong, healthy, well-functioning herd. This includes farm management approaches; the supreme importance of the health of your first goats; assessing breeder reliability; tricks and tips for managing herd size, gender balance, and introduction of new animals to support optimal herd dynamics; evaluating and choosing goats for their specialized jobs; and the strengths and weaknesses of using genetic data, pedigrees, and show awards as the basis for choosing goats.

That sounds like a lot, and it is, but as you read through this chapter you'll be able to sort out what is most important for you at this phase of your caprine career—whether you are new to goats or a seasoned goatherd. Remember that the holistic approach requires a big-picture awareness of all of the issues mentioned above. The details of this picture will come into focus gradually and evolve over time. Let's get started!

Making a Case for Organic and Extensive

Goats are raised in many different settings and on varying scales of production. The type of management you employ doesn't have to be one of extremes; most farmers apply different approaches as time, the realities of their land, and the animals' needs require. My purpose in writing this book is to support farmers in reaching the goal of the most organic and extensive management that is possible and practical. Not everyone is lucky enough to have the acreage to manage goats more extensively, nor fortunate enough to have organically produced feeds readily available—or affordable. I believe the most realistic approach to reaching this goal is to find a balance between what

Figure 2.1. The shiny coats of these beautiful Nubian does at Mama Terra Micro Creamery in Oregon are an indicator of their vibrant health.

you're currently comfortable with and what you can envision as the most organic and extensive management choices possible for your farm.

Conventional Farming

Conventional farming by today's definition implies, among other things, an acceptance of the routine use of antibiotics for preventing and treating problems such as mastitis; the routine use of synthetic (sometimes referred to as chemical) dewormers and antimicrobials for the control of parasites; and the use of feeds formulated from non-organic and genetically engineered crops. This conventional approach is historically young. Given the current wisdom, it might even be considered an experiment designed to increase production and animal population density while reducing labor costs and animal mortalities. And it's worked for a while. We are entering an age, however, of change. Antibiotic and antimicrobial resistance is rampant and seems to be increasing at an exponential rate. Herbicide-resistant weeds are popping up in fields and yards where long-term herbicide use has occurred. And the public's perception of farming increasingly favors organic production.

For second-generation farmers of my age (I'm in my midfifties) and older, administering antibiotics was often a part of the day-to-day operations on farms when we were growing up. We simply popped down to the local feed store and chose a bottle of penicillin, tetracycline, or other antibiotic. Many folks feel deeply concerned about new regulations that change the norms—making most antibiotics available only with a prescription from the vet. I've even heard some vets advise that producers stockpile antibiotics while they still can. What many seem to forget is that antibiotics have been widely available and easily accessed for only a short time in our history of farming—less than a century. The end of their easy accessibility in the United States won't mean an end to farming, but rather a return to what we did before these miracle drugs came on the scene—the embracing of alternative treatments, the attention to proper nutrition, and the purposeful selection of animals for increased hardiness. We should also remember that we will still be able to obtain these medications relatively easily; the only difference is that our veterinarian will have to approve their purchase and emphasize the importance of proper dosage and adhering to milk and meat withdrawal times. This is already true with many drugs that goat producers use regularly, including the vitamin E and selenium supplement Bo-Se, which most of us have in our remedy cupboards.

There are situations for which I recommend the use of antibiotics—remember, they can save an animal's life and should be used for that purpose. That's part of the reason I feel so passionately about *not* using them routinely. Let's keep antibiotics viable for when they are the only hope.

Organic Farming

Because the word *organic* has become a legally defined term encumbered by a lot of governmental language and oversight, *certified* organic production of any crop or animal looks different depending on what country you are farming in and what organization is providing the certification. To legally call your production organic in the United States involves paying fees and following many rules. This may or may not be advantageous to you if you are a commercial farmer, but remember, any farm can practice organic *principles* without engaging in the required process to earn the title of *certified organic*.

Even though I was raised by my parents to follow strict organic methods for growing produce, it took a visit to Pholia Farm from certified organic goat farmers and their sharing of how they dealt with mastitis to push me past my

> The end of antibiotics' easy accessibility won't mean an end to farming, but rather a return to what we did before these miracle drugs came on the scene—the embracing of alternative treatments, the attention to proper nutrition, and the purposeful selection of animals for increased hardiness.

The Challenges and Rewards of Humane Certification

Among the multitude of options for certification, those that focus on humane treatment of the animals are particularly appealing for many goat farmers who sell milk or meat products to the public. There are two main certifiers in the United States, Humane Farm Animal Care (HFAC) and Animal Welfare Approved (AWA). There are slight to significant differences in their standards (which can be accessed online) and requirements. They address all aspects of care, including medication, feed, housing, and euthanasia. The standards are lengthy and detailed, so it bears a side-by-side comparison to find out which best fits your farm and philosophy.

Redwood Hill Farm and Creamery in Sebastopol, California, became the first humane-certified goat dairy in the United States in 2005. Jennifer Bice is the farm's founder, and her brother Scott Bice manages the dairy, home to 250-plus registered milk goats. He also serves as producer liaison for the other goat dairies that supply milk to make the company's widely distributed goat yogurt and kefir. Humane Farm Animal Care also certifies all of those dairies. Jennifer notes, "Smaller farms that sell only regionally have an advantage: People can come visit the farm or meet the producer at a farmers market to get their questions answered. For us that isn't possible." For Jennifer, third-party oversight is an effective way to answer customer questions about how the goats in the Redwood Hill network of farms are cared for.

Organic certification was her and Scott's first choice. However, the geography and climate of their farm and that of many of the farms supplying milk to Redwood Hill Creamery often doesn't supply enough rainfall to support pasture, and access to pasture is a fundamental criterion for organic standards. Jennifer also points out, as emphasized in chapter 4, that goats thrive better on browse than on pasture, making the pasture requirement of organic certification more cow-centric than ideal for goats. Indeed, even AWA and HFAC have different requirements for access to pasture and browse. AWA stipulates, "Continuous outdoor browse and pasture access is required for all animals," while HFAC states, "When climatic and geographic conditions allow, goats must have free, voluntary access to pasture or an outdoor exercise area."

Both certifications require inspections by the organization providing its seal of approval, as well as extensive documentation by the producers. For Redwood Hill Farm, Scott says, "It wasn't much more work, as we were already doing and documenting much, just as a part of good, holistic management."

previous paradigm of the somewhat casual use of antibiotics for my goats. It really made me stop and think—if these folks could successfully manage a large, commercially viable herd of milk goats without using any antibiotics, other than in life-or-death situations, then why couldn't I? Why couldn't all of us change our approach? I believe that organic practices require more from the farmer, but in the end they will help not only the land and humanity, but also animals.

At the basic level, organic principles call on farmers to avoid using synthetic fertilizers, synthetic pest control, or antibiotics. Animals raised and certified organic must have feed and bedding that fits this broad definition. In addition, the largest percentage of the feed must come from access to organic pastures and paddocks—not stockpiled forage and grain. This alone ensures more extensive management practices, and it is a challenge for farmers without access to enough acreage capable of growing the right amount of feed. Organic regulations also place restrictions on use of other types of medications, such as dewormers and hormones.

Defining Scale

There are no widely accepted criteria about how to label the size of a goat farm. For the purposes of this book, however, it will be helpful if you know what I envision when I say small scale or large scale. There are no specific quantitative measures. For example, a farmer who keeps 100 goats on a single acre (0.4 ha) (I know someone who had a commercial dairy of this scope) will have management practices as intensive as those of a large-scale operation. However, a farmer who tends 300 goats on 100 acres (40 ha) is likely to have more extensive management practices than many small-scale operations do. I can offer these basic parameters of scale:

Micro or home scale: 2–20 goats
Small scale: 21–100 goats
Medium scale: 101–250 goats
Medium-large scale: 250–500 goats
Large scale: 500–1,000 goats
Mega scale: 1,000-plus goats

It's important to note that according to the regulations of the United States' body of rules, the National Organic Program (NOP), even under organic management animals *must* receive proper medical care, including antibiotics, if no other option is working to prevent suffering. The NOP regulations also stipulate that an animal treated with these drugs must leave the organic herd. In other words, the farmer must sell the animal to a non-organic herd, slaughter the animal (after the required drug withdrawal time), or retire the animal to another location apart from organic production. Regulations in other countries are a bit different. The European Union organic standards are governed by European Council Regulation EC 834/2007, and in Canada CAN/CGSB-32.310 provides oversight to organic production. In the EU standards (each EU member country is free to enact stricter standards) animals given antibiotics for the treatment of disease (as opposed to being administered as preventives at what are called subtherapeutic levels) are allowed to remain in the herd after twice the recommended withdrawal time (the time from the end of the drug's use to the time when the animal's milk can be used by humans or the meat eaten). The Canadian regulations allow the use of antibiotics only in life-or-death situations, after which the animal can remain in the herd providing withdrawal times are met. In my opinion the US NOP regulations are skewed in favor of larger-scale producers, who are more likely be able to afford to either send an animal to a conventional slaughter outlet or treat it and then sell it to a conventional herd. They are also more likely to have replacement animals available or the resources to procure them. The small goat farmer is likely to have more limited resources and might even place a higher financial and emotional value on each member of the herd, making the requirement of disposing of the treated animal a hardship and discouragement. In addition, with proper record keeping and longer meat and milk withdrawal times, there is no risk to the human food supply posed by animals that have been properly treated with antibiotics—meaning given the right dosage and duration of treatment. The burden of keeping proper records of drug treatments that pose a threat to the food supply is one that must be seriously addressed in any type of farming, not just organic!

Some farmers practice organics because their own lifestyle and philosophies demand a holistic approach to managing land and animals. However, there's an interesting phenomenon created by the increasing popularity of organics on the consumer end of the goat farm. Organic milk and meat fills a higher-priced niche

Biodynamic Farming

If you're interested in taking organic management one step, or really a few steps, further, biodynamic farming might be the answer. I have to admit that my first encounter with biodynamics left me a little skeptical—it involved the burial of a cow's horn and something about the cycle of the moon. Now that I've learned more, though, I see biodynamics as an attempt to truly bind the farmer to the nuances of nature—to the rhythms of the seasons and the invisible needs of the land.

Biodynamics began in the 1920s with the teaching of Rudolf Steiner, an Austrian philosopher and scientist who sought to help farmers improve the ecology of their farms through relating practices to the greater cosmic events. Biodynamics has since evolved into a comprehensive approach to farming that focuses on a "triple bottom line approach of ecological, social, and economic sustainability" according to the Biodynamic Association (established in 1938 with membership from North America).

Many people associate biodynamics with odd-sounding ceremonies such as burying a cow's horn filled with manure in the fall, digging it up in the spring, and using the contents to make a preparation, known as BD500, that is applied to the soil. Biodynamic farmers and gardeners use several such soil amendments. They also closely follow lunar cycles for planting guidelines. To some folks, these practices might seem a little out there, but defenders of biodynamics point out that following lunar cycles for farming is an ancient practice and the rituals surrounding things such as the cow horn preparations help connect the farmer to the land and the seasons in a way that benefits both.

Biodynamics does focus on the health of the soil and developing a farm that creates and utilizes its own cycle of fertility and health. Goats can certainly fit into this picture quite nicely, providing manure for crops and helping consume crop residue. If you explore biodynamics as a goat farmer, you'll still have to rustle up a cow horn, though, as the cow is seen as more earthbound than a goat and thus the most appropriate choice. For more on biodynamics visit www.biodynamics.com.

---

that is expanding to the mainstream. These market pressures are driving farmers to switch to organic, and new farmers to try organic. Whatever the reason for organics, the end result is of benefit to all. If you plan on practicing organic principles or becoming a certified-organic farm, then you will have to focus on strong genetics, nutrition, and other practices that more closely resemble the natural environment goats are meant to live in—in other words, holistic management!

Intensive Farming

Intensive farm management has many degrees of "intensity," from mega goat dairies where thousands of goats are housed indoors year-round to the small meat goat operation where the animals might live part-time indoors, be confined during birthing, and range out to rotational paddocks. To one degree or another, more goat farms than not are managed slightly toward the intensive side of center. As you might imagine, each step toward the extreme of intensive requires a concentrated focus on the health issues that will arise because goats are being separated from their natural way of living. These include parasites, manure management, stress from overcrowding, and artificial rearing of offspring. I imagine very few goat keepers reading this book are dreaming of one day building a 10,000-head goat farm, but even if you are, your chances of success will still be helped by understanding the natural needs of the goat and figuring out how to try to meet those needs even in the less-than-natural situation of a concentrated farm.

Richard Johnson and Mia Nelson
LOOKOUT POINT RANCH, LOWELL, OREGON

In the foothills of Oregon's Cascade mountain range, a brave experiment is being conducted by two stalwart and thoughtful goat farmers, Mia Nelson and Richard Johnson. I visited their extensively managed 500-acre (202 ha) ranch, Lookout Point, in the early fall, as they were preparing for an important matchmaking session—the artificial insemination of about 60 of their Kiko goats with carefully selected semen. (To assist them they are hosting BIO-Genics—see appendix D—which will be concurrently teaching an AI workshop at the ranch.)

Mia and Richard didn't start out seeking to be leaders in the meat goat world. Instead, they were simply trying to find the right livestock to utilize their mountainous, blackberry-shrouded forestland. After disappointing results with cattle and then beefalo, they focused on Kiko goats for their hardiness and history of thriving in environments similar to that of Lookout Point Ranch. In 2002 they started their herd with 10 Kiko does and a buck from Dr. An Peischel, then in California. Twenty more does and

Figure 2.2. These Kiko goats at Lookout Point Ranch in Oregon are extensively managed on over 400 acres (162 ha) of natural pasture and forestland.

a buck were purchased from her the next year, and in 2010 about 40 frozen embryos were hatched from Garrick Batten's original Kiko herd in New Zealand. Today the herd consists of more than 200 does and bucks, with the population swelling to over 400 after kidding season.

In many ways, Lookout Point is an admirable model for extensive management of meat goats, but few farmers would be able to successfully emulate the model in full. Indeed, Richard and Mia are quite open about the challenges they have faced, and are still facing. Their primary goal has been to physically challenge the herd and then selectively breed for strong, parasite-resilient, hardy stock that can thrive with minimal input of labor and costs—all in the moderate but less than naturally goat-friendly environment of the western side of the Cascade mountains, where rain is bountiful in the winter (with about 50 inches/127 cm of annual precipitation) and spring and the grasses brown and die by the end of summer. The winter temperatures are cold (dropping into the 30s F/-1 to 4°C) but not so cold as to help reduce parasite loads on the land. Such conditions mean that gastrointestinal parasites and reduced feed quality and quantity are concerns in the winter. In addition, the goats have done such a good job decimating the invasive blackberries and understory shrubs that Mia estimates the ranch can now only support about one to two goats per acre—meaning the animals have to work harder to sustain their nutritional needs.

The ranch is divided into large paddocks that Richard calls interior and exterior pastures. Each enclosure has a water supply (provided by wells and water storage tanks that dot the property) and a mineral feeder. Somewhat centrally located and surrounded by the interior pastures is "the corral." This is a well-organized management area where the couple can weigh animals, check hooves, perform AI, and load animals that have been sold to be shipped off the ranch. Higher up on the ridge are the exterior enclosures, where Richard says the goats "are almost feral." Seven adult livestock guardian dogs, one per paddock, provide security, and three herding dogs help move the goats to new pastures or into the corral. Richard says the LGDs and herding dogs have learned to tolerate each other's contrasting instincts. The only confrontation occurred when two intact males tried to work the goats at the same time, and a visit to the vet for neutering literally fixed that issue.

Until this year, the goats received no supplemental feed, and all the paddocks are sheltered only by large oak, pine, and fir trees. This year, as an experiment, Mia and Richard tried feeding grass hay to certain groups of goats, such as those being prepared for breeding, those for sale, and those in pens without enough native forage to meet their needs. All of their goats are horned, so Richard has created clever designs for mineral feeders (see figure 3.22) and hay mangers that allow animals safe access, horns and all. Richard and their son Arbor have even reengineered several motorcycle lifts—designed originally for mechanics to work on bikes—to serve as AI stands for the goats (see figure 8.4).

Mia says that the goal of breeding a population of goats that can grow and flourish through the winter, without the use of dewormers of any kind, has been an extreme challenge. They had been making progress, but then an influx of genetics that proved disappointing set them back. The farm's record-keeping program is admirable and thorough. They carefully document many details, including mothering ability—how well the doe cares for her kids. During kidding season, the does are checked frequently, and new litters are weighed and eartagged within hours after birth. They evaluate the potential of each buck by tracking information such as scrotal size, growth weights, and other genetic values. They are one of only a handful of meat goat breeders, and currently the only Kiko breeder in

Figure 2.3. Mia Nelson snuggles a pup from a litter of the farm's livestock guardian dogs.

the United States, who utilize the National Sheep Improvement Program's goat program for genetic improvement. In addition, they have a certified scrapie-free herd and are approved exporters of both live animals and semen.

Through their breeding program, Mia and Richard are attempting to speed up the evolutionary clock and nudge their animals beyond simply surviving to thriving on the marginal but beautiful acreage of Lookout Point. Their success will help other farmers with similar land and conditions to create productive businesses, rather than having to rely on expensive supplemental feeds and high management costs. You can learn more about Mia and Richard's research, challenges, and vision at their website, www.lookoutpointranch.com.

Extensive Farming

The most extreme example of extensive goat management is the feral goat population in Australia. These goats are mainly left to fend for themselves—ranging over vast tracts of open land, foraging for feed, breeding, and kidding. As it does for wild goats, this creates a natural selection process, ensuring that only the most thrifty, wily, and hardy survive. The feral goats might not be pretty, be of show quality, or have a fancy pedigree, but these are the animals that nature intended.

A more realistic way to shift toward extensive farming is by utilizing rotational pastures and paddocks, relying on a human or a guardian dog as a goatherd (to lead the goats to range farther afield), and feeding a diet that is not as rich in protein and carbohydrates as the typical confinement and high-production diet. Most important, extensive farming requires a disciplined culling program for animals that have difficulty kidding, aren't good mothers, have persistent parasite problems, or are difficult to maintain on a reduced-energy diet. Without a doubt the most difficult part of trying to mimic nature is getting rid of goats that you love, but know you shouldn't keep. I'm guilty of keeping some in that category myself. Luckily, culling can often be accomplished by sending that undesirable genetic package to a pet and non-breeding home, rather than the freezer.

Evaluating Your Land and Resources

If you plan on practicing organic and extensive principles, it won't come as a surprise that more land and space is better than less. If you are fortunate enough to own, lease, or otherwise have access to ample acreage, the goals of natural management will be much easier to implement. The reality for most farmers, however, is one of smaller lots or limited usage due to zoning, environmental concerns, or other restrictions. Whatever the size of your farm, it's important to assess how well you can manage your herd and meet the goals of

your program (including management approach and income production) before you invest in animals or changes in current management approaches.

Calculating how much land is the right amount for goats isn't a simple task, and there is no one-size-fits-all formula. Even if you don't have the ideal number of acres, you can manage your goats well through supplemental feeding. In chapter 5 I cover the important considerations regarding feeding goats, but here is a rule of thumb based on average goats (size, gender, and occupation) and an almost 100 percent browse-and-graze diet that is managed properly using rotational stocking: When given the choice between an excellent pasture and brush, goats follow their hereditary instincts and will select 60 percent of their diet from trees and shrubs.[1] This is advantageous for the small producer who has the right type of land (that readily grows these sorts of plants) as there is more feed per acre thanks to the height at which these plants grow.

New farmers should keep in mind that their ventures may come under scrutiny by neighbors and regulation by local government. Even if your neighbors are keen on the idea of having goats next door, you must make sure that zoning laws will allow for both the species and the number of animals you envision having on your farm. Unless your property is zoned for agriculture, it is important to verify that the laws support your farming goals. You can find out this kind of information by contacting your local jurisdiction's zoning department. Fortunately even many urban areas now allow for the ownership of a couple of small goats. No matter what the zoning allows for, neighbors can still be a source of challenge for any farmer. By doing your best to anticipate all of these possibilities you will help establish a stable farm for your enjoyment as well as the contentment of your herd.

Managing Herd Dynamics

The realities of herd dynamics—how goats interact and behave—are an important consideration before you establish a foundation herd or add new animals to an existing herd. You'll recall from chapter 1 that in nature, goats live in herds they were born into. Mother goats raise their own offspring, and young goats have many other kids as a peer group. Goat farmers rarely have the opportunity to start with an intact, complete herd that has developed in this fashion. Instead, farm herds usually consist of individual animals or small groups of goats procured from multiple farms and then plunked into a paddock together. Such a newly formed herd will take time to integrate and establish a productive dynamic, no matter how well you have sized your facility and planned your farm. Indeed, some animals may never function as well as they would have otherwise. This doesn't mean that you shouldn't start your herd in this manner, but expect possible drops in milk production. Some goats may also lose weight or experience other health challenges during the adjustment period.

To help limit the negative effects experienced during herd forming, I encourage you to attempt to start a new herd and integrate animals over a short period of time, rather than continuing to purchase and add goats over a year or more. (See The Pecking

The Pecking Order and the New Goat

It's normal for animals that live in herds to have a social order that can appear quite unfair to those of us who just want everyone to get along. This order (often called a pecking order after the fact that chickens enforce their hierarchy by pecking their way to the top) makes it difficult to introduce a new goat to an existing herd.

Here are some things you can do to make the transition easier.

- Choose young goats rather than adding mature goats to an existing herd. Adult goats will much more readily accept new juvenile animals than other adults. They might bite the new arrival's ears or butt a young goat that they decide is in their way. But the herd members won't gang up on the new ones and attempt to do serious damage.

- Introduce a group of adult goats that already know one another into a herd, and the transition should go more smoothly. (It's like getting to take your friends with you when you move to a new high school.) The social order within the new group is a comfort to those animals. And the sudden appearance of a group of new animals seems to somewhat confound the existing herd—as if with so many options, they can't decide whom to pick on.

- Avoid introducing new or young bucks to a group buck pen during breeding season, when the aggression level is much higher.

- Keep the new animal or animals somewhat separate from the main herd (which is a good idea as a form of health quarantine anyway) while they get to know each other through the fence. Even with this measure, though, there will be some fighting both through the fence and after integration into the herd.

- If you can access a buck, rub a cloth all over his most stinky parts (this creates what is known as a buck rag). Then rub the cloth on a new doe's coat, especially her head and shoulders, before you turn her into the pen with your herd. In most cases, the other does will find her perfume rather appealing and go easy on her.

Even if you take special care to help ease the transition of animals into a herd, the goats will have to work out their pecking order on their own. This order changes over time, too. So even goats that have known each other all of their lives may one day decide to spend the day duking it out—leaving each other exhausted, with sore, even bloodied, heads. You might be tempted to separate them, but they will usually return to their duel, and sometimes gang behavior, when reunited.

Order and the New Goat for tips on bringing new animals into the herd.) If your goal is to have a closed or mostly closed herd (where no new goats are added except the occasional breeding buck), then you can rather quickly grow your own herd, easily doubling the population every year, from the initial purchase of a handful of well-selected stock.

Starting with Healthy Animals

Avoiding goat disease problems is another important consideration when starting a herd or adding animals to your herd. Some diseases are communicable from goat to goat, and some can become a permanent part of your property, dooming future goats to infection. Others are even communicable to humans (these are called zoonotic diseases). Don't let this scare you away from raising goats, but be sure you enter into the world of livestock management with some knowledge of reality and how to best manage the risks.

In North America the most common diseases of concern are caprine arthritis encephalitis (CAE), Q fever, caseous lymphadenitis (CL), and Johne's disease. If you live in another part of the world, please consult with a knowledgeable veterinarian about which diseases you should be concerned about, as well as how to screen for them. The maladies of concern (and that can be tested for) in North America are all discussed in depth in part 5, but here's a summary:

CAE (caprine arthritis encephalitis) is a relatively common disease that is passed from doe to offspring at birth and likely also to uninfected animals later through direct or even indirect contact. This reportable disease causes arthritis in older goats and encephalitis in younger goats. Blood testing should be done after six months of age, as reliability increases after this time.

Q fever is passed via airborne particles at the time the goat gives birth. This reportable disease is also spread in raw milk. Q fever in humans causes persistent flu-like symptoms or worse, such as inflammation around the heart. It is becoming increasingly common. Testing is unreliable for the most part.

CL (caseous lymphadenitis) is a contagious infection of the lymph glands spread by direct or indirect contact. The bacteria that cause CL can also be present in the environment of a farm and infect future goats. Testing can be performed.

Johne's disease (paratuberculosis or MAP) is a wasting disease (meaning it causes those infected to lose body condition and literally waste away) that also occurs in cattle and wild ruminants. This reportable disease is spread through raw milk or feces, to kids before birth, and via contact with infected soil. Testing is also not super reliable or conclusive at this time.

Choosing and Working with Breeders

No matter what goat breed appeals to you, it is more important to find a seller who seems honest and trustworthy and has high-quality animals than it is to locate your dream breed. Breeders supply more than goats; they also typically become mentors and resources for other goat farmers. No matter what, do not buy your foundation breeding stock at a livestock auction—repeat that with me! Most auction animals are rarely anything other than unproductive and sickly animals or those intended for immediate slaughter for meat.

Even working with a breeder can be a slippery slope. It is very easy to talk yourself into believing what someone says when you want to believe them. I could tell you several stories of people who bought animals that they either assumed were healthy, felt sorry for, or believed were of superior quality (without any documentation), only to later discover that those animals carried a disease that required them to be euthanized, suffered from health problems that resulted in big veterinary bills, or were such dismal milk producers that they later had to be given away as pets.

How do you tell the difference between a good breeder and a good salesperson? One good way is

The Genetic Package

When you buy breeding stock, you are also purchasing a "genetic package." In her lifetime, a doe may pass on her genetics to anywhere from 10 to 30 animals. A buck may pass on his genes to hundreds of offspring. There's a good reason that a single buck is often referred to as "half of your herd." Because of this fact, you take more of a chance when buying a buck—so the purchase or use of a buck should be taken very seriously, with more investigation of genetics and usually more money invested in the purchase.

It's not uncommon for goat breeders to put a price on the value of the genetic package that they are selling rather than the animal as a single unit. By this, I mean that they decide the value of the goat based on its parentage. This is a legitimate way to value kids, providing they don't have any discernible flaws. As an animal grows, however, the value gradually shifts to how it looks and performs. Consequently, it's not uncommon for a kid to cost as much as an adult. While you may pay more for a kid than it eventually proves to be worth, you also may find your best deal, because that kid might grow into an animal with superior qualities. Had the breeder been able to foresee that, they might not have been willing to part with the kid. Once an animal is proven as good breeding stock, it's likely to be either unavailable for purchase or very expensive. (There are more details about selecting stock based on genetic and performance information later in this chapter.)

---

simply to examine the herd. First, educate yourself about goat health, and be sure you know what a healthy goat looks like. Then set up a time to visit the herd you are interested in. When you inspect it, pay particular attention to the older goats. Ask the age of the oldest animal in the herd that is still producing milk or raising kids (a good response is that the herd includes some productive goats near 10 years of age). If the breeder claims specific production numbers— how many kids a particular goat has raised or how much milk she has produced in her lifetime—find out how the breeder has determined this, and ask to see the records. (More on records later in this chapter.)

Another thing to ask about is disease issues in the herd. Many breeders regularly test their herd for contagious diseases. Don't simply take breeders' word for it, though—ask to see the documentation. They shouldn't mind sharing this information and should actually take it as a positive sign of how serious you are. If you are considering buying animals from a breeder who doesn't regularly test, it's wise to arrange tests for diseases that you are concerned

Figure 2.4. There's a lot to assess when choosing a good goat, including mild deformities such as the crooked, or wry, face on this doe.

Beware the Bulletin Board and Auction

Plenty of reliable breeders with great reputations market their goats through venues other than a website. Sorting out who these folks are and if the goats are a good choice for you can be quite tricky.

Alternative ways of selling goats include auctions, craigslist and other online venues, and bulletin board postings, often at feed stores. These venues are popular ways to sell lower-value animals, such as milk goats that didn't produce enough, goats purchased on a whim and then passed around to new homes as pets, and goats without pedigrees. You can find good goats through these outlets, but be very wary and do your research.

Figure 2.5. The Spanish meat goat is well suited for being extensively managed thanks to generations reared under tough conditions. *Jeannette Beranger, The Livestock Conservancy, provided by Ryan Walker*

about before taking possession of the goats. You will have to pay the cost of the testing, but ask the breeder whether any money will be refunded if the tests are positive.

Evaluating a Goat's Job Suitability

Many people choose their first goats based on the "cuteness factor." From coat color to eye color, size, type of ears, and behavior, different things appeal to each of us. You must resist the cuteness factor at all costs if you are going to be more than a hobby goat owner! Rather than adorability, the way the goat is built—its conformation—should match its main purpose. A dairy goat is designed to put her energy into making milk, a meat goat into putting on heavy muscle, a fiber goat into growing a lustrous coat, and a pack goat into being a sturdy, strong worker. In

addition, all goats share certain characteristics that help them remain healthy and active throughout their careers.

Learning to judge the quality of a goat takes time and experience. You can get comfortable with the required skills in several ways. The quickest is to spend some one-on-one time with someone who is already accomplished and have them show you differences in living goats—both visual qualities and those you must use your hands to detect. Going to goat shows can be helpful, but since goats are prepped and primped for shows, it's not the same as evaluating them on the farm.

It's fairly easy to assess an adult buck or doe. (See chapter 8 for tips on evaluating a buck's breeding potential.) But young kids are more difficult. An immature goat's body goes through several awkward stages (just like humans) that make it difficult to judge conformation. Bucks have a different build from does, with obvious differences that you would expect, such as the doe being more feminine and the buck more rugged. There are also other gender-based differences, such as bucks having a narrower pelvis than do does (also much like in humans). Dr. Susan Beal, DVM, wants producers to remember, "You have to pick the right animal for the job"—meaning that most livestock today isn't ready to be thrown into extensive management. They've been bred for too long to be managed more toward the intensive side, with supplemental feed, shelter, and assistance with giving birth, for example. It's critical to keep this in mind when choosing your goat's job suitability! Now let's go over some of the physical qualities that are consistent for all goats, as well as some specific characteristics of dairy, meat, and fiber goats.

General Appearance

Genetics and environment both help to determine a goat's physical structure. When you evaluate this structure, or general appearance, you're looking at the whole goat and how it is, quite literally, put together. General appearance starts with the underlying bone structure and alignment of the entire animal. A good goat should be built to hold up well over a long, productive career. This category also includes whether the goat fits the desirable characteristics of the breed. For example, Nubian and Boer goats should have long floppy ears, not long ears that that stick out to the side (often called airplane ears). Nigerian Dwarf and Pygmy goats have a maximum height limit, and Angora goats have coat characteristics specific for their breed.

It is easiest to judge general appearance of goats as they move about freely—without someone trying to make them look their best (as is done in the show ring and in most photographs). Watch how the animal moves as it walks: Do its legs move straight forward or do they paddle or swing? Does the goat limp? As its feet touch the ground and bear weight, do the pasterns (the part of the foot between the hoof and the first leg joint) support that weight well, or do they seem weak? Is the goat's back (the topline) flat and level while walking and standing, or does it dip or sway? Is the rump long and slightly sloped

Figure 2.6. Crossbred goats' general appearance often combines that of the parents. In this case, the breed character of this Saanen Nubian cross (sometimes called a Snubian) shows ears that are described as airplane ears—a result of crossing pendulous ears with erect ears.

backward when you look at it from the side, or is it short and very steep? When you look at the rump from the top, is it wide or is it narrow? In all these cases, as you may have guessed, the first option is desirable, the second is not, although a doe with a steep rump is likely to have an easier time kidding than one without any slope.

With the goat standing still, take a good look at its head. There should be no over- or underbite, and the muzzle should be moderately broad and not too narrow and pointed. A goat's muzzle is naturally narrower in comparison with a cow's or sheep's, but if it's too narrow, the nostrils will be pinched, restricting airflow. The goat muzzle is designed to squeeze between branches and seek out tender leaves in shrubs. A goat's teeth are difficult to assess, especially the back molars (see figures 4.2 through 4.6). The front teeth can tell a bit about the goat's age, and a good guide for this can be seen online (suggested websites are listed in appendix D). I personally have

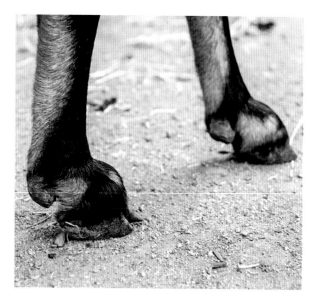

Figure 2.7. This goat's pasterns are slightly collapsed and appear weak, not upright and in line with the angle of the leg.

Figure 2.8. Evaluate the front legs and shoulders for problems by looking at them from the side and front. This 13-year-old doe's left front leg is rotated to the outside (note the placement of the kneepad) and bowed out at the elbow from arthritic changes.

never counted on looking at a goat's teeth to assess age; in fact, I probably wouldn't even consider purchasing a goat from someone who didn't have adequate records documenting the animal's age. But if you aren't confident of these records, then checking teeth is a good backup system.

Next, run your hands along the goat's body, feeling how the neck blends into the shoulders and how the shoulder blades sit against the body. Then run your hand down the front legs. There should be no swelling in the knees, and the bones of each knee should round slightly above the lower leg, not arch backward or forward.

The general appearance of the goat changes over time (also just like us), so do take age into consideration when evaluating the animal. I've seen goats that appeared to be headed in a less-than-desirable direction when they were young, but a problem never materialized. For example, as young animals, some goats of mine had pasterns that weren't strong and upright—meaning the legs might not hold up. Later in their lives, however, after years

of productivity and use, their pasterns remained at the same angle and held up quite well. (I'd like to be judged on how well I'm holding up, rather than on how I compare to a 17-year-old . . .). My point here is that appraising a goat's general appearance is important—but keep in mind that appearance is always subject to change.

Body Capacity

Body capacity doesn't refer to the number of people that can fit in a stadium; it refers to the build of the goat as related to working space in the body for digestion, breathing, and blood supply. Good body capacity allows a goat to effectively transform food into milk, muscle, and babies. A goat that is narrow in the barrel will not be able to process feed as well as one who has good body capacity. The chest is also where the lungs and heart reside, which relates to blood flow and oxygen supply. A narrow chest and rib cage will correlate with reduced efficiency, even if slight, of breathing.

It's easy to confuse body capacity with obesity or abdominal distention. We all have a goat or two that, due to her age and number of births, has a barrel so well rounded that from a distance people assume she is overweight. But there is a big difference between obesity and generous body capacity.

Stand in front of the goat and look at the space between the front legs. The goat should have width without looking like a Mack Truck. If the front legs are close together at the chest, you can bet that the rest of the body will also lack capacity. Next, put your hands on the spine and run them down the rib cage on either side. You should feel an ample outward arc, not a steep, flat drop. Often goats that are built too narrowly will have large, rounded abdominal walls that protrude farther back; it often looks a bit like the goat is hiding a beach ball in there somewhere. This happens when there is a lack of width in the barrel—both in the rib cage and farther down the back—to support the large size of the goat's rumen and digestive system.

Body capacity should increase over time. A year-old goat will look slender and a bit shallow compared with a mature goat of five or six years. Sometimes, though, you may see a young goat with the body capacity of a more mature animal. This might be an indicator that the goat will mature too early and not have as long a career as one that matures more slowly. This certainly has been the case in my experience.

The Mammary System

The qualities of a goat's udder are, not surprisingly, valued more highly if she's a dairy animal, but these same qualities are also assets for meat and fiber goat mothers that will be raising young. A well-designed udder is one that can provide ample milk, isn't prone to injuries, and will hold up to many years of service. Where the teats are positioned (called placement); teat size and ease of milking; the capacity of the udder (not just how big it looks); the texture of the udder (which is related to presence of milk production tissue); and the attachment of the udder to the body are all important things to evaluate. Let's take a look at each of these characteristics.

TEATS

In order for a goat to be easily nursed or milked, either by machine or by hand, her teats should hang straight down or slightly angled forward. If they are located a bit to the side or forward of the lowest part of the udder, then the milk will not as easily drain down into the teat. Goat teats come in a huge variety of shapes and sizes, unlike cow teats, which are typically uniform and easy to milk. A nicely sized teat is not the only goal; the size of the opening (orifice) at the end of the teat also plays a role in ease of milking. An orifice that is too tight can make extracting

How Parity and Age Affect the Udder

When analyzing the mammary system of a goat in milk, it is helpful to know the last time she kidded (had her babies); her parity (how many times she has kidded); and how many babies she has had, both in her last litter and over her lifetime. All of these factors will greatly affect how much milk she makes—and therefore the size of her udder. Most goat breeds produce the most milk, called the peak of lactation, about 30 to 100 days after giving birth. When the udder is past its peak, it is sometimes referred to as stale. (Don't worry, the milk isn't actually stale!) Depending on a goat's age and how many times she has kidded, her udder will reflect these influences. Generally the mammary system will look its best by the third or fourth freshening and age four to five. The number of babies the doe has had in each litter and in her career will also play a role in her milk production and therefore influence the size of her udder. A good doe will make enough milk to feed all of her kids.

milk take twice as long—for humans or goat kids. An orifice that is too open can leave the animal prone to udder infections, especially as the doe ages (when mammary infections become more likely for several other reasons as well). Extra teats and other teat deformities (discussed in more detail in chapter 8) are genetic possibilities for every goat breed and usually discriminated against more in dairy goats than other goats. No matter the goat's job title, though, it's best to avoid any deformity that interferes with nursing, milking, or udder health.

CAPACITY AND TEXTURE

When a goat's udder is full of milk it's difficult to judge its true capacity and texture. If you're choosing a goat in milk or evaluating her female relatives, it's helpful to see the udder both full and empty. The halves should be even, or nearly so, in size. If the udder looks large when full and not that much smaller when empty, it likely means that the udder is made up of more unproductive tissue than productive

Figure 2.9. The mammary system of the doe on the left is poorly supported, leaving it vulnerable to injury. The one on the right shows good attachment.

tissue. A healthy, productive udder should be soft and rather flappy when empty. Sometimes the udder on a doe who has just had kids will feel full even after being emptied. This can be due to inflammation and possibly not releasing all of the milk (called holding back), especially if the doe is feeding her kids. It is best to withhold judgment of udder texture until several weeks after kidding.

ATTACHMENT

The udder is mainly supported in the abdominal cavity by a ligament, a long band of connective tissue that runs down the middle of the udder separating it into its two halves. A highly productive *and* well-attached udder will be capable of producing milk for a much longer career than a dangly, how-low-can-you-go mammary. Goats are active creatures, and a poorly attached udder will be prone to injury from being stepped on, dragged across branches, and wear and tear.

The udder of a young doe should be much higher and tighter to her body than that of a senior doe. If a young doe starts with her udder too large and too low, by the end of her career or sooner (which may be shorter anyway) her mammary system will likely be at risk for injury.

Milk Production Characteristics

The characteristics that indicate an animal is more likely to be good at producing milk—or, in the case of the buck, making babies that will be good at giving milk—are referred to as dairy strength. This doesn't mean that an animal without these characteristics won't be a superior milk producer, though; they simply are usually associated with animals that do. When choosing dairy goats, an extreme degree of dairy strength is the goal. If you're choosing meat goats, a balance of meat and dairy strength is desirable in the females. They must be able to produce ample milk to feed their young but also pass on the traits that will increase the likelihood that their offspring will be good at putting on muscle.

When looking at an animal with good dairy strength, you will see a body that looks slender but

Figure 2.10. This dairy goat shows the angular, flat-boned structure associated with dairy strength.

Figure 2.11. The wrinkles on a buck's neck, thanks to loose, dairy-type skin, are considered a good indicator of his dairy strength.

not emaciated. This is often called free of excess flesh. The muscles are flat and long, rather than bulky and rounded, and there is a lack of body fat. The thigh of the goat is a good place to assess for excess flesh. An otherwise svelte dairy goat might become overweight and hide her dairy strength. This common transformation happens readily when an animal is not giving milk and/or hasn't been bred in some time. If you suspect that a goat is simply overweight (rather than lacking in underlying dairy strength,) assess the other characteristics associated with dairy strength but also be suspicious that her metabolism might be more suited to making fat than milk. (See chapter 6 for more on evaluating body condition.)

Good dairy strength gives the goat an "angular" rather than a blocky appearance. This may sound similar to the qualities of excess flesh, but it refers instead to the shape created by an animal's bone structure (it's evaluated separately from the bone structure I referred to under the general appearance category, though).

Dairy strength can also be felt in the goat's ribs and skin. There should be a decent amount of space between each rib; this is called openness. The rib bones should be flat, not rounded, and they should slope gently backward from the spine rather than straight down toward the ground. It's easiest to feel flatness of bone in the ribs, but you can feel it in the leg bones as well—again, especially when compared with a meat goat. The skin of a dairy animal should be soft, pliable, and "loose." The neck skin of a breeding-aged buck is often covered with wrinkles from the loose skin. You can check for skin texture at the same time you feel the ribs—just grab a handful of skin and pull gently out while you also rub it between your fingers.

Meat Production Characteristics

Meat animals, whether they be beef cattle, sheep, or goats, exhibit a different pattern of bone growth and muscling than their dairy counterparts. The desirable bone pattern in dairy animals is described as flat-boned. In the meat animal, bones are rounder and thicker. Muscling is short and pronounced, rather than long and lean.

Evaluate the bone pattern of meat goats by observing and feeling their lower leg bones and ribs. The ribs should feel rounded rather than flat and the legs rounded rather than oval. Evaluating the bone structure of the animal will help you decide if it has the potential to develop good muscling. If the animal is already well muscled, you can easily see the results, but if it's young or not in the peak of condition, bone pattern will still be something you can evaluate.

Muscling in the meat goat is obviously incredibly important—that's the meat, after all! It can be difficult at first to distinguish a well-muscled goat from an overweight goat. Muscling should be evaluated by

Figure 2.12. This young Boer wether, the grand champion at our county fair, shows the round bone pattern, broadness of structure, and muscling highly indicative of superior meat-production potential.

looking at the rear legs, shoulder, and loin and noting the shape of the muscles, which should be rounded rather than long and flat. Fat deposits can be felt as soft padding that doesn't change texture when the animal moves (whereas muscle will contract and relax) behind the elbow and on the sternum. Feel all of these areas when you are inspecting a goat for its meat potential.

Fiber Production Characteristics

The main quality to look for when choosing fiber goats is the quality of their fleece. Goats for mohair and those for cashmere are assessed differently when it comes to fleece quality. Both types will be difficult to analyze if the animals have been shorn recently. If the breeder has records of shearing and yield weights, these can be extremely helpful in choosing breeding stock. When evaluating a goat for cashmere, look for a summer coat with long guard hairs that are more coarse than the undercoat (making it easier to separate these hairs out from the desirable fiber after harvesting). Mohair, and to some degree cashgora (the fiber produced by the cross of an Angora goat and a cashmere-producing goat), is evaluated for uniformity of type and the density of coverage—how many hair follicles populate the animal's skin. Evaluating both characteristics well takes some time and experience—as with the other traits, learning to compare and assess good fleece quality is best done with the tutoring of someone who is experienced. The same loose skin on the neck that is exhibited by a goat with good dairy strength also helps ensure more fiber production, because 25 percent of the fiber growth occurs on the goat's neck—the more skin there is, the more fiber there is.

Interestingly, good nutrition, which correlates with increased fertility and reduced abortion rates, has a somewhat negative effect on fiber: There is an increase in volume, but also in diameter, thus lessening its quality. It's important to note, however, that an increase in nutrition will increase fertility and birth rate, and since kid fiber is of the highest value, it makes sense to focus on that, rather than the quality of the adult fiber. (See the sidebar Feeding for Future Fiber in chapter 8).

More Ways to Evaluate Herd Quality

When you're trying to build a strong herd or choose new animals to improve or grow a healthy, productive herd, it's helpful to have documentation that supports those goals. Several types of records can be very useful in this regard. Some of these are records that producers maintain themselves; others are generated through programs run by organizations that track pedigrees and performance. The most organized and broad programs are those offered by organizations associated with the dairy goat industry. This is simply because the dairy goat world has been organized far longer in the United States—and at one time dairy goats were far more numerous than any other type of goat. Today there may be far fewer dairy goats, but there are still more dairy goat shows, registries, and breed clubs, and several impressive data collection options that track milk production and genetic transmitting abilities, all of which help define the value of the animals.

Meat goat enthusiasts are moving toward creating equivalent genetic programs, with the American Kiko Goat Association and the American Boer Goat Association both evaluating them. But as of now, not as much in the way of centralized, large-scale programs is offered. Let's go over the options available and talk about their role in creating and choosing a healthy herd.

Producer Records

Although dairy goat producers generally have access to a wealth of data collection through milk production and genetic evaluation programs, some small dairy producers and all meat goat producers have limited options for accessing objective, broad data-based programs. Thus records of health and productivity created and maintained by the goat breeder are imperative if progress toward a stronger, more efficient herd is the goal.

All types of goat breeders should maintain health records, including data such as difficult kiddings; number of offspring and their viability; hoof and leg problems; and general health issues. (Appendix D lists some resources for finding software and record-keeping information.) At the minimum meat goat producers should maintain production records that include the weights of a goat's offspring at different target ages and the amount of feed required to attain those weights. Carcass yields and scores will also be helpful. Dairy producers should keep milk production records that include not only milk volume but also butterfat, protein, and especially somatic cell counts (indicative of udder health). Fiber goat producers will want to track fleece yield and quality. Producer records are only as accurate and honest as the person inputting the data, of course. For the records to be of use, it behooves the goat farmer to be as accurate as possible, something that should be reassuring if you are shopping for animals and evaluating these records.

Pedigrees

You can think of a pedigree as the animal's family tree. Depending on who maintains the pedigree—an official registry or simply the breeder—the document might contain different information and be more or less helpful to both the breeder and potential buyers. An accurate pedigree is one of the best ways to manage concerns such as line breeding, inbreeding, and crossbreeding (see appendix A for sample pedigrees) and to keep track of dates of birth, age, and identifying characteristics (such as color and tattoos).

Some breed organizations, especially in the dairy goat world, also include show ring championships, milk production awards, genetic-related scores, and genetic awards. Boer goats, too, have pedigree designation programs that honor animals with proven production records, but other meat goats do not. Official pedigrees are only updated with current information about awards when they are sent to the registry for revision, and in many cases this occurs only at the time of a transfer of ownership.

May I See Your Passport, Please?

Pedigrees are becoming valuable for another reason in the United States—the ability to transport goats across some state or any international borders. Regulations require official documentation of animals that match a unique identifier, such as a tattoo or ear tag, before they can be issued an entry permit. For unregistered animals that are ear-tagged with a USDA-issued tag, the paperwork that accompanies that process is also acceptable.

Show Ring Performance

Goats that do well in the show ring can help a breeder earn a good price when selling animals and can be a source of pride, but do prize ribbons and trophies mean much when it comes to evaluating animals as a buyer or breeder who is looking for strong genetics? Not so much. Show ring wins reflect only the beliefs of the person judging those animals, indicating that the winner was the best example of the breed present on that given day. In a group of outstanding animals the one placed at the bottom might be the winner at another show where the competition was less-than-impressive examples of their breed. For this reason I recommend all show wins, and losses for that matter, be taken with a grain of salt by both producers and shoppers. As the old saying goes, "Pretty is as pretty does." Shows are fun and a great way to interact with other goat breeders and learn how to evaluate animals. And of course it's satisfying to bring home the odd rosette or two to hang on your wall.

Milking Awards

Dairy Herd Improvement (DHI) is a nationwide program originally designed to help cow dairy farmers improve their milk production and genetics.

The dairy goat world initially adopted the program in 1939. It is not a government program, but the data collected is shared with the US Department of Agriculture (the USDA) and is available for all to see through their website at www.adgagenetics. org. When you are enrolled in DHI, it's colloquially called "being on milk test." You do not have to own registered or purebred animals to be on milk test; you don't even have to be a member of one of the goat registries. In addition, any goat you want to milk can be put on test, be it a Boer doe, an Angora doe, or a pet Pygmy doe. Those that do have animals recorded or registered with a major registry, though, can earn various types of awards and designations.

Each dairy breed has a set of criteria by which animals can earn what are called milking stars. The criteria are based on the amount of milk, fat, or protein produced by the animal. Depending on the age of the animal, the minimum amount that she must produce to earn her star varies. As you might guess, the minimum increases with age. The amounts needed to earn a star vary by breed. You can find this information in the registry's guidebook or online.

Herd Appraisal and Genetic Programs

In addition to the above programs, there are a handful of other options for goat breeders that are meant to reflect on the genetic potential or track record of a goat. As with the above programs, the most in-depth programs are currently available only for dairy goats, but there's growing interest and motivation to create similar opportunities for meat goat breeders. The American Dairy Goat Association (ADGA) has two programs that can not only help buyers choose stock and breeders choose new genetics but also provide a depth of information invaluable to a serious breeder—linear appraisal and superior genetics. Linear appraisal, usually an annual event, evaluates goat traits that are considered heritable. Height (also called stature), for example, is a highly heritable characteristic. Linear appraisal also scores individual animals based on four categories: general appearance, dairy strength, body capacity, and mammary system.

The second program, called superior genetics, gives onetime designations to animals that meet certain criteria for milk production and genetic traits. These designations are extremely useful when evaluating goats as objectively as possible. Ennoblement is awarded to pedigreed purebred and full-blooded Boer goats that prove themselves through a combination of show wins and offspring that all meet specific visual requirements. The specifics of the program are quite detailed, much like the superior genetics program, through ADGA. The goal is to reward those breeders working to improve the breed and provide incentive to all breeders to join the effort.

If you raise dairy goats, or even meat and fiber goats, I highly encourage you to participate in a linear appraisal session. No matter what the specialty

Linear Appraisal Category Scores

E = Excellent; at least 90% of ideal—
  Mature Program
EC = Extremely Correct; at least 90%—
  Young Stock Program
V = Very Good; 85–89%
G = Good Plus; 80–84%
A = Acceptable; 70–79%
F = Fair; 60–69%
P = Poor; less than 60%

Figure 2.13. At this linear appraisal session at Pholia Farm, veteran appraiser Eric Jermain and apprentice Ben Rupchis score one of our does.

of your goat farm, watching these sessions is a great way to train your eye to evaluate animals. Even if you are not able to bring your own animals, find a friendly breeder who'll let you observe an LA session. It is more informative and more objective than any goat show could ever be. And of course, when I say it is more objective, it still depends on humans, and you will definitely see that results will vary. But over time and repeated sessions, a balanced picture of your herd's quality does appear.

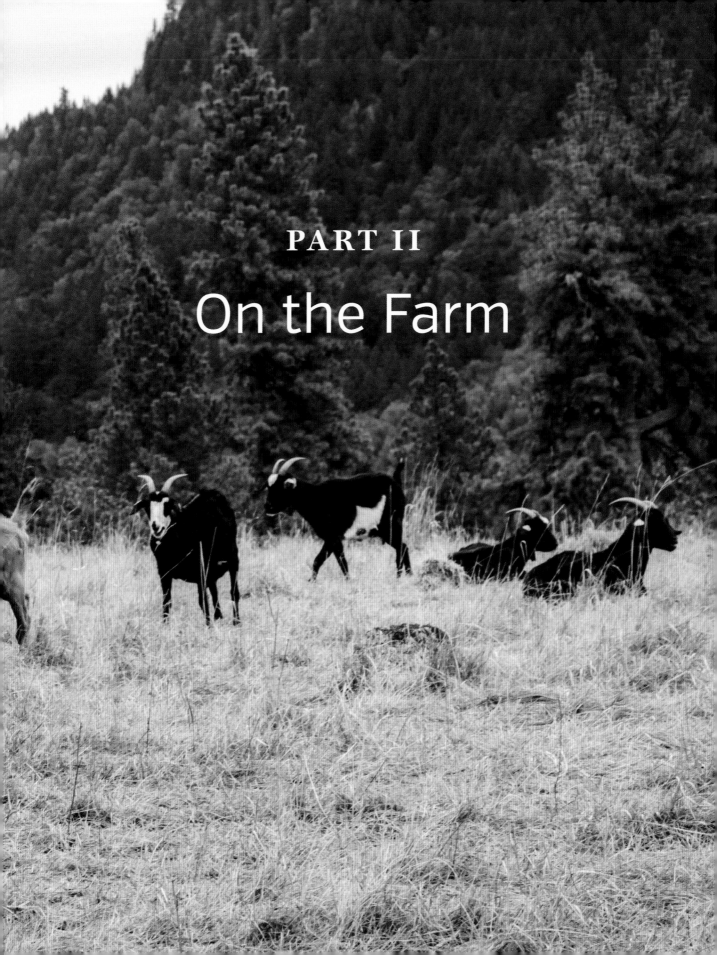

PART II
On the Farm

CHAPTER THREE

Farm Management

The best goat farm is one that attempts to mimic the ideal natural environment for goats. But since a farm can't precisely match a wild setting, farmers have to compensate for some of the things they can't provide, such as unlimited access to open grazing land. Any type of livestock farming involves some compromises to keep a balance between ensuring a thriving herd and maintaining a sustainable business and lifestyle for the farmer. The way each farm does this is likely to be different. I've tried many different management approaches over the years, and I will likely continue trying something new almost every season in the hope of not only improving herd health and productivity but also adapting to changing realities such as feed availability and quality. Staying flexible and being open to new management ideas is the best way to discover what works best for you and for your goats.

The core aspects of management are those related to your goats' basic needs for shelter, water, and food: enclosures, shelter from the elements, handling techniques, protection from dangers, and providing for the animals' nutritional needs. It's also important to figure out a system of identification for your animals, to consider how you will manage manure and bedding waste, and to be prepared for the reality of dealing with goat mortalities.

Goat Handling

One of the most fundamental aspects of goat farming is being able to safely and efficiently move goats from one place to another: from pasture to pasture, from pen to milking stand, and through gates and chutes. It's also a health-related concern, at least in terms of your herd's, and your own, mental health. If goats suffer fear and anxiety every time they are moved or handled, that can translate into stress, which in turn affects overall physical condition and productivity. In chapter 1 I explained how to respect a goat's instincts and modify your behavior when you want to approach and catch hold of them. You can also deliberately employ predator-like behavior to your advantage when you need to move goats from one place to another. For example, a dairy might need to move one group of milk goats out of the parlor so that another can take its place. A meat or fiber goat operation might need to bring goats in off the range and sort them for processing or shearing. In these cases you can use various "pressures" to humanely move the animals without inciting them to fear. Here are some tools that are very helpful for moving goats:

Rattle paddle: This is a long-handled plastic paddle that can be purchased from livestock supply companies. The wide paddle section is partially filled with metal pellets. The paddle serves as a physical barrier. When shaken, it makes a noise that slightly startles the animals.

Water pistol or sprayer: You can use goat's natural dislike of water to enforce behavior, such as getting them to move through a chute or not use a one-way gate in the wrong direction. I've even used a motion-activated Rain Bird–type sprinkler (designed to scare off deer) to keep goats away from fruit trees.

Warning sound: I like to condition goats to respond to a harmless warning sound, such as a whistle or clicking. At first the sound is followed immediately by a physical enforcement such as a spray from a water hose. After a time, the animal responds to the sound alone.

Fencing panels: You can use rigid fencing panels to create temporary funnels, chutes, and enclosures to sort and work with goats. It's important to set the height high enough so that fearful goats won't try to leap over—vaulting goats can end up hurting themselves. The panels can be temporarily connected to one another with carabiners (clips) or zip ties.

Electric net fencing: You can use flexible net fencing (temporarily disconnected from the charger) as a "herding net"—either by sequentially shifting it using its built-in stakes or by having several people hold and move the fence as the herd moves.

Herding dogs: Herding dogs can be used to move goats, though the decision to employ them must be made on a case-by-case basis. If a livestock guardian dog is present in the herd, for instance, it must be accepting of the predator-like behavior of the herding dog. Also, if some goats are tame and bold or have horns, they may choose to stand and defend the herd instead of submitting to being herded.

If your farm receives frequent visits by the public, consider how spectators will interpret what they see when goats are being moved from place to place. You might understand that hitting the ground behind the goats with a stick is a harmless way to encourage them to move forward, but non-farmers might think you're trying to hit the animal. This is just one example of why explaining and teaching is often an important part of managing an operation with high public visibility.

Collar Caution

It can be an alarming experience to lead a goat by the collar and suddenly feel the animal collapse and see it jerking spasmodically on the ground. If you haven't experienced this, you will someday! This can happen if the goat turns their head or stops suddenly, causing the collar to push up into the junction between the neck and jaw, putting pressure on the carotid artery or arteries and briefly stopping the blood flow to the brain. (This is similar to a maneuver sometimes taught in judo—or to the Vulcan nerve pinch, if you're a fan of *Star Trek* like I am.) Don't panic! Loosen your grip and pressure on the collar, and wait a few moments. The goat will recover quickly.

The first time this happened to me I was leading someone else's goat. I wasn't dragging the animal or being unkind, but the owner interpreted the outcome as due to my rough behavior, and I received quite a scolding. Some goats seem much more prone to this, perhaps due to the anatomy of their necks or the angle at which they carry their heads. The type of collar doesn't seem to matter; both plastic chain and nylon webbing can cause the response.

Fencing

Rare is the goat farm that won't need fencing, yet it's common to see farms where this part of the equation looks as if it were an afterthought. Goats have a reputation for escaping from fenced areas, and if you've been keeping goats for a while, you know it's true. I don't consider it *escaping*, though, because goats will just as quickly "escape" back into their pen if they so choose.

Goats are small, clever, and curious, as well as eager to venture out and find the foods or environmental stimulation they crave. To a goat, an inadequate fence is simply a challenge to overcome rather than a barrier to respect. Because goats are also wary of danger and of straying too far from the safety of the barn, once they're outside a fence they're unlikely to venture far. They may, however, do their best to make a culinary visit to your heirloom rose garden; check out the tap-dance suitability of your redwood deck; or play an energetic game of king of the mountain on the hood of your neighbor's classic car. To avoid these headaches—and, more important, the risk of injury to goats that get stuck while trying to push through or under a fence—it's important to design fences with goat behavior in mind.

Choosing the Right Style of Fence

Goats enjoy poking their heads into and through any space they think they can fit. Because of the triangular shape of their heads, they are often able to push their heads through an opening but then cannot extract themselves quite so easily. The situation is even more complicated when a horned goat gets its head stuck. (Grazing specialist Mark Kennedy refers to this situation as a "coyote fast-food restaurant.") Often, a fence has to be cut or taken apart to release the entrapped critter. Because of this, openings in fences should either be small enough to discourage goats from attempting to thrust their heads through, or large enough to allow goats to move their heads through easily.

Figure 3.1. While fences are designed to keep goats in their pens, there's always one that will test the limits, sometimes finding itself caught midway.

Figure 3.2. This sturdy welded wire fencing, also called a hog panel, keeps goats contained and is resistant to warping. Goats can reach their heads through to graze the other side, which reduces the need for mowing and edging along the outside of the fence.

Is Tethering Goats Okay?

There aren't many circumstances when tying a goat and leaving it to graze or browse is a good idea. Although a goat can be trained to be tied and be comfortable and competent moving about with a tether, it makes it quite vulnerable to predators or harassment. Dogs in particular might startle the animal and cause it to try to escape—in the process breaking its neck. If the goat is well trained to the tether and tied within a protected space, it's possible to make it work. But in general, I don't advise it.

Figure 3.3. I took this photo of a goat tethered along the roadside in Costa Rica, where vegetation is lush, fences are rare, and predators are unusual. Who knows if the breakfast-and-lunch sign is related to the goat? . . .

As you're making choices about fencing materials or assessing your existing fencing, consider whether there will be young goats in the mix. Young goats with their mothers are less likely than adults to try to leave an enclosure. But if they can easily fit through a fence, they might decide to come and go. Before long, they will be in the habit of squeezing through the fence openings at will, and as they grow, they will eventually be at risk of getting stuck. The bottom line is, no matter how well you have designed your fences, it's also a must to keep a proper set of wire and/or bolt cutters on hand!

In addition to sticking their head into any hole through which they think it might fit, goats also love to use any type of resilient vertical surface as a scratching post. Chain-link fencing and woven wire fencing (also called field fencing) are perfect surfaces to rub against. If used in an area where goats spend a good part of their time, the fencing will soon become stretched, worn, and weakened. Goats also like to place their hooves on fencing so that they can stretch their necks up to eat low-hanging branches or simply so they can see farther. (Because of their cloven hooves, goats are quite adept at standing on narrow surfaces.) This places a lot of strain on the parts of the fence bearing their weight. If goats don't ruin a woven wire or chain-link fence by rubbing on it, they will by standing on it.

Setting Fence Height

Goats are good jumpers, and the younger the goats, the more they like to leap (both disposition and body mass have something to do with this). In areas where jumping is likely, such as over a fence that is meant to protect a garden or to separate bucks from does in heat, make sure the fence is tall enough to prevent the goat from even attempting to jump over. A 4- to 5-foot (1.2–1.5 m) fence is satisfactory for most goats. A 3-foot (0.9 m) fence is usually tall enough to contain adult Nigerian Dwarf or miniature goats. To be safe for all ages and types of goats, I recommend a 5-foot (1.5 m) fence, particularly in areas where they spend a lot of time loafing, rather than foraging or grazing.

If a fence is too short, a goat might not only jump but—even worse—jump and get a back leg hung up in the fence, snapping the bone. I've witnessed this kind of accident, and believe me, it's a sound you won't forget. Even with your best-laid plans, however, every once in a while a goat will be born that seems to be part deer and able to leap out of almost any pen, especially when it's young, light-bodied, and full of mischief.

The best fences for smaller goat pens and paddocks are panels made of rigid steel rods welded into place, creating a structure that can hold up to all of the abuses and challenges goats provide. Hog panels are one example of this type of fencing. These panels have a grid pattern with smaller openings toward the bottom, and they are about 3 or 4 feet (0.9–1.2 m) tall. Another example is security or horse panels, which are 5 feet (1.5 m) tall and have openings that are only 2 inches wide by 4 inches tall (5 × 10 cm). Of course, both these types of fencing are also the most expensive, but it's likely you'll never need to replace them.

Fencing Large Areas

In larger permanent pastures and enclosures you can use woven wire, high-tensile hot or electric wire, or a combination with woven wire below and hot wire above. You won't see multistranded barbed wire in use very often for goats, because they are not quite as pain-motivated as sheep and cattle and are therefore less likely to respect barbed wire and more likely to challenge it. When choosing perimeter fencing, think about the following:

- Is the space large enough, with enough feed and activity options to keep goats busy so that they don't need to challenge the fencing? If so, then the fencing can be less substantial.
- Are there other animals (including pet dogs) that might enter the pen and frighten the goats so that they might run through the fencing? If so, then barbed wire or hot wire that they could get tangled in is not a good choice.
- Are there predators that you can keep out by choosing the right fencing? If so, then use that style of fencing, even if it costs more, to avoid the financial hit and the heartache of losing animals to predation.
- Are there livestock guardian animals (llamas or donkeys) that need to be kept confined with the goats? If so, the fencing will need to be taller than one for goats alone.

The guidelines for openings in the fencing and height of the fence stated earlier also apply to perimeter fences. It's important to monitor the integrity of pasture fences. They should be inspected regularly for compromises from downed tree limbs, large animal damage (in our part of Oregon, elk are notorious for their ability to plow through almost any fence), and even vandalism by people, including those hiking or on horseback and looking for easy trail routes.

Electric Fencing

The most affordable and effective choice for fencing of rotational and temporary pens is almost always electric net fencing. This wonderful option provides incredible flexibility when the fences are used and maintained properly. For permanent paddocks, woven wire and high-tensile hot-wire fencing are also good choices.

There's a bit of an art to choosing the right setup. I found this out the hard way by spending money on several rolls of electric net fencing only to discover that most of the year our soil is too dry and our goats too lightweight to properly "ground" themselves and feel any shock from the fence. It's alarming to see your fence being towed up the driveway by a couple of goats who have become entangled (fortunately without being harmed). You have several options when it comes to making sure that electric fencing is effective.

Pick a time when the ground is wet for training goats to respect an electric fence. That way, they will be sure to ground easily and feel a shock. Once the animals have had that experience, then they will rarely try to challenge the fence in the future, even at times when it isn't providing adequate shock.

Choose the right charger with enough power to provide the right amount of shock. Ask an electric fencing dealer for help in this decision, because there is danger to unwary humans from hot-wire fences and increased liability with increased fence strength.

Consider "positive/negative" fencing (also called wire return or ground wire circuit), which includes grounding (negative) wires in the fence. Most net-style electric fences are made with positive wires; the earth serves as the negative "ground." With a positive/negative fence, the animal will receive a shock anytime it touches a positive and negative wire on the fence at the same time, even if the animal is far from the fencing ground rod. Net or strand fences can be configured as positive/negative.

If you're using high-tensile (HT) fencing, provide enough wire strands to also create a physical barrier (nine strands over a 4-foot/1.2 m span is usually enough for goats). Five strands is sufficient for adult goats in situations where there is no risk of predators attempting to enter the enclosure or kids trying to leave the pen.

Space the strands properly, with the wires closer together toward the bottom. For example, wires in

Figure 3.4. When installed properly, five-wire (or more) high-tensile electric fencing is a very popular and effective system to contain standard-sized adult goats. The goats kept in this pen have access during the day to lush pastures farther down the hill.

a six-wire fence would be spaced (starting from the ground up) 5, 5, 6, 7, 8, 9 inches apart (13, 13, 15, 18, 20, 23 cm).[1]

Energized wires should be placed so that two are close to the ground (where animals are more likely to challenge the fence), one is at the top, and one is in the middle. The remaining gaps can be filled with non-energized wires.

Make sure that fences are not being negatively affected by contact with weeds, grasses, or branches. These can causes fences to lose efficiency and can drain the battery of battery-powered types.

There are many combinations of fencing and energizers, and it's important to choose the right combination when investing in electric fencing. Take some time to go over those options with a representative from one of the companies selling these supplies. Be sure to describe fully the conditions of your land, what you'll be using the fencing for (permanent paddocks or rotational grazing), and your expectations. See appendix D for sources of information on electric fencing.

Gates

There are two great secrets for designing gates in a goat pen. First, set up frequently used gates to swing inward only, if possible. This allows you to "sweep" goats away as you enter. It also makes it very difficult for the goats to force a gate open outward if it is accidentally left unlatched.

Second, if you frequently move goats into and out of a certain holding area, be sure to include more than one gate into that area. Reserve one of the gates for human traffic only. Remember that goats are creatures of habit. If their habit is to go through a certain gate to visit the lush pasture, then every time you open that gate, they are likely to assume they should pass through it. This creates problems if you need access to the pen to bring in a tractor or wheelbarrow, for example. As soon as you approach the gate, the goats will run to it, blocking your way. However, if there's a gate they never pass through, they are unlikely to show interest when you approach that gate. American poet Robert Frost said, "Good fences make good neighbors." I say, "Good fences and good gates make good goats."

Paddocks and Pastures

As with so many facets of holistic goat care, keeping goats' evolutionary habitat in mind is important when designing or analyzing open areas where they spend their days lounging, romping, exercising, or eating. The most idyllic way to meet their needs would be a large space with natural browse, rocks to climb on, and no threats from predators. The reality for many farms will be less than picture perfect but

> American poet Robert Frost said, "Good fences make good neighbors." I say, "Good fences and good gates make good goats."

still suitable. Options include loafing sheds, dry-lot paddocks, pastures, and open range.

Loafing sheds are covered spaces with open sides, often paved with concrete, where goats can stretch their legs and interact with the rest of the herd. They are covered to protect the animals from sun, rain, or both.

Dry lots are fenced areas without any forage plants. The ground is usually bare, composed of dirt, gravel, or sand. Windbreaks and trees for shade might be a part of a good dry lot. In contrast, pastures are large vegetated fields either enclosed with a single perimeter fence or divided into rotational grazing paddocks. (Pasture is discussed in more detail in chapter 5.) A dry lot or loafing shed or barn can be used in conjunction with pasture as a holding area when pastures are too wet to be grazed (because parasite risks are higher) or have been overgrazed, or when conditions are otherwise not ideal for grazing.

Open range is any large area that is not farmed (not purposefully planted with crops) to which the

Figure 3.5. At Lazy Lady Farm in Vermont, dairy goats are grazed in rotational paddocks using three-strand hot wire. Pastures are only grazed when they are over 8 inches [20 cm] high; they're mowed twice per year, and browse is fed two times per week.

Figure 3.6. Goats enjoy a stimulating environment. This tire has been hung in the pen of young bucks to give them something to tussle with.

goats can have access. In some cases this will be full-time access; in others, such as at Pholia Farm, it'll be part-time in the company of a goatherd. (I get to experience my inner Heidi.) In some parts of the United States, large tracts of open-range public lands are grazed (through lease agreements) by cattle, sheep, and occasionally goats. Sometimes goats are being sought to help restore lands that have been overgrazed by these other livestock.

No matter what type of enclosure you supply for your goats, you can enhance their physical and psychological health by supplying activity options. A hanging tire, climbing structures, stumps, a downed log—almost anything will do. Goats like change, too, so consider rearranging their "furniture" once in a while—suddenly everything is new again!

It's also important to consider the placement of items in an enclosure that goats will climb on, such as stumps or play equipment. If they are too close

Figure 3.7. Stumps, removed after a tree has died, are often hard to dispose of. Using them as goat enrichment tools is a great option!

to the fence, goats will use them as a launching pad for a leap. Depending on the motivation of individual goats, this distance should be at least twice the length of their bodies. I once saw a wonderful video online of an ingenious goat so determined to reach the leafy branches of a tree within its pen that it wedged itself between the trunk of the tree and the fence, then alternately used its hind legs on the trunk and its front legs on the fence to work its way to the top of the rather tall fence. At that point it could have easily jumped out of the pen, but that wasn't the little goat's goal. Those leaves must have been delicious!

Barns and Shelters

Unless you live in a spot where the land and climate are ideal for goats, you'll need to provide some type of shelter, even if you plan on managing your herd extensively. This might be as simple as a portable field shelter or as elaborate as the goat barn version of a Hilton hotel. Shelter can do many things: It protects the goats from climate and weather extremes that challenge their health; it provides a place to protect them from predators; it's a place to perform exams and management tasks such as trimming hooves and weighing animals; and it's a place to store supplements, feed, and tools. To determine the ideal design for your barn or shelter you'll need to consider these factors: weather, ventilation, slope of the land, herd size, predation risk, and the desired purpose.

Weather

Goats that are healthy and acclimated to the weather of an area can make do with very little shelter. Even when they are adjusted, however, you'll need to provide shelter for times of extreme climate conditions. Humidity and wind are particularly important factors to consider when you're making housing decisions. Protection from cold winds and dampness but with plenty of fresh air exchange is of the utmost priority for goat health. It's essential to avoid dampness from wet bedding—whether the moisture is due to urine, rain, or condensation.

Figure 3.8. Field shelters that can be moved to different paddocks are an important part of a rotational grazing system. This shelter at Sospiro Goat Ranch, North Carolina, is about 10 feet (3 m) long and 6 feet (1.8 m) tall; the framework is welded wire panels (each is 16 feet/4.8 m long) that have been curved to form an arch.

Goats need ample shade during hot summer months, and when goats are on pasture, a simple field shelter such as the one shown in figure 3.8 can fill the need. See appendix D for links to plans for shelters like this one.

In the winter, however, the goats will crowd to the first spot of sunlight striking their enclosure. Knowing where the sun will hit each part of the structure and pens throughout the year is very helpful. If the main doors into a barn face north, it's unlikely that the pens inside will receive any direct sunlight. If barn doors open to the west, the pens will receive afternoon sun, which could be scorchingly hot in some regions. Deciduous trees (protected from the hungry, searching lips of the goats, of course) can be used to provide shade for the barn or pens in the summer, and they won't block the sun in the winter. Shade cloth, which can be purchased at most hardware stores and greenhouse supply companies, can be hung in a variety of ways—as a curtain from an overhang or an awning supported by poles—to keep the temperature lower in the summer. Goats with peers and friendly herdmates will stay close together to take advantage of one another's warmth in the

Will and Caroline Atkinson

HILL FARM DAIRY, SOMERSET, ENGLAND

Figure 3.9. Will and Caroline Atkinson and their daughter, Kitty, at their goat dairy in Somerset, England. *Photo by Angie Hannon, courtesy of Will Atkinson*

When I visited Hill Farm Dairy in the beautiful rural foothills of southwest England in 2012, I was extremely impressed with the planning that had gone into the operation, from the barns to the dairy building to the streamlined fashion in which owners Will and Caroline Atkinson turn excess buck kids (usually regarded as a farm problem) into an asset.

The farm's 100-plus registered goats, which include Anglo-Nubians, British Toggenburgs, and British Saanens, graze the 20 acres (8 ha) of pasture and 10 woodland acres (4 ha) that surround the 16th-century farmhouse in which the Atkinsons live. As can be guessed from the name, the property is quite hilly, but the Atkinsons used this to their benefit by designing and placing buildings to take advantage of gravity. For example, their milking parlor is at a level above their processing rooms, which allows them to transport the milk by gravity flow, instead of pumping—a process that is always more or less damaging to the milk. In another example, one of their main goat barns is built in steps along a slope. The top tier is hay storage, which allows for hay to be off-loaded at ground level, and then dropped down into feeders at the next level as needed. The ergonomics of these designs are helpful to the famers, the animals, and the quality of their cheeses.

In another terrific example of efficiency, Hill Farm doe kids not kept for the farm are sold to other dairy farms, and their billy kids (in England the does are often called nannies and the bucks billies) are collected and raised on another farm to be a part of the meat supply. The goat meat company Cabrito UK collects buck kids from many English dairies and processes them into high-end goat meat products for chefs and restaurants. It's a wonderful model of sustainability that is helpful to the farmer, the meat processor, and the public alike.

I asked Will what advice he has for people considering building a goat farm. He summarized

Figure 3.10. This well-ventilated barn at Hill Farm Dairy has been built with slatted sides that can be opened in the summer and closed in the winter, while the ends remain open year-round for air circulation.

the couple's successful approach. "Buy your goats from one, good source. Plan your site if you can, to allow all the various jobs to flow easily, and to allow for expansion. Let the goats graze on healthy pastures. Factor help into your financial analysis, to allow you to have some breaks. Treat your animals like human beings with different habits; and remember that you are demanding milk from them twice a day to help you realize your dream—respect that, and treat them accordingly. Spend a lot of time in close proximity to them so they get to know you—but also have a spot somewhere in the barn so that you can sneak in and sit and watch them without them knowing you're there. Keep bedding immaculate at all times, and give them really good hay to eat."

Figure 3.11. This hoop-house-style barn at Prairie Fruits Farm in Indiana helps keep goats warm in the winter and cool, with the help of hanging fans, in the summer.

winter. As long as you provide the right options, goats will moderate their comfort quite well.

Ventilation

Good ventilation is essential for herd health. It's easy to view a snug, tightly sealed barn as cozy, but remember that the inside of the barn is living room, bedroom, and bathroom for your goats. As their bedding becomes soiled, it contributes contaminants to the air they breathe. The ammonia fumes produced by urine can be damaging to the lining of their respiratory tracts, making them prone to pneumonia and general ill health. A barn should have adequate

openings to allow fresh air to flow in and stale air out. It's also important to time the frequency of removal of soiled bedding from the pens to limit the buildup of fumes. How can you tell whether the ventilation in your barn is adequate? Go into the pen, crouch down to goat level, and take a deep breath. If you don't like how that feels, well, neither do your goats.

Slope

The slope of the land will influence the drainage of rainwater runoff into a barn and wastewater runoff away from a barn. Water runoff into a barn is unacceptable because bedding will be damp and fumes will accumulate. Water runoff away from a barn or an outdoor goat pen is also of concern if creeks, streams, or even wells are located downhill: Runoff from the goat area is likely to pollute this water. Many states require well-planned and -overseen management of

Table 3.1. Animal Welfare Approved Goat Housing Standards

Type of Goat	Minimum Indoor Bedded Lying Area
Adult goat (buck/doe)	16 square feet (1.5 square meters)
Kid	4 square feet (0.4 square meter)
Doe with one kid	22 square feet (2.0 square meters)
Any additional kid	4 square feet (0.4 square meter)
	Minimum Additional Loafing Area
Adult goat (buck/doe)	27 square feet (2.5 square meters)
Kid	5.4 square feet (0.5 square meter)
Doe with one kid	33 square feet (3.0 square meters)
Any additional kid	5.4 square feet (0.5 square meter)

water runoff from animal pens. In the United States farmers and land managers can consult with the local office of the Natural Resources Conservation Service, a branch of the Department of Agriculture (USDA), which can provide a great deal of assistance, at no charge, to those who are striving to manage all their land resources optimally.

Space Requirements

The number of goats that can comfortably and humanely occupy a space depends on several factors, including breed, size, and disposition. Whether the animals have been reared together is also important, as is the philosophy of the herd manager. For example, a herd that is stable in social interactions (meaning they all know one another) can safely and without stress share a much more crowded space than a herd made up of animals that have been recently introduced. Because of this, it can be advantageous to buy or rear animals in peer groups, also called contingents. This can be accomplished, to one degree or another, on farms of all sizes. These groups move through the system, from birth on, staying together for the majority of time. When the younger groups are integrated with the main herd, they will find equilibrium sooner and be overall more productive and less stressed.

There are various guidelines for the minimum space to allow per goat based on gender and age. Tables 3.1 and 3.2 provide guidelines from two humane certification programs in the United States; the guidelines are similar but not identical. Visit Humane Farm Animal Care's program at www.certifiedhumane.org and the Animal Welfare

Figure 3.12. Goats adore shelves and platforms for lounging and romping. Shelves also expand the square footage of available space. The bottom shelf of this unit is sloped so that most of the manure pellets roll off.

Approved program at www.animalwelfareapproved .us for more information. I think both of these programs allow adequate space for holistic care, especially when you take into consideration herd dynamics as described above.

Predation Risks

Closing animals in a barn at night is one of the easiest ways to ensure their safety. If this is the method of predator protection you opt for, then the area in which you enclose your animals must provide ample space for movement and access to feed and water. I discuss other options for predator protection later in this chapter.

Table 3.2. Humane Farm Animal Care Goat Standards for Bedding Areas

Type of Goat	Weight of Goat	Space Allowed When Straw Bedding Is Supplied
Adult doe	Up to 230 pounds (105 kg)	18 square feet (1.7 square meters)
Young kid, up to 5 months old	8–75 pounds (4–34 kg)	8–10 square feet (0.7–0.9 square meters)
Buck	165–265 pounds (75–120 kg)	30–40 square feet (2.8–3.7 square meters)

Note: Total area required per goat should be 1.5 times the bedded area.

Farmer Efficiency

Don't forget about yourself when designing your barns and shelters! Truly, the farmer's needs should be given as much consideration as the goats', because the efficient and content farmer is better able to care for the animals. Design adequate feed and tool storage; hoof trimming and procedure areas; bathroom facilities; and anything else you can anticipate will make your work easier and more rewarding.

Flooring and Bedding

When it comes to the surface that your goats stand and lie on in their barn or shelter, there are no perfect answers. Goats are indiscriminate poopers and pee'ers: They go when they have to go. Their small size and ability to move around in tight spaces makes it impractical to keep them in dairy-cow-style loafing stalls—which are bedded or padded with mats but keep the cow in position such that she must urinate and defecate over a gutter or grate just outside the comfy stall. Also, because goats dislike rain so thoroughly, they spend more time under shelter than cows do, so urine and feces accumulate more quickly. Goats also seem to show a preference for urinating in dry bedding, so much so that you could almost, just almost, litter-box train them—for that one event, anyway.

Common choices for flooring are dirt, decomposed granite (a fine-grained granite that is much coarser than sand), or concrete (sometimes overlaid with rubber horse stall mats). The usual practice is to cover this flooring with layers of bedding, such as wood shavings, sawdust, or straw. Periodically, the bedding has to be removed. Depending on the scale of your operation, a tractor with a small bucket loader can work well for this task—as can, of course, a pitchfork and strong back. In the winter this type of bedding, sometimes called deep litter, can provide heat from the decomposing bedding below. Moisture in the air, though, will be high, so good ventilation is a must. When the bedding is removed, the underlying surface should be treated to a straight vinegar spray to neutralize ammonia fumes or a dusting with hydrated lime powder before adding fresh bedding. Wear a dust mask when spreading lime, because it can be irritating to the respiratory system. Keep the animals out of the barn if possible until the lime dust settles and either soaks up a bit of moisture from the floor or is covered with other bedding. Note that

Figure 3.13. This raised slatted floor (*left*) at Monte Azul cheese dairy and resort in Costa Rica helps keep goats clean and greatly reduces bedding needs. At Pholia Farm, a retrofit of our barn included movable sections of raised flooring (*right*) that offer the same benefits.

organic regulations do not allow use of hydrated lime for this purpose.

Goats love to sleep on wooden platforms, and those made of spaced lumber, like an outdoor deck, are just as appealing to them. The manure and urine falls, for the most part, between the slats, keeping the goat cleaner and drier and eliminating the need for bedding. Most raised flooring is built over a concrete surface that allows the manure to be removed, usually by water or scrapers. Raised slatted floors are more typical in tropical or subtropical climates—I hadn't seen any here in Oregon, where winter temperatures can drop to well below freezing (of course other parts of the United States are even colder). Where winters are severe, cold air infiltrating under the decking is a concern because animals can become chilled.

Since our barn was already built, adding raised slatted floors would have meant a major rebuild. Instead, I decided to experiment with installing decking raised only 6 inches (15 cm) off the ground. I built it in sections sturdy enough to stay in place but light enough to be periodically tipped up for cleaning. I used pressure-treated boards for the parts that touched the ground and regular 2×4 lumber for the top slats, spacing them ½ to ⅝ inch (1.3–1.6 cm) apart. There is minimal draft under these low floors, and the manure below creates some heat. Our goats seem to be quite comfortable standing or lying on them even during our coldest of nights, when temperatures drop to the low 20s F (about -5°C). The initial cost of materials (and my time) was moderately high, but over the years the lack of need for bedding, the incredible time savings on labor devoted to waste removal, and the reduction of volume of waste that needs to be managed have all made the investment well worthwhile. In fact, we reduced our bedding use to the point that we no longer need our large composting area (described later in the chapter).

If possible, a barn should include other spaces for special housing needs. One of the most important is an isolation or hospital pen where a sick goat can be taken care of and kept away from the stress of the main herd and also where new animals can be quarantined. Depending on how you manage kidding, you may also want special maternity areas and kid pens (see chapter 8 for more about this).

Feeders

Feeding goats well is a critical part of holistic goat care. In fact, it's so important that I devote two chapters to the topic—chapters 4 and 5. Those chapters provide in-depth information on goat nutrition and the types of feeds that are good for goats, including pasture, hay, and grains. Here the topic is feeders, the logistics of the structures you'll set up for providing access to hay and other dry feed. When designing, building, or buying feeders, keep the following in mind: safety, feed cleanliness, feed waste, and accessibility for workers.

Feeders on goat farms are often inefficient and wasteful. The wily nature of goats makes it difficult

Figure 3.14. Hay feeders for goats present challenges for both the safety of the goat and the retention of the feed, thanks to the acrobatic and determined nature of caprines.

Figure 3.15. A slanted bar feeder helps keep goats from pulling feed out. This one is on the farm of Fran Brown, Oregon.

to design the perfect hay and forage feeder. Goats are amazingly ingenious at finding or creating ways to squeeze into a feeder—and their inventive contortions can result in injuries. When herds include young goats, the challenge is even greater, because kids can squeeze into spots an adult cannot.

Keyhole feeders, or any types where the goat must raise their head to access or exit the feeder, prevent feed from being wasted or dirtied. Such a feeder works fine when it has more openings than there are goats and the animals using the feeder are all part of a peer group. The downside is that injuries can occur when a goat who has their head in the feeder is rammed from the side by another goat.

Slanted bar feeders work well for a group of large, mature goats and when you're offering short, chopped feeds. The drawback is that smaller goats may attempt to shove their way through the bars. Also, goats can pull long-stemmed hay through

Figure 3.16. Feeding goats in a wide alley accessible by a tractor or cart is an efficient way to manage dry forage. This alley feeder is at Hill Farm Dairy, Somerset, England.

Figure 3.17. At Pholia Farm this alley between feeders set up with welded wire fencing panels allows us to access the feeders without entering the pens. The design also keeps excess feed from being pulled out onto the ground, and allows a munching goat to see when another, more dominant, goat is headed her way.

the bars, dropping some, which is wasteful. When this style of feeder is placed along an alley, it provides easy access for workers.

Freestanding rack feeders are versatile because they can be moved and repositioned as needed. They tend to lead to a great deal of waste, especially if tall goats bypass the rack and pull the hay directly from the top, so they should be designed with a lid or should be tall enough to prevent access from the top. You may find it awkward to get to refill one that's located in the middle of a pen. Young animals in the pen are likely to climb or jump into the top of the feeder, putting them at risk of becoming stuck or injured, and of course soiling the feed.

Welded fencing panels work well for creating both temporary and permanent feeders for goats. If you use welded wire for fencing, you are likely to have a few pieces lying about, so it's convenient and economical to use these to construct feeders. Pieces of paneling can be connected to the inside or outside of an existing fence using zip ties, hog rings, wire, or twine to create a V-shaped receptacle for hay.

Such feeders are more or less efficient, depending on the size of the openings in the panel. If the feeder is designed with open ends, goats will pull hay through the openings and even climb into the feeder (often with unfortunate results). Panels can also be used to form a bit of a barrier that goats can still put their heads through. This can help limit hay loss a bit and also allows an animal a chance to see if a more aggressive goat is about to ram her.

Water and Mineral Stations

Water and mineral stations don't get the respect they deserve on some farms. We all know goats need them, but it is rare to see as much attention to placement, maintenance, and availability as you will with feed stations. If you want your goats to have healthy, productive, long lives, then clean, pure water and access to the right minerals is critical (I cover the topic of choosing minerals in chapter 4). So even if you are an experienced goat farmer, it won't hurt to occasionally reassess these systems on your farm.

It is crucial to provide enough water stations to accommodate the number of animals present and place them so that even the lowest-ranking herd member can access the trough without stress. It's hard to give an actual number, but the smaller the waterers, the more you will need. For example, if a water nipple is the source, there should be at least one for every 10 to 20 goats. If the water source is a large trough, one should be adequate for 20 to 40 goats, providing it's easy to access.

Positioning a trough or water nipple for easy access—and easy escape—is important. Place the waterer where goats that are further down the herd pecking order can get to it without being cornered by

Figure 3.18. A water nipple, more commonly used for dogs and hogs (and often found through suppliers of equipment for these animals), can work great in a temperate climate where freezing isn't a concern. When water is supplied this way, it ensures that each sip of water is clean.

Figure 3.19. Heated water troughs help prevent water from freezing. The heater for this setup is inside the red housing atop the trough. The mildly warmed water is more palatable to the goats, too, which helps keep water intake high. This trough is ready to be cleaned of the bits of floating hay.

a pushing herdmate. Weather conditions are also a factor when you're figuring out how many waterers you need. When the weather is hot, goats will drink more water. When it's cold, consumption will go down naturally. Thus, be sure to plan for sufficient access even at the hottest time of year.

During cold weather, goats may have trouble getting enough water, not because of too few water stations but because the water is too cold or the water surface is frozen. If you live in an area with cold winters, you'll need to design your system so the water is in a protected area. You may need to use

heaters, or otherwise supply warmer water, both to prevent freezing and to increase palatability so that the animals will drink their fill. Water nipples, such as the type designed for dog kennels or hogs, can be readily used for goat waterers, but only in areas where freezing of water supply pipes or the nipple itself is never a concern. Wrapping pipes with heat tape can help to prevent problems if temperatures drop to freezing.

Clean water is more flavorful for the goats, which encourages them to drink their fill. Equally important, dirty water is a health problem—disease can spread

Figure 3.20. Using a repurposed ice chest as a water trough helps keep water clean and cool in the summer, and unfrozen in the winter.

through contaminated water. (A bit of natural biofilm and a bit of algae isn't usually a problem, but manure and other debris is.) Whatever type of waterer you use, keeping the water clean can be quite a challenge. Even water troughs placed outside a goat pen can somehow end up the target for a pooping goat. All troughs should be regularly cleaned. It's not a bad idea to follow the cleaning with a sanitizing step using something like chlorine bleach, boiling water, or other food contact sanitizer diluted to the proper strength. This is especially helpful if some parts of the water container cannot be scrubbed, such as ball float valves in automatic waterers. Another tactic for helping keep water clean is to clip the beards—which serve no helpful purpose—from all of your goats so that the beard doesn't soak in the water as an animal drinks.

The best way to ensure clean water is to place water troughs so that they are easy to inspect several times a day. If they are easy to inspect, they will also be easier to clean.

If you raise fiber goats, take care to design waterers so that their necks stay dry, which helps maintain the quality of their fleece. Water nipples are great choices for this.

Goats prefer to drink water that is not too cold or too hot. On a cold day, goats will relish water that is quite warm—warmer than their body temperature but not scalding hot. In hot weather, cool but not icy-cold water encourages consumption. At Pholia, our cool well water refills the automatic waterers in the summer; in the winter we fill a separate trough with hot water for the does to drink after being milked. Finding ways to support these preferences isn't always practical, but it's good to aspire to, for the sake of reduced stress and a healthy level of water consumption. And when you watch your goats enjoying water that's just right, it makes you feel good, too.

Stations for minerals, salt, and baking soda should be easy to access and easy to clean, just like water stations. Having more stations is usually better than

Figure 3.21. Mineral feeders made of PVC piping are a popular way to provide a supply of loose minerals to goats. The white PVC has become discolored from contact with the goats' muzzles, but that doesn't deter goats from using the feeder.

Figure 3.22. This mineral station is sheltered from the weather and from the shenanigans of goats at Lookout Point Ranch, Oregon.

having too few. (The importance of minerals, salt, and baking soda for goat digestion and overall health is explained in detail in chapter 4.) As a rule of thumb, plan to set up from half as many mineral stations as you have water stations, on up to an equal number. It's a good idea to periodically observe your herd to confirm whether all of your goats are getting a turn at a station, and add more if needed. Keep in mind that even if your herd composition remains the same, the dynamics can change so that goats that used to have higher status and easy access can suddenly be excluded. I especially see this happen as some of the herd members age.

One of the biggest mistakes made in regard to mineral feeders is overfilling them. Intake will drop, simply because the supplements will become stale or dirty. You may get a false sense of security if it seems like your goats aren't bothering to visit mineral stations, but the reality is likely that the goats aren't eating the supplements simply because they don't find them palatable. As with water troughs, place mineral stations where you can easily inspect and refill them. Put out small portions that will stay fresh until they're used up. I recommend putting out only enough to last two to four days. If your goats aren't consuming that amount in that time period, try putting even less out.

Identification Options

Visitors at many small to midsized goat farms are often surprised that the owner can identify most of the goats by sight—even if all the animals are the same color. For most farmers, it is also practical to have another form of ID for the goats, one that doesn't rely on recognizing animals visually. This is helpful when others must identify an animal, such as for milk testing, veterinary work, shipment, or sales.

ID systems can be temporary or permanent. Temporary methods include name tags and numbers on a collar or an ear tag. Permanent ID includes tattooing or microchipping. Microchipping is uncommon in goats, but when a chip is used, it's injected into the goat's tail webbing (unlike dogs and cats where the chip is injected into the neck skin).

Regulatory Issues for Goat ID

Many states have regulations that require a method of permanent identification for any goats (and other livestock that is commonly used as food for humans) being transported into that state. Most of these regulations have to do with concerns with the contagious and zoonotic reportable disease scrapie (transmissible spongiform encephalopathy). ID methods include a tattoo that can be matched with a registration or other official document from a recognized organization, a microchip, and an ear tag with a number issued by the US government. If you plan on selling animals out of state or country, it is a good idea to consider how you will deal with possible regulations—before you are caught off guard and potentially lose valuable sales. A link to a state-by-state list of identification requirements can be found in appendix D.

Figure 3.23. Ear tags are the ID of choice for goats on extensive management farms. The tag must be placed in the center of the ear, not too close to the edge, where it could be ripped out. *Photo by Willee Cole/ BigStock.com*

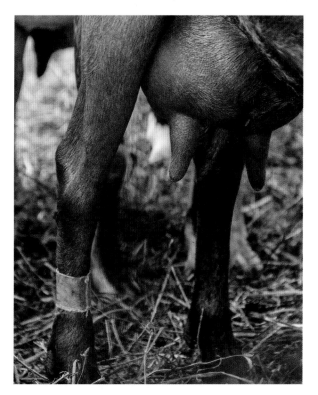

Figure 3.24. Velcro leg bands aren't used as permanent ID but can help mark a goat that needs special attention, such as for medication or withholding milk when medications have been given.

(See chapter 9 for more about tattooing and identification for kids.)

Plastic numbered tags on collars are easy to use but also easy for goats to remove. They become a great target of oral investigation. Tags can also get caught in feeders and fences, and when the goat tugs, the tag may be yanked off the collar, or the collar itself may snap. Goats can even slip out of their collars entirely. Collars made of nylon webbing, such as those that dogs wear, are not as easy for goats to remove as chain-type collars are, but web-type collars can also be a hazard if they catch on a protrusion—leaving the goat dangling. Once, a very young goat at Pholia Farm jumped off a crate, and as she descended, her collar caught on the crate's upper latch. She hung there and might have died had I not been standing nearby and able to quickly rescue her. Goats are just good at getting into predicaments. I prefer plastic chain collars because they're well priced, it's easy to select colors for color coding, they will break if a goat is hung up, and they don't leave an indentation on the goat's neck. Colored plastic chain for making collars can be purchased online in a huge selection of hues.

Ear tags are commonly used for meat and fiber goats but rarely for dairy goats because of aesthetic concerns (as many dairy goats are also show goats) and the risk that one goat may pull out another's tag in the more confined environment of a dairy. The USDA has accepted the unique tattoo that is given to most dairy goats as a form of official identification. For the tattoo to be accepted by the USDA (and other countries when exporting from the US), it must be identical to that noted on an official registration or recordation paper. These papers are provided when the goat is entered into the herdbooks of the American Dairy Goat Association or another officially recognized registry.

Protection from Predators

Very few locations are free from concerns about loss of animals to predators. Goat predators include roaming domestic dogs, coyotes, bears, feral hogs, lynx, bobcats, mountain lions, and even eagles (which usually attempt to take only kids). In some parts of the western United States, including the county where I live, and in parts of Europe, wolf packs are returning and are becoming a potential predator for ranging livestock of all kinds.

Figure 3.25. Mother goats that naturally raise their young practice "planting." They leave their kids tucked away in a sheltered spot and go off to browse and graze. This behavior must be considered when predator issues or other threats are present. These Boer kids are planted near the shelter of an old stump at Sospiro Goat Ranch in North Carolina.

Protecting Predators

Most of us focus solely on protecting our livestock from predation, but predators are an important part of ecosystems. A holistic farm must learn to balance its needs with those of the biosystem of which it is a part. For some producers, especially those with large, biodiverse farms, it might be helpful to become Certified Predator Friendly. This program ensures that farmers have the information and knowledge to coexist with predators and provides a stamp of approval that might be useful from the standpoint of marketing and sharing your vision of sustainable farming. For more information visit www.predatorfriendly.org.

Be sure to fully analyze this risk before you purchase goats. The form of predator protection you choose will depend completely upon the threat and what is manageable for your property and neighborhood. Choices include containment, livestock guardian dogs, and guardian llamas and donkeys.

Containment

One of the best ways to keep goats safe is to make it impossible for a predator to reach them. Just what these security measures look like will depend on the type of predator threat. Roaming dogs and coyotes will go over and under fences, and they prowl at all times of the day and night. Containment against this threat might consist of adding a strand of electric fence at the bottom outside the fence and another along the top—even jutting outward from the top of the perimeter fence. This makes it more difficult for a predator to jump over the fence.

Bears, at least common black bears, are much more opportunistic. They rarely invest a lot of effort into breaking through fencing. If bears are the risk, your strategy should be to prevent easy access to the goats and to prevent the goats from escaping. Mountain lions are typically dawn and dusk predators; to help avoid them, keep goats closed inside a barn or covered pen from sundown to sunrise. Lions have been known to easily clear tall fences with prey in their jaws, so if you decide to rely on fencing alone to protect the animals, it must be quite tall—at least 6 feet (1.8 m)—and should have an electric wire, set to a strong shock, along the top.

Livestock Guardian Dogs

Several breeds of livestock guardian dogs (LGDs) have been bred to live with and guard sheep and other livestock from a variety of predators. The most common breeds date back hundreds of years and include Pyrenees, Komondor, Maremma, Akbash, Anatolian, and Tibetan mastiff. These are not herding dogs—they will not help move animals from paddock to paddock—but instead have a natural instinct to protect whatever animals they live with.

In places where large predators, such as mountain lions, are common, a good livestock guardian dog

Figure 3.26. One of our livestock guardian dogs, Milo—a Pyrenees Maremma cross—stays with the herd when they are out free ranging.

is worth its weight in gold. Working with livestock guardian dogs requires a different mind-set than does typical dog training, because LGDs eventually need to think independently of their owners. This involves a very different approach and set of expectations from puppyhood. If you decide that LGDs are the right choice for protecting your herd, it is crucial that you learn a great deal about working with them before you acquire your first one. In fact, a good breeder of LGDs will be unlikely to sell you a puppy if you are not experienced with such breeds. Don't be offended if that is the case. The book I learned from is called *Livestock Protection Dogs: Selection, Care, and Training.* (See appendix D for this and other sources of LGD information.)

Caution: In areas where wolves are a potential or real risk, an LGD may also be at great risk. Wolves are themselves canines and will usually intentionally attack, punish, and mentally destroy an otherwise great working dog.

Guardian Llamas and Donkeys

Llamas (not to be confused with their smaller relatives, alpacas) and standard-sized donkeys can help

Figure 3.27. This guardian llama is on duty at Mountain Lodge Farm in Washington State.

protect against losses from small predators, such as dogs and coyotes. Llamas' ability to help is due to their natural curiosity and their size—most can meet the average-sized human eye-to-eye. When a disturbance occurs, such as a person approaching a fence at a spot where humans don't usually appear, most llamas will come charging, their necks stretched out and their long coats billowing. The boldness of this behavior is often enough to intimidate all but the most desperate predator. If confronted with an extremely aggressive large predator, however, llamas are likely to be the first victims. They can be used in conjunction with an LGD to help provide comprehensive coverage (sounds like car insurance).

Donkeys are naturally curious animals, too. They also have some impressive defensive abilities, including biting and very accurate kicking. Neither llamas nor donkeys are likely to bond as closely with the herd as a dog will, but the goats might sense their protective abilities and stay near, especially when danger threatens. Only females or castrated males should be used, as intact male llamas and donkeys can become quite aggressive with the goats, especially during breeding season. An advantage to both of these guardian animals is that they can usually eat all or most of the same feeds that the goats are eating.

Manure Management

Anticipating the ongoing need for management of manure, bedding, and feed waste is critical in order to efficiently keep the barn environment healthy and prevent high nitrogen runoff, which can negatively affect the land—and subsequently the animals'

Figure 3.28. Windrow composting is a way to store manure and other wastes, allow them to break down, and then spread them on crops and fields with minimum labor and time. This windrow is at Prairie Fruits Farm, Illinois.

health. The amount you need to manage will depend on how your herd is managed, the climate of your region, your approach to pen maintenance, and the quality of your feed and how you offer it. Herd goats that spend a significant part of their day on pasture or browse will generate less manure in their pens because they'll spread it on the land during the day as they move about. In regions with less goat-friendly weather, more bedding is likely to be needed, which in turn increases the amount of waste generated. As noted earlier in the chapter, good feeders help reduce feed waste. Of course, the higher quality the feed, the less picky and wasteful the goats will be.

Manure waste can be sold or given away to gardeners and crop farmers or spread on your own land. You'll probably need to pile and store the material for some period of time, and ideally you'll compost it during that time. The more bedding and feed waste that's present, the more it will benefit from composting. At our farm we sometimes spread uncomposted waste directly on our land, and other times we utilize a large compost pile. (Or we did until very recently when the addition of raised slatted flooring greatly reduced the amount of bedding we use.) The uncomposted matter makes great weed prevention mulch between garden rows and around trees. Over time it breaks down and can be worked into the soil the next season. Our large compost area serves as a place to stockpile waste when not needed elsewhere or when the weather doesn't allow for spreading.

To compost manure and bedding waste properly requires that the pile have the right amount of moisture and balance of carbon-rich (bedding) and nitrogen-rich (manure) organic materials. I'm a fan of low management needs for composting, simply because there isn't time on our dairy for more chores. We built our compost pile by first forming a ring about 16 feet (5 m) in diameter using horse stall panels and one gate panel. We lined this with chicken wire to hold the bedding in place. We pile up the bedding as it is cleaned from the pens, wet it with a sprinkler or allow a convenient rainstorm to do the work, and then keep it covered with heavy black tarps (I like the kind used by big-rig trucks, which I buy from a custom tarp company that sells them used). Covering the pile accomplishes two things: We keep additional rain from washing nutrients out of the pile, and we keep in the heat generated while the nutrients break down. The pile is usually full by the end of winter; by the fall it has broken down to one-eighth of its original volume. That's when we spread it on the fields.

Be sure to plan for manure management before designing your farm, and be ready to adapt a new system as your herd grows or shrinks. Even simple things such as locating piles downhill from the barn—so that transporting waste is easier and runoff doesn't contaminate waterways—must be considered. (See appendix D for sources of more information on these topics.)

Mortality Management

Every goat farm will have to deal with mortalities, both unexpected and planned. It's important to know how to deal with this issue both from a practical standpoint and for the sake of herd health and safety (including not attracting predators). In 2012, according to the USDA, mortality rates for both meat and dairy operations in the United States averaged 13.8 percent for kids and 7.2 percent for adult goats.[2] These numbers included goats that died from predation or illness or were euthanized. Commercial meat goat operations should plan on losing 10 percent of kids and 5 percent of adults.[3] Dairy goat breeders who don't run intensive operations typically expect a 5 percent loss of kids (not including stillborn kids) and 2 percent adult losses. On our farm we average about a 2 percent loss of kids and less than 1 percent of adults from the working herd. We also have a small retired herd, but we don't factor the death of those animals into the losses—whether they pass away naturally or I have to provide that service to end suffering.

In years past, a rendering plant (where carcasses are processed into products such as tallow, lard, and bonemeal) was the farmer's usual choice for dealing with carcass management; it's what we used when I was young. Very few rendering plants still exist, and those that do may not accept cattle, sheep, or goat carcasses because of concerns over zoonotic diseases such as scrapie and bovine spongiform encephalitis (mad cow disease). The other legal options are landfilling, burning, burying, and composting. Landfills don't always accept animal remains, and those that do may charge a fee. And I consider it a waste to send to a dump what could be returned to the earth as valuable nutrients.

Burning a carcass is not a simple task—it must be done in a proper incineration facility for both health and nuisance reasons. Only very large-scale farmers are likely to make this kind of investment. Burying is an option for many farmers as long as they have the right equipment and ample land. Often local ordinances will dictate where and how animals can be buried on your property. Regulations may stipulate minimum setbacks from waterways and property lines, for example. There are also depth regulations, because burying either too shallowly or too deeply can create problems. In the United States you can contact your local extension office regarding local and state regulations for burying mortalities.

Composting is becoming the most effective and economical choice for mortality management. It may sound a bit unsavory, but the conversion of an animal's body into compost can be clean, efficient, and economical, and it allows for the return of nutrients to the farm. At Pholia, we bury adult goats, but we compost stillborn and kid mortalities. We utilize the same pile where we compost bedding waste and manure. Compost setups for goat carcasses must be built properly to ensure the rapid breakdown of the animal without undesirable odors or attracting vermin or predators. Regulations regarding composting of carcasses must be followed if they apply. (See appendix D for links to websites that offer more details.)

CHAPTER FOUR

Understanding Nutritional Needs

Other than genetics and environment, the most important factor influencing a goat's health is correct nutrition. The official word (for English speakers anyway) on what a goat needs to consume is *Nutrient Requirements of Small Ruminants*. This academic, research-based volume is produced through the auspices of the National Research Council, and it is the bible of goat nutrition. Veterinarians, nutrition specialists, farmers, writers, and others rely on it to obtain reliable information. It was one of my primary references as I wrote this chapter. However, not everything is known regarding goat nutrition. This should really come as no surprise, given the fact that

even for humans, what's healthy to eat and what's not continues to change as scientists make new discoveries.

If even the experts aren't certain, how can producers feel comfortable making dietary choices for their herds? First we can learn what is known, theorized, or believed at the moment, and second, and more important, we can focus on goats' natural attributes and physiology and apply what we do know about these traits to feed choices (discussed in detail in chapter 5). To begin, let's look at how nature has designed the goat for a specific type of diet, and then relate that to mineral and vitamin requirements. You'll learn how all of these topics are intertwined and interdependent in a delicate balance that supports every aspect of the animals' health and productivity.

Understanding the Ruminant

Many animals are herbivores, designed to eat only plant material. But not all herbivores are equal when it comes to digestion. For example, even though goats and horses eat similar foods, such as grain, hay, and grass, they process those foods in different fashion. A horse has one stomach and is called a monogastric, but a goat is a ruminant—with a multi-chambered "stomach." In addition, not all ruminants are created equal—sheep, cows, and goats all have digestive

Figure 4.1. Fresh browse and graze offer vitamins, minerals, exposure to sunshine, and exercise, all important aspects of goat health.

systems consisting of four compartments, but the systems work somewhat differently because they are designed for different types of diets. Goats are the most versatile. I call them "growsers" (grazers/ browsers) because they consume a variety of plant material, including browse, roughage, and grass. They are considered an intermediate type between cows and sheep in one group and animals such as deer and elk in the other. Let's start by taking a chamber-by-chamber look at the upper digestive tract in order to understand how goats use their feed to meet their nutritional requirements.

Mouth

A goat's muzzle and mouth are structured perfectly to selectively choose leaves, stems, and seeds and nip them from a plant. Compared with the muzzle of a sheep, the goat's is more pointed, just for that purpose. Cows and sheep press their muzzles parallel to the ground and tear off grasses. All ruminants lack top front teeth; instead they have a thick, tough area called a dental pad, which is designed to protect the gums from the sharp lower front teeth as they bite, as shown in figure 4.2. Scissor-like molars at the back of the mouth make the tough cuts and are also important for chewing. Goats use their sensitive lips to select the desired forage and then work the plant material back toward the molars. More tender nibbles are cut off the plant using the sharp lower front teeth. As the goat ages, its teeth show increasing wear and spacing, as shown in figures 4.2 through 4.6.

Once food is inside the mouth, a goat chews rapidly, but only partially, and then swallows. You can think of the first ingestion as the goat simply collecting, compressing, and storing the feed for processing later. This allows it to quickly collect feed when the opportunity arises. In fact, it is this proficiency that leads many to believe that goats can process poorer-quality feed more efficiently than do sheep—it's really about efficient foraging. A goat that doesn't have a properly formed muzzle—for example, one that is too narrow or too broad—or properly placed and healthy teeth will be less than successful as a competitive browser.

Figure 4.2. Between the ages of one and two, a goat's first two permanent incisors appear, replacing the small, sharp baby teeth.

Figure 4.3. By age two, four permanent incisors are present (note the difference in depth where they emerge from the gums compared with the others).

Figure 4.4. By age four, all eight permanent incisors are present.

Figure 4.5. This 11-year-old doe has two broken incisors, which makes foraging more difficult.

Figure 4.6. This 14-year-old doe has kept all of her incisors. Note the wear and spacing of the teeth.

The Rumen, the Reticulum, and Rumination

The first two compartments of the goat's upper digestive system are the rumen and the reticulum, more properly called the reticulorumen, because the two compartments are only partially separated and work together as a unit. They differ in the texture of their lining. The rumen has little finger-like protrusions called papillae, which have a great capacity to absorb nutrients, and the reticulum has a distinctive honeycomb pattern and serves to gather reticulum contents and help push them upward, either to be chewed as cud or to move on to the omasum.

The reticulorumen occupies almost the full left half of the abdominal cavity in the adult. Newborn goats, on the other hand, have a very small, non-functioning reticulorumen. Milk bypasses the non-functioning organ thanks to a reflex, stimulated by suckling and by contact with milk proteins, that forms a tube-like diversion called the esophageal fold or groove, which lets the milk flow directly into the fourth compartment, the abomasum. By about 3 weeks of age, the rumen begins functioning, and by 12 weeks it will have reached close to full proportions and functionality. (For more about the development of a young goat's rumen and its special feeding needs, see chapter 9.)

The reticulorumen is essentially a large, continuously functioning fermentation container. Inside, a vast microbial population, which includes bacteria,

Figure 4.7. A goat's muzzle (*left*) is narrower and more pointed than a sheep's muzzle (*right*), which is designed more for eating grasses than leafy browse.

yeasts, fungi, and protozoans, is responsible for breaking down the otherwise indigestible parts of the plant. To assist with this, the goat regularly regurgitates wads of rumen contents and rechews them. The process of chewing cud, called rumination, not only breaks down the plant material but also mixes in a great deal of saliva. The saliva is important for several reasons, including keeping the rumen acid pH level from getting too high—a condition called *acidosis*. The goat's saliva, with a pH of about 8.1, contains many buffers—compounds or substances that can absorb and neutralize the excess acid produced by fermentation.[1] When a goat stops ruminating, due to illness or other problem, acidosis is a risk, because of both the loss of saliva and the likelihood of a change in rumen bacteria. Offering free-choice baking soda or other buffering substances to your goats is a good practice. They will consume some of the buffer when needed to keep their rumens pH balance.

> The goat's saliva, with a pH of about 8.1, contains many buffers—compounds or substances that can absorb and neutralize excess acid produced by fermentation.

In addition to buffers and some other nutrients, goat saliva contains mucin, a type of protein capable of binding with tannins—which are found in oak leaves, acorns, and many other plants. Tannins can prove toxic to sheep and cattle, but because of mucin and their proportionately larger salivary glands, goats can safely dine on high-tannin plants.[2]

Rumen bacteria also produce many of the vitamins the goat needs in order to thrive, such as B vitamins and vitamin K. A goat's diet greatly influences the balance of bacterial types and the health of the bacteria in the rumen. A diet high in grain can produce too much acid. This is because grains ferment more rapidly than fiber and have a higher sugar content, meaning more acid is quickly produced. When this happens the goat's buffering system can't keep up. This results in a change of the type of bacteria that inhabit the rumen, with

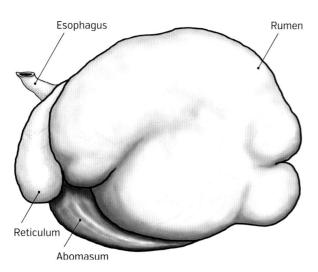

Figure 4.8. A view of the goat's stomachs from the exterior left side.

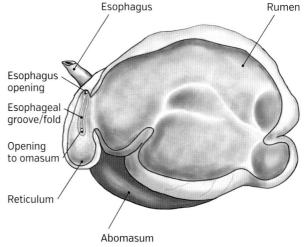

Figure 4.9. A cross section of the reticulorumen as viewed from the left side. Note the honeycomb pattern of the reticulum and the esophageal groove that closes in the nursing kid, diverting milk to the abomasum.

more acid-tolerant varieties thriving. The result is chronic or acute acidosis. Although scientists are still not fully in agreement about the behavior of rumen and reticulum contents, the current popular theory (and one that makes sense to me) is that when roughage and fibrous feeds first enter the rumen, they float on the surface of the liquid contents, forming what is called a *feed or fiber raft or mat*. This raft is thought to capture small particles of grain and slow their processing. Therefore, it's beneficial for a feed raft to be present in the rumen before a goat ingests grain, so that the fermentation of the grain will occur more slowly—optimizing its nutrients and reducing the risk of acidosis. About 10 to 20 minutes of access to forage before you feed grain is adequate.

The wall of the rumen and reticulum moves in an undulating fashion and returns a wad, or *bolus*, of partially processed feed forward and toward the esophageal fold. A contraction then pushes the bolus back up into the mouth (along with a belch of rumen gases, which aren't too pleasant to be near) for more processing through chewing and the addition of saliva. When the goat swallows the bolus, it's diverted to the omasum by the esophageal groove. It's estimated that healthy goats will chew their cud for almost eight hours per day, with most of the rumination occurring during the night. Special collars, sometimes called smart collars, are available for cows that monitor rumination time 24 hours a day. This can greatly help you assess animals' overall health, as a decrease in rumination will correlate with decreased health. I would love to see something like this available for goats.

The Omasum

The third compartment, the omasum, is a situated on the right side of the abdominal cavity. The omasum is lined with many small folds or pleats of tissue that absorb water and allow feed particles to move on to the abomasum. The goat's omasum, like the rumen, is smaller than a cow's. The omasum is the least understood of all of the chambers in the ruminant's upper digestive system.

The Abomasum

The last compartment, the abomasum or true stomach, is similar to the stomach of other non-ruminating animals, including people. The abomasum produces hydrochloric acid and enzymes that further break down nutrients. In adult goats, the abomasum is proportionately larger than the stomach of a cow, partially because the wall is thicker. It is also thought likely that more digestion takes place here, compared with the cow's digestive process.

Energy: The Primary Nutritional Need

Every living creature has its own built-in power plant. This power plant creates heat to keep the body warm, electricity to run the nerves, and power to move the body around, perform maintenance functions (including waste removal), and process more fuel. Energy is provided through the breakdown of starches (carbohydrates), proteins, and fats (lipids). When

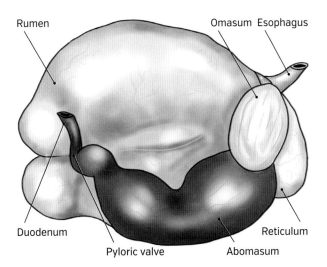

Figure 4.10. A view of the exterior right side of the goat's stomachs.

The Lingo of Labels— Energy

These are some of the more common terms and abbreviations describing energy-producing nutrients that you might encounter when reading feed labels, feed test results, or narrative about feed.

Gross energy (GE): The total amount of energy in feed.

Digestible energy (DE): The amount of energy available to the animal through digestion.

Total digestible nutrients (TDN): All of the digestible nutrients in the feed, including fiber, fat, protein, and carbohydrates.

Metabolized energy (ME): The actual amount of energy utilized that is used by the animal. This is calculated by subtracting the amount of energy from the feed that isn't absorbed (it remains in the feces and urine or is lost in gases) from digestible energy.

The Lingo of Labels— Carbohydrates

These are some of the more common terms and abbreviations describing carbohydrates that you might encounter when reading feed labels, feed test results, or narrative about feed.

Starch: The main carbohydrate in feed, minus fiber.

Water-soluble carbohydrates (WSC): Carbohydrates that can be extracted in water, including simple sugars.

Ethanol-soluble carbohydrates (ESC): Carbohydrates that can be extracted in ethanol, mainly simple sugars and sucrose.

Neutral detergent fiber: Three structural carbohydrates—hemicellulose, cellulose, and lignin.

Acid detergent fiber: Cellulose and lignin.

Lignin: A structural fiber that isn't digested.

people provide food to their livestock, or themselves, the need for energy is usually the first consideration, rather than minerals and vitamins (a bit like when we reach for a chocolate chip cookie instead of a salad). This is quite understandable, because energy needs must be met on a continual basis.

Goats, like other ruminants, are amazingly adaptable power plants. The way they process some feeds and meet their body's needs is quite amazing. Let's go over the three types of substances from which goats obtain fuel and also consider the importance of fiber.

Carbohydrates

Carbohydrates provide most of the energy needs of goats. Carbohydrates are present in plants and in digestible form range from simple and easy to process to complex, also known as structural, and

slower to process carbohydrates. Thanks to the amazing fermentation ability of the rumen, goats can process the structural carbohydrates cellulose and hemicellulose, complex carbohydrates found in great quantity in the leaves and stems of plants. Humans can't digest these carbohydrates but ruminants can. Rumen microbes are able to ferment and break them apart into digestible carbohydrates.

Fiber is an indigestible part of plants. *Lignin*, a strong, structural fiber part of the plant, is indigestible even by goats. Lignin is the strongest of the plant fibers and found in the highest quantities in thick, coarse stems—such as those of woody plants. Because it isn't digestible, the percentage of lignin within a feed is often noted in a feed analysis. It's important to check lignin content in order to decide whether goats will be able to utilize a feed well.

The energy produced in the rumen is first used to feed the microbes (after all, they are doing most of the work), which need it not only to fuel their work of fermenting more starch, but also to process protein, make vitamins, and so on, for their own metabolic needs, including making more microbes. The rest of the energy is then used by the goat. Grains contain more simple carbohydrates, while forages contain more complex carbohydrates. This is why grain can ferment too quickly in the rumen, with the unwanted side effect of producing too much acid.

Protein

Protein processing in the goat is complicated and pretty amazing. As in the case of carbohydrates, a portion of the protein contained in the plants and grains that goats ingest is first used by the microbes. Some of the protein escapes utilization in the rumen, or cannot be processed by the rumen microbes, and continues on through the other compartments. This type of protein is called *rumen bypass protein* and is processed later in the digestive tract by the abomasum and small intestine—or passes out of the goat in the manure.

The microbes in the rumen are themselves on the menu as a major protein source for the goat's body. Microbes that die or simply pass through into the lower compartments provide a direct source of protein for the goat. What a great symbiotic relationship!

Protein that doesn't bypass the rumen and isn't used directly by the rumen microbes is changed in the rumen into ammonia, absorbed into the bloodstream, and transported to the liver. In the liver it's converted into a protein called urea, and then put back in the bloodstream, where it's removed and used as an ingredient of the goat's saliva. During rumination, the goat is not only rechewing the contents of the reticulorumen but also sending recycled protein, as urea, back into the digestive system, where it can be utilized by the body and by rumen microbes. This system of urea recycling is one way goats can continue to meet their protein requirements even when dietary protein is limited. It's also another reason why when goats stop ruminating, their health is in danger.

> During rumination, the goat is not only rechewing the contents of the rumen but also sending recycled protein, as urea, back into the digestive system, where it can be utilized by the body and by rumen microbes.

The Lingo of Labels—Protein

These are some of the more common terms and abbreviations describing protein that you might encounter when reading feed labels, feed test results, or narrative about feed.

Crude protein (CP): Total protein in the feed.
Adjusted crude protein (ACP): Crude protein minus the portion that isn't usable.
Digestible or degradable protein (DP): The amount of protein available for digestion.
Metabolizable protein (MP): The actual amount of protein ingested that is used by the animal. This is calculated by subtracting the amount of protein that isn't absorbed (it remains in the feces and urine or is lost in gases) from digestible protein.

Lipids

Lipids are a large group of compounds that include what we all know as fat, but also other compounds that aren't exactly fats, like fatty acids and waxes. Goats obtain lipids in their feed and use them as an energy source (fat provides more energy than the same amount of protein or carbohydrate can).

Some fatty acids are taken in through diet, while others are manufactured in the goat's body. Both types are stored in the muscles and fat and released in milk. Several of these are called *essential fatty acids*. Humans and goats must ingest these fatty acids through food. The amount of essential fatty acids, specifically *conjugated linoleic acid* (CLA) and *omega-3 fatty acid*, consumed by a goat are greatly influenced by diet. Natural browse and pasture help produce a much higher content of these valuable nutrients. Although fats are an excellent source of energy for the goat, they cannot be effectively added in any significant volume to a goat's diet because they act as an antimicrobial and reduce the rumen microbe population and efficiency.

The Importance of Minerals

We all know that minerals are an important part of what we eat and feed our animals, but just what are they? Simply put, minerals are inorganic (something not associated with a living organism) substances, or compounds, found in nature. As an example, moss is an organic organism that you might find growing on a rock (inorganic), which is made of minerals. Many are necessary for the health and survival of all living (organic) things, including us, goats, and moss. Why are minerals so important? They play many roles in metabolism. In a nutshell, the goat "power plant" that feeds on energy can't complete its job without the assistance of minerals.

Minerals can be divided into two categories. The macro minerals are those needed in relatively large amounts (compared with micro minerals); the trace, or micro, minerals are needed only in very tiny amounts. Macro mineral requirements are usually expressed as a percent of the daily diet or grams per day, because this is a good way to measure something that is needed in larger quantities. Requirements for trace minerals, which are needed in very small amounts, are often expressed as parts per million (ppm) of the daily diet or in milligrams per day.

The macro minerals are calcium, phosphorus, magnesium, sodium, potassium, chloride, and sulfur.

Table 4.1. Suggested Average Macro Mineral Requirements

Mineral	Percent of Diet
Calcium (Ca)	0.3–0.8
Phosphorus (P)	0.25–0.4
Sodium (Na)	0.2
Chloride (Cl)	0.2
Potassium (K)	0.8–2
Sulfur (Su)	0.2–0.32
Magnesium (Mg)	0.18–0.4

Note: The amounts listed in this table are based on averages. Actual needs vary greatly depending on gender, age, parity, production volume, and stage of lactation. Complete requirements based on body weight and other factors are available in *Nutrient Requirements of Small Ruminants* (National Research Council).

Source: Adapted from Steve Hart, *Introduction to Goat Nutrition* (Langston University, Langston, OK), accessed at www2.luresext.edu/goats/training/nutrition.html#macro.

The trace minerals discussed in this chapter are cobalt, copper, iodine, iron, manganese, molybdenum, selenium, and zinc. There are other trace minerals, too, but whether they play a role in health is not known.

Determining Mineral Requirements in Goats

Recommendations for goats' minimum mineral requirements continue to evolve. According to the *Merck Veterinary Manual*, "Requirements for minerals have not been established definitively for goats at either maintenance or production levels."[3] It is thought that goats probably have a greater overall mineral requirement than other livestock thanks to their having evolved as browsers of forbs (leafy plants that are not grasses), shrubs, and trees, whose root systems are better able to draw minerals up from deep in the earth than can grasses and cereal crops. If the best livestock nutritionists have yet to

The Lingo of Labels—Mineral Measurements

If you have tried to read feed and supplement labels, you might have found them difficult to decipher or to compare with other labels. You are not alone! Besides the confusing nature of the abbreviations and acronyms, the way the quantities of minerals are listed varies from label to label. Let's go over some of the most common ways quantities are listed: percentage (%), parts per million (ppm), and milligrams per pound (mg/lb).

The word *percent* is the combination of the words *per* and *cent*. Put that way, it makes simple sense. It is quite literally "for every 100." So percentages always deal with dividing something into 100 equal parts, or *parts per hundred*. This works fine for some substances, but not those that are present in very, very tiny quantities. That's where parts per million (ppm) comes in. In fact, some substances are quantified in parts per billion or trillion, but luckily those don't appear on feed labels. Parts per million, like percentage, is not an actual weight—it's a comparison. In fact, if you like decimal places, you can write ppm as a percentage. It would look like this: 1 ppm = 0.0001%.

Macro minerals will be represented as percentages on feed labels. For example, if a 100-pound (45 kg) bag of minerals contains 1% calcium, then it contains 1 pound (0.45 kg) of that mineral. Nice and simple! Micro, or trace, minerals will be listed as ppm, mg per kg, or mg per pound. (This is an example of the odd time when metric and imperial measurements get together.) Because the amount of the mineral is so small, there is no way to reasonably express it in ounces or pounds. For example, if a 100-pound bag of minerals contains 1 ppm selenium, it would take 10,000 bags (that's 1,000,000 pounds) to supply 1 pound of selenium. In some cases, micronutrients are listed on labels as real measurements, such as 1 mg/pound or kilogram.

Even if this explanation doesn't completely demystify feed and supplement labels, it may help you gain some perspective on the vast differences in the amounts of mineral nutrients needed in a goat's diet and the precision that goes into formulating supplement and feed rations.

---

pin down just what goats' nutritional needs are, then what is the producer to do? I suggest that you start by learning what the current recommendations are and then doing the following:

- Learn the signs of mineral deficiency and toxicity, and constantly observe your herd for signs of problems.
- Investigate your region's known soil mineral deficiencies, as well as those of the source of your feed.
- Be aware that there are likely breed and individual differences in needs for minerals, as well as seasonal and stage-of-life differences.
- Have forage tested for minerals, especially those minerals that may be excessive or deficient. (See appendix D for more on testing.)
- When it's possible or as needed, have a blood or even better a liver sample from one of your goats tested for mineral levels. (If the animal is healthy and lives with the main herd, its test results are likely to reflect the rest of the herd.)
- Consider rotating use of different brands of mixed minerals to provide long-term balance.

Offering Individual Minerals Free-Choice

A few years ago I picked up a book called *A Holistic Vet's Prescription for a Healthy Herd* at an Acres USA conference. When I read the section that promoted the use of a free-choice, or cafeteria-style, mineral feeding program, I was intrigued. I was already offering my herd several individual minerals; namely,

Figure 4.11. Free-choice mineral buffets: At my farm, I use plastic bins (*left*) that I divided by cutting plastic cutting-board sections and hot-gluing them into place. Oats and Ivy Farm in California sets out minerals in small containers (*right*) called quarter pans that are typically used by caterers. *Photo at right courtesy of Amanda Nuñez*

sulfur, copper sulfate, and dolomite (for calcium), in addition to a high-quality complete mineral mix (with a blend of all the macro and micro minerals) and plain salt. I'm always looking for new techniques to experiment with that might help promote herd health, so I ordered the starter goat mineral kit from Advanced Biological Concepts.

While many nutritionists and veterinarians shake their heads or roll their eyes at this approach (and believe me, I've seen it) there are even more producers, both cow and goat, whose anecdotal stories suggest it works. In my opinion it does. I believe in doing your best to prove your approach, and to that end, I annually send a liver sample in for testing to double-check whether my herd is suffering from any deficiencies or toxicities (details on how to take a liver sample are in chapter 7).

KEEPING TRACK OF CONSUMPTION

I document daily what minerals have been consumed and need refilling. I have seen interesting patterns over time. The most obvious one occurs when the goats are on forest browse—their consumption of copper almost ceases. A day or so without browse, and it goes back up. Another friend reports that only one of her goats consumes the copper supplement but has done so consistently for many years. In the late winter, phosphorus consumption among the bred does increases, presumably to provide for the demands of the growing fetuses.

Two other companies also offer some individual minerals: Nutritional Ag Products in Oregon (limited product line and availability) and Free Choice Enterprises in Wisconsin (see appendix D for contact information). Each mineral is supplied in a mix, so it isn't strictly a single mineral. Rather, it is the most prominent one.

If you decide to try this approach, document your goats' intake and note what minerals are used throughout the year. It's likely that you won't need to continually supply all of the minerals, because the animals will ingest a sufficient supply of some through their feed.

COPPER SULFATE– USE WITH EXTREME CAUTION

Taking the free-choice feeding approach a step further, some producers, myself included, have chosen to offer straight copper sulfate instead of a copper mix. For good reason this causes concern on the part of veterinarians. A small amount of copper in this form can cause toxicity and death in an animal. I know people who have lost goats in this manner. Yet many of us have been doing this for years. So what's the secret, and why take the risk? Copper sulfate is relatively easy to source and is inexpensive compared with mixes. Because it doesn't contain salt (sodium chloride), it doesn't attract moisture from the air, so it stays fresh and dry. When first offering copper sulfate to goats, it's important to do the following:

First find out whether you live in a region where soils are naturally high in copper, copper-deficient, or have an excess of a copper antagonist such as we have here in southern Oregon, with high molybdenum levels. Second, it's critical that you already have been providing a good-quality mineral mix that includes the recommended levels of copper. Third, start by putting out only an extremely small amount of copper sulfate, and take care to observe that *every* goat in the herd is testing or noticing it. If any animal consumes more than the tiniest lick you should remove the copper sulfate. The following day, repeat the procedure. Once the transition is made, only a small amount, say 1 tablespoon per 40 goats, should be offered per day. And finally, it's imperative that you confirm liver levels of copper on a periodic basis.

Even when you approach this with caution, you have to understand the risk. No one is sure why some animals will "choose" to consume too much of this extremely unpalatable mineral, but it happens. *Even with the advice I'm offering here, there are no guarantees of safety.*

Mineral Interactions

The most important thing to remember about all minerals is that they are codependent upon other minerals (and vitamins, too) as they play their role in metabolism. Mineral metabolism can also be inhibited by certain minerals, such as molybdenum blocking the absorption of copper. A delicate balance and a precise nutritional dance occurs that can be quite easily upset by too little or too much of any one mineral, as shown in figure 4.12.

With the understanding of the complex interactions of minerals in mind, let's look in more detail at individual macro minerals and micro minerals: their role in metabolism, problems that result from deficiency, and sources of the mineral in different types of feed.

The Macro Minerals

The macro minerals are those usually found in abundance in nature—in feed and in the earth. They are also needed in the greatest quantity in goats' diet. The macro minerals, listed here from those needed in greatest quantity to least, include calcium, phosphorus, sodium, chloride, potassium, sulfur, and magnesium.

CALCIUM (CA)

Bones and teeth contain almost all (99 percent) of the calcium in the goat's body. Most of us think of bones as hard, dead structures, thanks to seeing skeletons and old bones in museums. But in living animals, bones are active structures that are constantly growing and "shrinking," absorbing and releasing minerals. (I explain this in greater detail in the section on milk fever, hypocalcemia, in chapter 10.) Calcium also plays a role in the transmission of nerve signals, muscle contractions, blood vessel function, blood clotting, and hormone function.

Calcium requirements change greatly throughout the goat's life, depending on its occupation, such as putting on muscle, producing milk, or growing fetuses. Fortunately calcium is usually readily available through feed, including from milk consumed by the young goat. Legume feeds, such as alfalfa (lucerne) and clover, are particularly high in calcium, while grass hay and grains are lower. The common feed options and the ratio at which they are used need to be considered when you're choosing a mineral mix for your goats. Most mineral mixes formulated for goats contain a 1:1, 2:1, or 1:0 calcium-to-phosphorus ratio. The first, with equal parts Ca to P, is intended for animals on a high-calcium legume diet, such as dairy goats; the second, with twice as much Ca as P, for those on a light-grain or high-grass forage diet, such as meat and fiber goats; and the last, with only Ca, for those on a very high-grain diet (never recommended, but common in feedlots and concentrated operations).

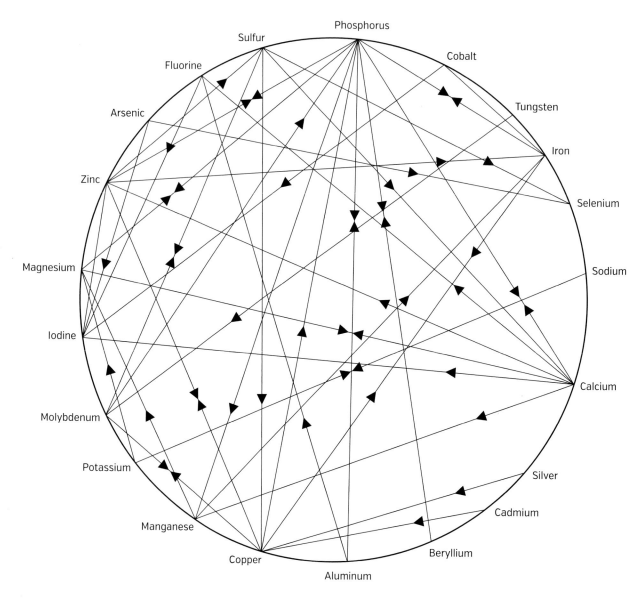

Figure 4.12. In this chart, arrows indicate an interaction with another mineral. The mineral toward which the arrow points is the one affected by the first. For example, sulfur points toward copper. Too much sulfur can cause a deficiency of copper. Some arrows point in two directions, indicating two minerals that can affect each other equally. *Reproduced with permission from Advanced Biological Concepts*

Young animals that are deficient in calcium can experience rickets, which is poor development of bone leading to enlarged joints and bowed legs. Osteomalacia (sometimes called adult rickets) has similar symptoms, with the addition of potential fractures and lameness. Hypocalcemia (milk fever)

in milking animals is also due to a sudden calcium deficiency in the blood. Other mineral and vitamin deficiencies can cause these same disorders, as discussed later in this chapter. How crops are fertilized (many synthetic fertilizers are very high in phosphate, and even manure used as a fertilizer

The Need for Liver Biopsies

If you are concerned about a deficiency or toxicity of copper, or any other mineral for that matter, in your goats, I suggest planning for the harvest of a liver sample from an otherwise healthy member of the herd—one that lives with the main herd and is therefore subject to the same diet and conditions— whenever the opportunity arises; for instance, if a healthy goat dies during delivery of her kids or you decide to butcher a goat that was living with the main herd and in good health. I purposely keep an animal or two just for this purpose, usually a yearling milker that is a disappointment in milk production. Once a year, I harvest the goat for meat and also take a liver sample, which I send to the lab for testing for all pertinent mineral levels (most are stored in the liver). It is a worthy sacrifice for the greater good of the herd. For a detailed description of how to take a liver sample, refer to chapter 7.

can be too high if it's collected from animals being fed too much phosphorus) greatly influences the calcium-to-phosphorus balance in the animal's diet; thus, the feed source needs to be considered when choosing a mineral mix.

Calcium levels can be monitored through a blood serum reading. On feed labels calcium is usually listed by the minimum and maximum percentage— typically as "not less than" and "not more than."

PHOSPHORUS (P)

Like calcium, most of the phosphorus in the goat's body is contained in bones and teeth (about 80 percent of the total), where it supplies structure and strength. It also plays a role in energy metabolism and is contained in saliva, enzymes, cell walls, nerve tissue, and genetic material.

Phosphorus deficiency is the most common goat mineral imbalance and is seen as reduced appetite, suppressed weight gain and growth, bone deformation, fertility issues, pica (cravings that cause an animal to eat things it would otherwise not, such as dirt or bones), and reduced milk production. Goats whose diet contains a major portion of browse and/or legumes typically are not as likely to be phosphorus-deficient as those consuming grass pasture or grass hay diets. Cereal feeds that contain a lot of grain are typically high in phosphorus.

Excess phosphorus can lead to urinary stones (calculi) and softening of the bones (due to the imbalance of calcium to phosphorus). Too much phosphorus in the diet can be influenced by the soil pH where the crops are grown, the use of high-phosphorus fertilizers, and improper manure application or waste runoff that leads to excess phosphorus in feed crops, as explained below.

Phosphorus is the most expensive of all of the minerals, so mineral mixes with high calcium-to-phosphorus ratio (such as 2 parts Ca to 1 part P) may be less expensive than a mix that is 1:1. Excess phosphorus in the diet can lead to manure and urine high in the mineral—a potential environmental concern. For example, the waste from feedlots where animals are on a heavy grain diet can contain more phosphorus than is ideal, leading to a higher level in the manure. Water runoff from this waste can greatly damage the biology of water systems. The same manure used as a fertilizer can lead to higher-than-ideal phosphorus levels in crops.

Phosphorus levels can be monitored through a blood serum reading. On feed labels phosphorus is usually listed as "not less than."

SODIUM (NA) AND CHLORIDE (CL)

Sodium and chloride are two separate minerals but are most commonly available in combination as sodium chloride—or common table salt. While all minerals are "salts," the word *salt* most often refers to sodium chloride, probably thanks to its abundance in nature and the fact that all members of the animal kingdom naturally crave sodium chloride (there's a good reason pretzels and peanuts are served at happy hour).

Sodium is critical to many functions in the goat's body, including transport of glucose (a sugar) and amino acids (proteins) to body cells, nerve conduction, temperature regulation, and muscle activity. Chloride plays a role in the digestion of proteins and the absorption of minerals and amino acids: Chloride is one component of hydro*chloric* acid, an important acid in the stomach. Chloride also balances the osmotic pressure within blood vessels to keep them from leaking, and helps keep the blood pH in balance.

When given free access to salt, goats will regulate their own intake to the proper levels. In the wild, salt (often along with other minerals) is consumed at naturally occurring salt licks, where erosion has exposed mineral deposits or salt has been concentrated by evaporation. Most feeds are naturally low in sodium, unless grown in soil where salt concentrations are high. Because animals naturally seek out salt, it is easily added to the diet via supplements or mixed into rations. Salt is sometimes used in mineral mixes for two opposing reasons: first, to entice the animal to consume a mix containing minerals that are otherwise not tasty, and second, to limit overconsumption of an otherwise toxic mineral needed only in minute quantities.

When salt is available to goats, they will not suffer sodium deficiency. In most types of animals chloride deficiency has not been observed. If salt is unavailable, animals will often experience pica and resort to eating dirt or licking any object that might have a salty residue.

> Salt is sometimes used in mineral mixes for two opposing reasons: first, to entice the animal to consume a mix containing minerals that are otherwise not tasty, and second, to limit overconsumption of a mineral needed only in minute quantities.

Sodium and chloride levels can be monitored through blood serum levels.

POTASSIUM (K)

If you've ever suffered with leg cramps, you may have been told to eat a banana or a sweet potato to increase your potassium levels. Potassium, along with sodium and chloride, serves as an electrolyte in the body. Electrolytes play a role in many functions within the body, including osmotic pressure, blood pH balance, nerve impulses, fluid balance within the cells, muscle contraction, and hormone release.

Unlike some other minerals, potassium is not stored in the body and must be consumed daily to maintain the proper supplies. Fortunately most feeds, especially leafy greens, have adequate potassium content. Winter squash also is a good source (incidentally, goats also like bananas and their peels). Diets that are too high in grain (which is lower in potassium than forages), or forages grown in soils that are low in potassium, may lead to deficiencies in the goat. Signs include reduced growth, reduced appetite, stiffness, weakness, pica, and even muscle paralysis.

Higher-than-ideal potassium levels in feed lead to excess potassium in the manure. If that manure is spread on agricultural fields, crops grown there may end up with high potassium levels. While excess potassium in the goat's diet doesn't typically lead to health problems, it might be linked to grass tetany (see the Magnesium section on page 86) and milk fever (hypocalcemia).

Potassium levels can be monitored through blood serum readings.

SULFUR (S)

Sulfur is an essential part of many amino acids. Hair (which is made of protein) has a very high percentage of amino acids containing sulfur (that's one

Michigan State University

DCPAH

Diagnostic Center for Population and Animal Health

Report of Laboratory Examination

Owner:
Ginnie Doe

Rcvd Date:	10/16/2015 1:40 pm	**Animal:**	NONAME	**Breed:**	Goat, other
Admitted By:	Buck, Dr.	**Species:**	Caprine	**Gender:**	Female
Ordered By:	N/A	**Age:**	20 months		
Encounter:	01234567				
CR#:	AP				

MINERALS

Collected Date/Time (If Provided)	10/04/2015 13:41:00		
Procedure		**Ref Range**	**Units**
Specimen Tissue	Liver		
Dry Weight Fraction*	0.349H	[0.260–0.340]	
Arsenic, Tissue	<0.07	[<=9.00]	µg/g dry
Lead, Tissue	<0.07	[<=3.00]	µg/g dry
Mercury, Tissue	<0.36		µg/g dry
Thallium, Tissue	<0.07		µg/g dry
Cadmium, Tissue	0.08		µg/g dry
Selenium, Tissue	1.29	[0.70–5.00]	µg/g dry
Iron, Tissue	126.86	[125.00–800.00]	µg/g dry
Copper, Tissue	378.51	[20.00–650.00]	µg/g dry
Zinc, Tissue	101.92	[90.00–370.00]	µg/g dry
Molybdenum, Tissue	4.27	[2.40–5.90]	µg/g dry
Manganese, Tissue	7.83	[5.80–23.00]	µg/g dry
Cobalt, Tissue	0.29	[0.11–0.71]	µg/g dry

Figure 4.13. These lab results for a liver sample of a two-year-old doe show normal ranges for important minerals, although her iron level is on the low side of normal. When having liver samples tested, it's important to find a lab that uses reference values based on ideals for goats, not sheep or cattle.

reason hair stinks when it burns). For fiber goat producers, sulfur is an important mineral for ensuring high-quality fleece. Sulfur is also used by many of the microbes within the goat's rumen.

Deficiencies might be seen as hair loss and reduced body condition and growth. High levels of sulfur can lead to copper and selenium deficiencies (sulfur interferes with the absorption and utilization of both of these trace minerals). Sulfur is acidic—it's even used to lower the pH (raise the acid level) of soils to grow plants that like acid, such as blueberries. Because of its acidity, an excess of sulfur, often from high-sulfur feeds such as grain, in the rumen can make the acid level too high and lead to an imbalance that is linked to many problems, including polioencephalomalacia, also called goat polio. (See the Polioencephalomalacia entry on page 300 for more information on this disorder.)

Sulfur is found naturally in water and feed. If the soils and water are from volcanic sources, sulfur levels may be high, often resulting in an unpleasant odor due to the presence of bacteria that convert the sulfur in the water into the smelly gas hydrogen sulfide. Sulfur is also available to the goat in feeds containing amino acids that themselves contain sulfur. If you have feed analyzed for its nutrients, you may see that the *methionine* content is often listed. Methionine is an amino acid that contains sulfur. It is also a prime source of sulfur for many of the compounds that the goat needs. Goats eat plants that contain methionine, and their bodies use the sulfur to create other amino acids. That's why you will see the methionine percentage included on feed labels and nutrient analyses.

In areas where water and soil sulfur levels might be low, sulfur can be supplied through a yellow salt lick that includes the mineral along with regular (sodium chloride) salt. It can also be provided in a loose form. A few years ago I provided a sulfur block in the hope of helping one goat improve her coat condition. She did indeed utilize the block, as did many of the goats. Intake declined over time to the point that the block now seems to sit unused. I wipe it off periodically, in case someone wants a clean taste.

Whether or not the supplement was of help, the doe's coat improved permanently.

MAGNESIUM (MG)

Magnesium is found in bones (about 70 percent of it), in cells (29 percent), and in extracellular fluids (1 percent). It is used in hundreds of metabolic functions and in skeletal development. An example here of codependency is that magnesium is involved in the use of calcium, phosphorus, and potassium (which affects nerve impulses and muscle contraction).

Magnesium deficiency leads to tetany (which literally means "spasms") with convulsions. Accompanying symptoms include loss of appetite, excitability, and staggering; death may result. A common time for this risk is in the spring when grasses grow quickly and are low in magnesium or too high in potassium. In this case it is often called grass tetany or staggers. Magnesium toxicity in goats is unlikely.

Magnesium levels can be monitored through blood serum readings.

The Micro or Trace Minerals

The micro or trace minerals are those needed in tiny amounts compared with the macro minerals. They include cobalt, copper, iodine, iron, manganese, molybdenum, selenium, and zinc.

COBALT (CO)

Cobalt is a trace mineral, but it is not used directly by the animal; instead it's necessary for the rumen bacteria to manufacture vitamin B_{12}. Deficiencies of this vitamin can cause serious health problems (see the Vitamin B section on page 91). Fortunately most feeds provide plenty of cobalt, and deficiencies and toxicities in goats are almost unheard of.

COPPER (CU)

Copper is one of the most discussed minerals when talking about goat nutrition. Copper deficiency is often blamed for many issues with goat health, many times with good reason. Worldwide, copper is second only to phosphorus when it comes to

Dairy One

FORAGE LABORATORY

DATE SAMPLED	LAB RECEIVED 10/19/15	DATE PRINTED 10/21/15	LAB USE .886

ADDITIONAL DESCRIPTIONS
ALFALFA HAY

ENERGY TABLE – NRC 2001		
	Mcal/Lb	Mcal/Kg
DE, 1X	1.58	2.97
ME, 1X	1.15	2.55
NEL, 3X	0.66	1.47
NEM, 3X	0.70	1.55
NEG, 3X	0.43	0.95
TDN1X, %	64	

KIND DESCRIPTION LEGUME HAY	CODE 100	LAB SAMPLE 22107980

DESCRIPTION 1
ALFALFA HAY

ANALYSIS RESULTS

COMPONENTS	AS SAMPLED BASIS	DRY MATTER BASIS
% Moisture	11.5	
% Dry Matter	88.6	
% Crude Protein	20.6	23.3
% Available Protein	19.3	21.8
% ADICP	1.3	1.4
% Adjusted Crude Protein	20.6	23.3
Soluble Protein % CP		47
Degradable Protein % CP		78
% NDICP	3.1	3.5
% ADF	22.6	25.9
% aNDF	29.1	32.8
% Lignin	6.2	7.0
% NFC	28.8	32.5
% Starch	.6	.7
% WSC (Water Sol. Carbs.)	8.8	10.0
% ESC (Simple Sugars)	7.1	8.0
% Crude Fat	2.2	2.5
% Ash	7.87	8.89
% TDN	59	67
NEL, Mcal/Lb	.63	.71
NEM, Mcal/Lb	.61	.68
NEG, Mcal/Lb	.37	.42
Relative Feed Value		196
% Calcium	1.41	1.59
% Phosphorus	.21	.24
% Magnesium	.33	.37
% Potassium	1.91	2.16
PPM Zinc	31	35
PPM Copper	9	10
PPM Molybdenum	< 0.1	< 0.1
% Sulfur	.30	.34
% Chloride Ion	.48	.54
IVTD 30hr, % of DM		79
NDFD 30hr, % of NDF		36
kd, %/hr		4.53
% Lysine	1.05	1.18
% Methionine	.32	.36
Horse DE, Mcal/Lb	1.08	1.22

Figure 4.14. Forage tests can provide information about a variety of nutrients and concentrations of components such as moisture and lignin (insoluble fiber). For this test I selected options that would reveal the presence of minerals I was particularly interested in, including calcium, potassium, molybdenum, and copper.

Table 4.2. Suggested Average Micro Mineral Requirements

Mineral	Parts Per Million
Cobalt (Co)	0.1–10
Copper (Cu)	10–80
Iodine (I)	0.5–50
Iron (Fe)	50–1,000
Manganese (Mn)	40–1,000
Molybdenum (Mo)	0.1–3
Selenium (Se)	0.1–3

Note: The amounts listed in this table are based on averages. Actual needs vary greatly depending on gender, age, parity, production volume, and stage of lactation. Complete charts based on body weight and other factors are available in *Nutrient Requirements of Small Ruminants* (National Research Council).

Source: Adapted from Steve Hart, *Introduction to Goat Nutrition* (Langston University, Langston, OK) accessed at www2.luresext.edu/goats/training/nutrition.html#macro.

Figure 4.15. Loss of tail hair, giving a look sometimes called a fish tail, is often associated with copper deficiency but can also occur from rubbing or pulling.

mineral deficiencies. However, in a few parts of the globe, copper excess is the problem. In these areas, a deficiency in a mineral that naturally inhibits copper often exists, and errors in feed calculations have led to goat deaths. Because toxic levels of copper are reached suddenly and death is the most likely result, copper in the diet should be carefully monitored. That being said, the recommended copper levels for goats (and cows) continues to change as more information comes to light. The National Research Council has not established the maximum tolerable copper level for goats as of this writing.[4]

Copper plays a crucial role in the working of many enzyme systems in the goat. It also is essential for red blood cell production, bone formation, the development of connective tissue (such as ligaments and tendons),

> Research indicates that copper needs vary by breed, with the likelihood of some being more sensitive to toxicity than others.

skin quality, and hair pigmentation. Copper at high but safe levels is shown to stimulate growth and feed efficiency and may reduce nematode parasites in goats. Copper oxide wire particles (COWP) can be given to the animal orally in a gelatin capsule, or bolus, where they then lodge in the lining of the rumen and release copper slowly. This product was developed to help reduce parasite loads and is now commonly used by goat producers.

Copper deficiency often manifests as changes in color (lighter) and quality (rougher) in the hair coat. Weight loss, diarrhea, infertility, decreased immune function, improper and uneven bone growth, anemia, and connective tissue disorders are also possible. Kids from copper-deficient mothers may show rear leg stumbling or discoordination, called *ataxia*. The disorder is also called swayback from the animal's appearance, which is a result of poor spinal cord development.

Copper deficiencies are common in many parts of the world. Many soils are not only lacking in copper but may also be high in molybdenum, which binds with copper, making it unavailable for use by the animal. It is believed that the copper level in soils and feeds must be four times that of the molybdenum level to be in the proper balance for goat health. It's important to find out the mineral status of your soil (or the soil where your feed is grown) before making decisions about copper supplementation. High iron levels in water or feed will also limit copper absorption.

Research indicates that copper needs vary by breed, with the likelihood of some being more sensitive to toxicity than others. Breed differences also exist with Nubian and Boer (meat goats) crosses, which are more tolerant of copper than are sheep,[5] and Angora goats may be more sensitive than other goat breeds.[6] Pygmy goats may also have a greater need for copper.[7] From my observations, I believe that Nigerian Dwarf goats may be as well. (It is interesting that all of these breeds originated in different parts of the world from the standard European and North American dairy goat breeds.)

Copper is stored in the goat's liver and released as needed into the bloodstream. A liver sample, taken live by biopsy or from a freshly deceased goat, is the best means of checking for copper levels. A measurement of copper level in the bloodstream is not a good indicator of the animal's true status, unless it is low, which indicates a true crisis, because it can occur only when the liver's copper supplies have been depleted. Acute liver toxicity is a crisis of a different type. When the liver reaches the limit of its capacity for storing copper, it releases copper to the bloodstream, and the animal suddenly experiences a body-wide rupture of blood cells—a *hemolytic crisis*—and dies.

Ideally, copper supplementation will occur only after you have determined whether your feed or soil have deficiencies or excesses (in minerals such as molybdenum or sulfur) that might require

supplementation. (See the Determining Mineral Requirements in Goats section earlier in this chapter.)

IODINE (I)

Iodine is used in the goat's body for the manufacture of thyroid hormones. The thyroid gland (located at the front of the goat's neck, near the base of the jaw) contains about 80 percent of the iodine in the goat's body. These hormones regulate fetus development and metabolic rate and have an effect on digestion, muscle function, the immune system, and the seasonality of reproduction.

Iodine deficiency causes the enlargement of the thyroid gland, called goiter. Other symptoms include fertility issues, a weakened immune system, skin and hair disorders, and reduced milk production. Some areas of the world have soils that are iodine-deficient. Also, some types of plants contain substances called goitrogens that interfere with thyroid gland activity.

Figure 4.16. This Nubian buck owned by Gail Swanson was diagnosed with a goiter caused by iodine deficiency. The goiter had been growing for years; he had also suffered from fertility problems. The owner administered iodine by placing iodine solution on the skin on the underside of the buck's tail. Within weeks, almost miraculously, the goiter had nearly disappeared.

Deficiency can also occur when a goat eats too many of these plants, such as broccoli, kale, and other cabbage-family plants.

Iodine can be obtained in the diet through supplementation and feeding kelp meal. One remedy for a deficiency and goiter (see figure 4.16) is to paint iodine tincture, such as that used as a scrub, wound cleaner, or navel dip, on the bare skin underneath the animal's tail. From there it will be readily absorbed into the body.

IRON (FE)

Iron, the most common trace mineral found in the body, is a component of hemoglobin, which transports oxygen and carbon dioxide in the bloodstream to and from the body tissues and the lungs. It is also involved in the building of connective tissue; it's among the body's antioxidants and therefore helpful in improving resistance to disease. Iron is stored in the liver. When liver supplies are used up, anemia results.

Goats can be deficient in iron before changes in blood cells can be detected through blood work. Symptoms of iron deficiency include loss of appetite, reduced growth rate, tiredness, pale mucous membranes and rapid breathing. (See the Inner Eyelid Color and FAMACHA section on page 126 for more about judging the color of mucous membranes.) Because iron is important for disease resistance, high mortality rates can be indicative of a herd-wide deficiency.

Iron also interacts with other vitamins and minerals. Iron toxicity can result as a consequence of vitamin E deficiency. High levels of iron in feeds, due to high levels in soil or water, or oversupplementation, can lead to copper deficiency. Iron levels can be measured in the liver, and anemia can be measured using a blood sample.

MANGANESE (MN)

Manganese serves several enzymatic needs and is involved in bone formation and reproduction. Deficiencies in goats are rare when diets consist of pasture and forage. Toxicity is also rare. Manganese levels are measured in the liver.

MOLYBDENUM (MO)

Molybdenum (*moh-LIB-deh-num*), besides being fun to say, is essential for its role in enzyme function. Deficiencies are rare, but high levels cause copper deficiencies. For this reason, molybdenum can be used to treat copper toxicity if this is caught in time. As noted in the section on copper earlier in this chapter, many areas of the United States have soils that are too high in molybdenum when compared with copper. Even in areas where the soils are not naturally high in this trace mineral there can be an imbalance thanks to the addition of molybdenum to some fertilizers. Molybdenum levels can be measured in the liver.

SELENIUM (SE)

Selenium is present in several important enzymes that serve as antioxidants, protecting cells from damage via oxidation. Selenium is involved in growth, immune support, and reproduction. In some regions soil levels are low; levels in other areas are high. It is important to know which area your feeds are from before determining selenium supplementation. Deficiencies result in infertility, poor growth rate, and sudden death by heart attack, but the most common result is white muscle disease (linked also to low vitamin E intake). Toxicity is rare in goats, but a subject of concern nonetheless. Symptoms include abnormal gait—lameness—and loss of hair. Selenium is often given to goats in deficient areas through an injection that combines selenium and vitamin E. Oral supplements with or without vitamin E are also used. Selenium levels can be measured in the liver.

ZINC (ZN)

Zinc plays a role in many enzymes that are involved in everything from protein, carbohydrate, and essential fatty acid metabolism to appetite and gene expression. Deficiency symptoms include loss of appetite, immune system dysfunction, skin problems, and bone and reproductive issues. High levels of zinc can cause rumen dysfunction and an imbalance in fatty acid production.

Vitamins

Vitamin is the term used to describe a group of organic (remember, that means something associated with living things) compounds required in the body for the performance of many essential metabolic functions. Vitamins are needed only in tiny amounts. They are grouped into two categories, fat-soluble and water-soluble. Interestingly, vitamins were named alphabetically in order as they were discovered, beginning with vitamin A. Are you wondering what happened to vitamins F, G, H, I, and J? At one time they did exist, but they have since been reclassified either as other vitamins or some other type of substance.

Vitamin A

Vitamin A is a fat-soluble vitamin available in feed as vitamin A but also as a previtamin, sometimes known as a provitamin (which sounds like the opposite of an amateur vitamin to me) or precursor. Provitamins are converted by the body into the namesake vitamin. Carotene, including one variation called beta-carotene, is the precursor of vitamin A. Carotene is best known for its presence in orange vegetables, especially carrots. But it is also present in leafy greens, browse, and leafy green hays. In the cow very little of the carotene that is consumed is converted into vitamin A. Goats, on the other hand, convert most of it. The presence of carotene in cow's milk gives it a more yellow color than goat's milk. Some literature has made health claims about goat's milk, citing its greater amounts of vitamin A as a reason for being superior to cow's milk. But our human bodies are quite happy with carotene in place of vitamin A. Vitamin A, like most fat-soluble vitamins, is stored in the goat's liver. Because it is present in milk, increased milk production levels might be associated with greater dietary requirements. Carotene levels in hay and stored feed decrease over time; it's estimated that all carotene disappears within six months after storage. A lack of vitamin A in feed may or may not be a problem. If a goat has stored adequate supplies of vitamin A in its liver during the fresh grazing season, then consuming older feed should not result in a deficiency. If you time kidding and milk production for the off season, though, the vitamin A requirements of the does may be higher than the stored feed can provide. You can address this by providing supplements or feeding other crops high in carotene. Carrots, winter squash, and winter kale are all good options.

Deficiencies can lead to symptoms that include loss of appetite, weight loss, and an unkempt look. Other more specific symptoms are reduced vision, nasal discharge, urinary stones, and reduced fertility, including abnormal sperm. Deficiency can bring about an increased susceptibility to eye problems including conjunctivitis (also known as pinkeye).[8] I've found that feeding carrots is a simple cure for a goat suffering from nutrition-related conjunctivitis. Kids that are low in vitamin A are more prone to parasites, respiratory problems, and diarrhea. Herds with coccidiosis issues have increased vitamin A requirements. (There's a good reason that sick goats crave those leafy greens and browse.)

If the intake of plants high in carotene is high, animals don't experience an excess of vitamin A. But if the intake of vitamin A, usually through vitamin A supplements, is too high, then problems related to skin quality and lameness might result.

Vitamin B

This water-soluble group of vitamins includes cyanocobalamin (B_{12}), folic acid, niacin, pantothenic acid, pyridoxine (B_6), riboflavin (B_2), and thiamine (B_1). All are produced in the rumen by microbes. In the healthy goat, there's no need to supplement these nutrients in the diet. Deficiencies can occur when the trace mineral cobalt is deficient, because cobalt is necessary for B_{12} production. Signs of B_{12} deficiency include reduced appetite and milk production, eye discharge, sensitivity to light, decreased

> I've found that feeding carrots is a simple cure for a goat suffering from nutrition-related conjunctivitis.

immune function, and increased parasite problems. A thiamine deficiency can also easily occur when something happens that changes or upsets the balance of microbes in the rumen. Acidosis from too much grain or other conditions favors the growth of unwanted bacteria that produce an enzyme, called *thiaminase*, that blocks the production of thiamine. This can rapidly lead to the development of so-called goat polio, also known as polioencephalomalacia, PEM, or stargazing. Thiamine utilization is also subject to inhibition by several deworming medicines (anthelmintics), in particular a coccidiostat called amprolium (a common brand is Corid). High levels of sulfur in the diet can also inhibit thiamine. Several plants that may be present in pastures, woodlands, and baled forage contain thiaminase including western bracken fern (*Pteridium aquilinum*), fresh fescue, and Italian ryegrass.

Vitamin C

Vitamin C, ascorbic acid, is a water-soluble vitamin, and it is produced naturally by goats (humans, guinea

> For goats, as for humans, stress and sickness increase the need for vitamin C.

pigs, primates, and fruit bats are the only animals that don't make their own). It's believed that kids, like calves, start producing vitamin C sometime after three weeks of age. That's why raw milk contains a bit of vitamin C (it's destroyed by pasteurization so should be supplemented in kids' diets when pasteurized milk is fed). For goats, as for humans, stress and sickness increase the need for vitamin C. It can be offered as a supplement in the form of ground-up tablets or ascorbic acid powder, or by feeding leafy greens. There is no concern for overdose.

Vitamin D

Vitamin D (fat-soluble) is known as the sunshine vitamin. It is produced by goats' skin, thanks to the influence of the ultraviolet light provided by sunlight. Another version of the vitamin is produced in leafy hays that are cut and then cured in the sun, with the production taking place in the dying leaf exposed to UV light. Vitamin D plays an important role in the proper usage of calcium and phosphorus by the

Vitamins and Feeding Kids

When you're thinking about the vitamin needs of newborns and young kids, keep in mind that their needs differ from those of adult goats. Adult goats produce vitamin C and the B group in the rumen, but kids that haven't yet developed a functioning rumen do not. However, they still need their vitamins! So where do they get them? First from colostrum and then from milk. But the B vitamin supply in milk begins to dwindle until it's all but gone—the timing roughly coincides with the age at which the kid should now be ruminating. When kids are not being dam-raised, you must consider what you are feeding them and determine if it supplies these vitamins. Vitamin C (which the kid will start making much later than the B vitamins) is destroyed by pasteurization. If you are feeding mixed milk from multiple does, pasteurized or not (and I always recommend pasteurizing comingled milk—more on that in chapter 9), vitamin B may not be present, because the does are likely at different stages in their lactation. Milk replacers will contain these nutrients, but as a whole, milk replacers are less than satisfactory for several reasons (also more on that in chapter 9). You can provide a supplement of both vitamins, without concern of an overdose, by adding them to the milk.

goat's body. Deficiencies in kids result in rickets—bowed legs, enlarged joints, and stiffness—along with reduced growth rate and poor body condition. In adults deficiency can cause osteomalacia, softening of the bones, or osteoporosis, brittle bones. An excess can cause calcium deposits in tendons and skin and other soft tissue.

High milk production is likely to increase vitamin D requirements, due to its presence in milk. Kids obtain their initial stores from their mothers, both from milk and before birth through placental transfer. Kids who are weaned early or housed inside (with limited or no UV light) may need more supplementation. Sunlight is the most effective form of vitamin D supplementation—something to be considered when you design housing for both adults and kids.

Excess levels will not occur naturally—too much sunlight will not lead to too much vitamin D. But feed supplements can be overconsumed, leading to symptoms of excess.

Vitamin E

Vitamin E, a fat-soluble vitamin, is an important antioxidant—a substance that helps prevent the oxidation, or breakdown, of cells. Earlier we talked about its importance for the absorption of the trace mineral selenium. In fact, it is hard to separate vitamin E and selenium when it comes to considering deficiencies of either. Vitamin E is present in many feeds, especially leafy greens and seeds, milk, and colostrum. As with vitamin A, its presence decreases in aging hay. Newborn kids are at the greatest risk of deficiency; the most common symptom is white muscle disease. In adults, low levels can lead to retained placentas and, interestingly, oxidized flavors (sometimes described as a plastic, burnt hair, or medicinal flavor) in milk. Increased vitamin E may help reduce somatic cell counts and stress on the mammary gland during dry-off (when high-producing animals might stay engorged and stressed for

longer times than lower-producing does). There is no occurrence of toxicity with E.

Supplements can be given, usually in the form of an injection combined with selenium (Bo-Se is a common one) that is available through veterinarians. Green leafy plants and many seeds, such as sunflower and pumpkin seeds, are great sources of vitamin E.

Vitamin K

Vitamin K is the only fat-soluble vitamin made in the rumen. Its main role is as an *antihemorrhagic*—it promotes blood clotting and the prevention of bleeding. Supplementation is only necessary when a goat has eaten something that inhibits vitamin K. Some plants naturally contain a substance called coumarin that can be converted to a vitamin K inhibitor. Examples include sweet white clover (*Melilotus albus*), mullein (*Verbascum thapsus*), and sweet vernal grass (*Anthoxanthum odoratum*). All of these plants have flowers that have a sweet vanilla-like aroma, indicating the presence of coumarin. Other antagonists to vitamin K include sulfa-based antibiotics and some toxins produced in moldy feed.

Water

Water is an obvious need for livestock, but it's a mistake to assume that any water will do, from that stored in algae-lined rubber tanks to chlorinated public water. The quality of the water your goats drink can affect their health in a variety of ways.

Whatever the source of your water, including private well, municipal supply (sometimes called city water), spring, or reservoir, it's wise to take samples and have them tested for mineral content and the presence of bacteria by sending them to a lab. A mineral analysis can reveal the presence of minerals that might inhibit metabolism of other minerals. Some types of bacteria found in water samples are harmless, but others, especially coliforms, indicate

> Sunlight is the most effective form of vitamin D supplementation —something to be considered when you design housing for both adults and kids.

contamination from organic matter, including soil and possibly feces. Although most coliforms are harmless, some originate from fecal sources and can be pathogenic. Water pH should also be checked. A pH of 7.0 to 6.5 is ideal because it's close to optimal rumen pH.

Water intake will vary depending on the ambient air temperature and the activity level and status of the goat (a goat in milk requires the most). A rule of thumb for making sure that you provide enough water—say, if the goats are being rotationally foraged away from their usual water supply—is as follows: At 50°F (10°C), an average adult goat needs 1.5 gallons (6 L) of water per day, but at 90°F (32°C), the same goat would need 3.5 gallons (13 L) per day.[9]

Water from a well will typically have a higher mineral content than city water. Water from a reservoir or surface spring, unless treated, will likely contain bacteria that may or may not be harmless. Some bacteria convert minerals from one form to another, such as sulfur to hydrogen sulfide, as described in the section on sulfur earlier in this chapter. Water from a public supply may contain chlorine, and that may negatively influence how much goats will drink.

CHAPTER FIVE

Feed Choices for Health and Vitality

Left to their own devices and in the right geography and climate, goats will make the right nutritional choices and keep themselves in great health. As captive domesticated creatures, however, they count on us to anticipate these needs and provide the right nutrition. In mega-agriculture, where animals are confined and all their food is served to them, scientists and feed specialists design a *total mixed ration* based on what is currently believed to be true about the animals' nutritional needs. The holistic farm, whether it be extensively managed or some combination thereof, must rely on other means to anticipate these needs. I believe that a combination of education and intuition is all that is needed to properly feed goats.

This chapter covers options for feed and supplements, keeping in mind that it's impossible to cover every possibility for good feed choices. Farms and homesteads have different natural feed resources as well as varying access to purchased feeds. Fortunately there are many creative combinations of the scenarios I describe that will help any goat farm provide the right nourishment to help their caprine friends maintain homeostasis.

In general, a goat will consume 3.5 to 5 percent of its body weight each day. For example, a 150-pound (68 kg) goat would eat between about 5 and 7.5 pounds (2.3–3.4 kg) of feed. This calculation is based on dry matter—in other words, the wetter the food, the more total poundage an animal will need to consume in order to take in the nutrients it needs. If you're interested in a formulaic approach, I recommend visiting Langston University's interactive (and free) feed formula calculator (see appendix D for more details). Many other factors affect this range, including the quality of the feed, the activity level of the goat, whether the goat is producing milk or pregnant, and the age of the animal.

Dining Out

There's a good reason that goat milk and meat are the most consumed animal products in the world, beyond the fact that they are tasty and packed with nutrition—goats can thrive on terrain and natural feed options that would not sustain beef or dairy cattle. If you're fortunate enough to own or have access to a decent amount of land on which to manage your herd, then likely your goats are in for a treat. Most types of terrain and plant growth can make great feed sources—pasture, forests, chaparral, and savanna. Let's go over these options for providing nutrition when the goats are browsing and grazing away from the barn.

Knowing what plants to eat and which ones to avoid is part instinct and part learned behavior. I've had the opportunity to observe this firsthand. When we moved to our current land from the smaller lot

Figure 5.1. Wild rhododendron, which is highly toxic, shares a paddock where goats often graze at this North Carolina meat goat farm, and yet it remains uneaten. In other situations farmers aren't as lucky, with unwary and naive goats eating the plant and sickening or dying. Browsing is very much a learned behavior.

where we lived when we bought our first goats, the herd had to "learn" how to browse. Some plants they took to naturally; others they had to be encouraged to eat—for example, in order to get them to nosh on poison oak, I withheld their evening feeding of hay and instead filled the feeders with poison oak (fortunately it doesn't affect me). Before long, they were motivated by hunger and tried the plant, and soon everyone was enjoying it. Now, kids imitate their mothers, trying everything they see them eat. We have poisonous plants as well, and I've seen our goats rush up to them with their mouths open, only to stop just before biting into them. In that case, the plant must be exuding some sort of compound that signals the goat not to eat it. Keeping this combination of learned behavior and instinct in mind is important when you're first browsing or grazing a herd of goats that aren't familiar with what is growing.

Pasture for Goats

The ideal goat pasture will include a blend of grasses, legume, and *forbs* (herb-like broad-leafed plants), many of which are known as weeds. One person's weed is another's feed! The diverse cross section of plants in healthy pasture provides a great spectrum of nutrition, in good part thanks to the varying depth of the roots—the deeper a plant's roots, the more minerals, in both variety and quantity, are brought up from the earth and stored in the plant matter. Goats like to eat many types of grasses and will prefer some types of plants at a young, succulent stage and others when they are more mature. Legumes, such as clover and vetch, serve both as forage for the goats and as fertilizer for the other plants. Legumes, in association with certain types of soil bacteria, have the ability to capture nitrogen from the air (nitrogen fixation) and transform it into compounds that plant roots can absorb. Forbs are a key part of a nutritious pasture because many of them are very deep-rooted and bring up minerals from the subsoil. For the most part, grasses have shallower root systems. Forbs also provide variety of flavor and help goats meet their natural need for a varied diet.

Chicory (*Cichorium intybus*) is one of my favorite forbs. This now widespread plant native to Europe is a perennial that readily reseeds itself and has a deep taproot. Forage chicory seed can be purchased online or through catalogs. Other desirable weeds that can be sown or found naturally in pastures include plantain (*Plantago major*), curly dock (*Rumex crispus*), pigweed (*Amaranthus* spp.), horseweed (*Conyza canadensis*), lamb's-quarters (*Chenopodium album*), and even star thistle (*Centaurea solstitialis*). In fact, goats are one of the best organic remediation methods for the troublesome star thistle.

Growing and managing healthy pasture is an in-depth topic that is worth studying on its own. If pasture is to be a main part of your goat's diet, then I highly recommend reading the classic *Fertility Pastures*, by Newman Turner, and *The Art and Science of Grazing*, by Sarah Flack (see appendix D). Some important aspects to consider regarding grazing goats on pasture include ideal plant height, the presence of desirable plants and grasses, the distribution

Which Plants Are Toxic to Goats?

If you search online for a list of plants that are poisonous to goats you will encounter a lot of contradictions—you may find yourself more confused than when you started. Some plants are certainly deadly, but many lists include plants that goat owners know from experience don't bother their animals. The way in which plants negatively affect an animal varies. Some deposit toxins inside the goat's digestive system. Some plants are harmless at times but become toxic after exposure to frost, or after wilting, or during drought. Still others contain no toxins, but they do contain substances that block the production of vitamins.

Western bracken fern (*Pteridium aquilinum*) is a great example of the confusion. It is listed on many websites and in books as poisonous to goats. Bracken grows prolifically in the part of Oregon where I live. My goats, as well as those of other area goat farmers I know, consume it on a regular basis, but by no means as a primary feed source! Instead, they take a nibble or nosh whenever they walk through a patch. In cows, sheep, and other livestock, bracken is known to cause thiamine (vitamin B_1) deficiency and can lead to polioencephalomalacia (stargazing or goat polio) and subsequent death. It is also linked to severe bleeding disorders in cattle. However, symptoms in goats are not well documented,[1] and *Goat Medicine* authors Smith and Sherman advise that "currently goats should be considered *potentially susceptible* to bracken fern toxicity." What is known is that for all livestock, bracken fern will not cause a problem until it is consumed in a certain volume and over a sufficient period of time. Thus, if other feed choices are too limited, bracken is more likely to be a problem. You don't need to prevent your goats from browsing in the forest and fields or spend unlimited amounts of time trying to remove the bracken, but you must be sure that the herd has plenty of other feed options.

To help you learn more about plants in your area that you may need to be vigilant about, in appendix D I have included a list of plants named by Cornell University Department of Animal Science as toxic to goats. But rather than going by lists alone, I highly recommend consulting other local goat farmers and knowledgeable veterinarians for their experiences with the real or potential dangers of local plants.

Figure 5.2. Goats relish chicory after it passes its lush, leafy spring stage and starts to flower. This chicory has been repeatedly grazed throughout the season and continues to flourish even without any irrigation or rain.

of manure as the animals graze, and the importance of soil fertility.

No matter a field's plants and nutritional status, the height of the growth is the first aspect to consider. Grasses and other plants should be dry and taller than 6 inches (15 cm)—about the length from your fingertips to your wrist. Observing this rule helps ensure that goats are unlikely to ingest parasite eggs and larvae as they graze. Parasite eggs are shed in manure. The larvae hatch and then move on to leaves of pasture plants. They prefer moist plants and stay relatively close to the soil surface, where it's cool and sheltered. Fortunately goats prefer to eat at their own eye level as well, as choosing dry grasses over wet. If the goats have limited access to

Figure 5.3. Goats not only love forbs and briars, but gain valuable nutrition from eating them. Here a Boer yearling nibbles on blackberry at Sospiro Goat Ranch, North Carolina.

Figure 5.4. Goats prefer to eat closer to eye level and do a good job eating seedheads in tall pasture.

other good feed sources, however, they will eat wet plants and graze close to ground level. In that case, not only are the animals at more of a risk for poor nutrition, thanks to the limited feed choices, but they're also at a greater risk for parasite problems. In tall pastures, goats graze farther away from ground level and from manure that would otherwise be accidentally ingested, increasing the chances of parasite problems, as well as other diseases and illnesses. When given the choice, goats will not eat feed that has been trampled or dirtied, but if feed options are limited, they will eat whatever they have to in order to survive. To avoid overgrazing of pastures, consider the number of goats on the pasture, as well as options for management such as rotational and mob grazing.

If you are planting new pasture or refurbishing existing pasture, the mix of grasses you choose is likely to be based primarily on your growing zone, the type of soil, and the availability of irrigation. Purdue University's Forage Identification website has very helpful photographs and information on many common pasture forages (see appendix D). A good place to start (in the United States) for learning more about your region is a local state university extension agent.

Mowing can also be an important part of pasture management. Mowing the fields might seem counterintuitive, but mowing helps add organic matter and minerals to the topsoil over time. Mowing also helps control some unwanted weeds that the goats might not be polishing off. If you have a seasonal or dryland pasture, mowing also helps to reduce wildfire risks. If a field has grown up very tall, you may need to mow it simply for practicality's sake, because goats tend to avoid entering into growth that is over their head—it's likely that their sense of security diminishes when they can't see what's happening around them.

Forest, Chaparral, and Savanna

If your farm includes rangeland such as forest, chaparral (shrub growth), or savanna (dry grassland with some trees), then a wealth of nutrition and food supply is available to your goats. Harvesting this supply, however, takes more than just a perimeter fence and an open barn door. You will have to consider the possibility of predators, protecting trees from damage by your goats, and getting them to forage away from the safety of their barn.

For those farmers determined to track their goats' diets—from total digestible nutrients to protein percentage—browsing presents a conundrum. There's no easy way to quantify the nutrition obtained by

Grazing Management Terms—More than a Mouthful!

There are about as many terms describing how animals are managed on pastures as there are burrs and stickers in your socks after a summer hike in socks and sandals. Indeed, the American Forage and Grassland Council lists 32 different pasture management terms.[2] *Mob grazing, strip grazing, cell grazing, first-last grazing,* and *mixed grazing* are just a few examples. You'll find practitioners, authors, extension agents, and other proponents passionate about one or more of these approaches (many of which are quite similar or variations on a theme).

Choosing a management approach should be based on what your land can offer that fits your animals' needs and what your own time allows for—not just the hot new grazing trend. Rather than choosing an approach based on name only, and then trying to conform to that approach, spend some time reading more about the fundamentals of grazing. I recommend *The Art and Science of Grazing* by Sarah Flack, and see appendix D for other references, too. Here are some quick definitions of the most commonly used terms:

Rotational grazing: Utilizing multiple paddocks through which animals are rotated. Each paddock is allowed a recovery time to regrow to the height you desire.

Interspecies grazing: Rotating difference species of livestock in a series that optimizes the crop and reduces parasites (by limiting species-specific parasites' access to a host).

Strip grazing: A length of movable fencing is placed to allow a long, shallow strip of a crop to be grazed. The fence is moved farther into the mature crop as each strip is consumed. A useful approach when grazing fodder crops.

Explaining grazing management principles in detail is beyond the scope of this book, but if grazing is your goats' primary source of nutrition, it's vital that you take time to learn about the basics of good grazing practices. Simply turning your animals out onto pasture without knowledge could have unanticipated unhappy results for your herd's health and productivity. You don't need to be an expert as long as you understand how to monitor plant density and height and how well your animals are thriving on what they consume. It's really all about observation and continuing to learn as you go. Attending a workshop on grazing management can be well worth the investment of time and money, too.

Figure 5.5. Forest browse offers great nutrition for goats.

freely browsing goats. Instead, learn how to interpret the nutritional status of your animals by careful observation—the animals will show you if their needs are being met through their body condition, coat condition, and brightness of eye and behavior. Chapter 7 covers such observation techniques, which are windows into the nutritional status of your herd.

A mature forest will have much for goats to eat: fallen leaves and understory shrubbery; bark (from trees of all ages); moss and lichens; and various herbs and annuals. Mature trees with thick bark, like conifers and oaks, are usually safe from destruction,

Figure 5.6. Trees must be protected if they are to survive the appetites of goats. In the picture on the left wooden pallets have been arranged and secured to create a perimeter that protects the trunk and lower parts of a young tree. In the image on the right, wooden slats, woven through wire and then wrapped with hardware cloth, surround the trunk of a mature tree.

Figure 5.7. Goats are well adapted to arid land and the nutrition that its plants offer. These sleek, healthy-looking Nubian dairy goats are enjoying juniper at Black Mesa Ranch in Arizona's high desert.

but goats can easily strip the bark off thin-barked adult and all young trees. If they remove the bark in a complete ring around the trunk—called girdling—it can destroy the tree. So a plan for forest management must be part of your planning before you browse goats in the woods.

Chaparral-type land is natural in some regions, while in others it's the result of overlogging, fire, or other means of deforestation (unmanaged goats are great at turning a forest into chaparral). Native or not, this type of rangeland can be a feast for your caprines. Goats love many types of otherwise noxious plants often found in these types of land, such as poison oak (*Toxicodendron diversilobum*) and poison ivy (*T. radicans*) (which erroneously appears on some lists of plants poisonous to goats), star thistle (*Centaurea solstitialis*), Himalayan blackberry (*Rubus armeniacus*), and even the bane of the southern United States, kudzu (*Pueraria montana* var. *lobata*). This taste for what are considered annoying weeds by us, but are quite nutritious to the goat, is a terrific opportunity for providing feed and accomplishing organic weed abatement—all at once. Many other shrubs and vines desired by goats, including wild (and domestic) grape-vines, are quite difficult to destroy and will regrow anew after being eaten. By selectively browsing your goats, you can treat these types of plants as crops.

Open savanna is also a bountiful banquet for goats with a combination of the above-mentioned types of plants. Susceptible mature trees should be protected from girdling so that they can provide shade and leaf drop for the goats.

Dining In

No matter how much browsing and grazing acreage you have, there will be times of the year when the goats cannot, or will not, go out to dine. The sustainable goat farm needs to consider all options for providing feed. Familiar feeds for goats include hay, silage, and grains. There are plenty of other options, too, including garden crops, fresh fodder, trees and shrubs grown for feed, and foraging for feed. This section will help you assess the options and provide a balanced diet for your goats, especially when they are confined to shelter.

Hay or Fodder

The terms *hay* and *fodder* refer to any grass, legume, grain-bearing plant, leafy plant, or combination that's grown, cut, dried, and then stored for use later. (In the United States, the term *hay* is most commonly used.) The quality of hay is directly related to the quality of the soil, the skill of the hay grower, management practices such as pest and weed management and storage, and a bit of luck with the weather. The quality of a hay crop from a particular field can vary greatly from year to year, with weather influencing not only the growth rate of the crop but also the timing of the harvest. A good hay grower will be not only excellent at managing all of these issues, but also honest about the quality of the hay produced.

Organically grown hay is the ideal, but it may not be available or affordable in your local area. No-spray hay might be equivalent, as long as you can trust the farmer, and it should cost less than the certified-organic equivalent. Remember, chemical sprays are expensive, so if you can convince a farmer that a few non-toxic weeds are not a deterrent to your purchase, you might make some headway. But also remember that the farmer is equally concerned about who will buy the hay after you stop. Committing to no-spray, without the added value of organic certification, is a tough change for some farmers.

Hay prices depend on all of the above conditions as well as demand. In some areas entire crops are pre-purchased by large dairy co-operatives or companies. In the Pacific Northwest of the United States, a good deal of hay is exported to the lucrative market in the Far East. Your best value will come from making a large single purchase, by the ton or cube of hay instead of the bale, rather than picking up a bale or pickup load as needed. If you don't have a barn, I recommend investing in construction of a hay storage structure.

GRASS HAY

Grass hays vary quite a bit based on region. Common types include orchard, timothy, rye, and fescue. The best grass hay for goats should have flexible, tender stems and fragrant green leaves (blades). It should be free of dust or molds. Check for signs that the grass went to seed before being cut, such as thickened dry stems, yellow to brown leaves, and seedheads visible in the hay. This hay will not be as nutritious as green hay. Grass hay might consist of only one type of grass, but a mixed grass that contains a variety of other "weeds" can be the most desirable for goats (as long as the weeds are not toxic, of course). In most areas, two cuttings of grass hay per year is the norm. The first cutting, from growth that grew slowly over the winter, is usually less desirable for most goats. The stems are thicker, there is less leaf, and overall nutrients are lower. Second cuttings usually have finer, tenderer stems—meaning less waste and higher digestibility.

LEGUME HAY

The most popular hay for dairy goats is without a doubt the legume alfalfa, also called lucerne. Other types of goats can be fed alfalfa as well, but its high

Figure 5.8. Goats relish high-quality orchard grass and alfalfa hay like these examples. They're excellent sources of nutrients.

Figure 5.9. Wet brewers grain stored in a water trough will last a week to 10 days if cared for properly. It offers an extremely high-protein, high-fiber feed option for goats. Dogs like it, too!

protein and calcium content (a particular need for goats in milk) must be taken into consideration. Because of legumes' ability to improve soils through nitrogen fixation, they are often planted in rotation with other crops (including grass hay and grain crops) and are also planted as a part of a blended pasture mix. Other examples of legume hays include clover, vetch, pea, and bird's-foot trefoil. Typically a legume hay of equivalent quality to a grass hay will have a bit more digestible energy and calcium, making it a great feed for high-producing dairy animals. Alfalfa stands (growths) will grow for several years before they need to be tilled under and replanted. During a single good year they can be harvested up to four times. The first cutting after the field is planted is much different from subsequent first cuttings, since the plants will be small and tender. Usually the first cutting of alfalfa has similar qualities to that of first-cutting grass hay—the stems are coarse, and there is less leaf than in subsequent cuttings. The fourth cutting, when available, has the greatest proportion of leaf, and the stems are smaller and more palatable to goats.

GRAIN HAY

Grain hay is simply a grain-producing plant harvested as fodder instead of grain. When grain crops such as wheat, rye, barley, oats, and triticale (a hybrid of wheat

and rye) are harvested as hay, the seedheads usually aren't fully developed or dry, and the stems are still green. Grain hays make a wonderful feed for goats, but can also attract rodents in the hay barn. If the hay is high in grain seedheads, it might also be a bit too high in phosphorus for feeding to bucks and wether goats (it would increase the likelihood of urinary stones).

Fermented Feeds

Most feeds for livestock, including hay, grains, and other plants such as corn, can be chopped, fermented, and stored for later feeding. This type of feed is called silage (originating from the fact that it was forage stored in a silo); feed made this way is referred to as ensiled. The terms *haylage* and *baleage* refers to hay bales harvested with a higher moisture content than if they were intended to be stored dry. The bales are encased in plastic to ferment and be stored. A commercially produced and widely distributed brand of non-GMO alfalfa haylage in the United States is Chaffhaye. All fermented forages offer some improvement of digestibility, because bacteria will have already partially broken down the plant material (just like what happens in the rumen). For the hay farmer, ensiling provides the opportunity to take a crop that cannot be made into high-quality dry forage and create a high-quality silage. A hay

Hidden Health Threats—Mycotoxins in Feed

As I was finishing writing this book, a goat farmer and cheesemaker friend from Maine contacted me about a problem she was having with the feed for her goats. A goat had become sick, and the farmer suspected mycotoxins from grain as the cause. The farm feeds organic grain, but organic grain has the same risk for contamination from mycotoxins as does conventionally produced grain. Mycotoxins are produced by molds, so hay is also a possible source, but grain is the more likely culprit. When animals consume such feed, a disease called mycosis can be the result. Depending on the type of mycotoxin, the symptoms are different and myriad. Abortions, mastitis, diarrhea, retarded growth, liver and kidney damage, and death are among the outcomes of mycosis.

Some feed companies test for mycotoxins—and there are advisory levels for the major ones—but as of now there are no definitive "safe levels." This is due to "lack of research, sensitivity differences by animal species, imprecision in sampling and analysis, the large number of potential mycotoxins, and integrations with stress factors or other mycotoxins."[3] My friend had her grain tested, and results showed levels just below the advisory level for the aptly named vomitoxin (5 ppm). Not all feed companies test their feed, and the only recourse for the consumer is to attempt to prevent further mold growth and to discard moldy feed.

The best option to prevent problems with mycotoxins, other than limiting the amount of grain, comes in the form of substances capable of binding mycotoxins in the feed and therefore blocking their absorption in the animal. Substances that are easily available to farmers and currently known to bind mycotoxins include activated charcoal (also known to bind with ingested residues of the herbicide glyphosate, as explained in the Glyphosate Remediation Products sidebar on page 113), and sodium bentonite (which comes in a powder and can be purchased from many livestock and feed supply companies). I keep charcoal on hand for treating digestive upsets and offer a free-choice tub of bentonite in our lineup of individual minerals. The goats consume it as needed. In studies both bentonite and charcoal reduce mycotoxins in the animal, but bentonite is more effective at reducing mycotoxins in the milk, so it's a great choice for lactating goats.[4]

crop that cannot be harvested with an ideal moisture content or dried properly can still make excellent silage. Some silage is made using added formic acid as a preservative. Probiotic and other fermenting bacteria may be added to speed fermentation and provide beneficial bacteria to the livestock.

All silage-type feeds include possible risks. If the acid level is not high enough or soil is present in the silage, *Listeria*—the bacteria responsible for causing listeriosis (an often fatal infection)—might also be present. Feeders where silage is fed should be cleaned frequently so that any leftover feed does not mold and spoil, which may present risks from mold spores and other unwanted bacteria.

Wet brewers grain (BG), a by-product of making beer, and distillers grain (DG), a by-product of making distilled alcohol (both are also called spent grains), have value for a dairy goat's diet. Distillers grain has a tainted past in the history of the dairy cow (it was the sole feed given to cows confined in filth and without health care at the turn of the 19th century), but both types of spent grain can be used thoughtfully as a part of a larger nutritional picture. Each is high in protein and low in starch (which has already been fermented). The ratio of calcium to phosphorus varies but is similar to that of other grains—about 1:2 or greater. This should be taken into consideration when you're feeding animals that might be prone to urinary stones, such as bucks and wethers, where a ratio of 2:1 or 1:1 is important.

Larger breweries and distillers are likely to dry their spent grains and then have them marketed and sold, but small craft producers often give away the grains for free to eager livestock farmers. We have been feeding wet BG for about 10 years, thanks to a local brewery (that also includes our cheese on its menu). Wet grains can only be stored for a week or so and must be inspected before each use for mold and spoilage. Their aroma should be pleasant, a bit "beery," and like that of cooked cereal. Most guidelines suggest feeding no more than 15 percent of the total diet as spent grains.

Grains and Seeds

Grain—corn, oats, barley, wheat—can play an important role in providing energy for certain goats, such as heavy milk producers unable to maintain their body weight and any other animal unable to obtain enough energy from fodder and forage alone. The energy available in grains is quite concentrated compared with leafy forage. Grains for feed are also called *concentrates*. Grains ferment differently in the rumen from forages (as discussed in chapter 4) and when fed in excess can cause acute or chronic acidosis. A small amount of grain can be used to reward animals for coming into the milking parlor (but don't ever let the goats teach you that they need more in order to remain calm!), and more can be given to individual animals who may be having trouble maintaining good body condition. (See chapter 6 for more about body condition scoring.)

Goats cannot completely digest dried whole grains because of their density and hardness, as well as the faster speed at which they move through the rumen compared with fibrous plants. For that reason grains are usually lightly processed by one of several methods that create smaller pieces or expose more surface area. Common treatments are grinding, crimping, steaming, rolling, and flaking. Steaming usually precedes crimping or rolling. Grains can also simply be soaked, or even germinated, to make them easier for goats to chew and digest. At Pholia Farm we

Figure 5.10. At Pholia Farm we feed a very small amount of sprouted barley to the milking goats—about 1 cup (225 g) per milking (parts of the year we milk once a day, parts twice) for every 100 pounds (45 kg) of animal. Since the sprouted barley is bulkier than dry, this would be equivalent to about half that amount if fed unsprouted.

germinate whole feed barley (as opposed to seed barley) by wetting it down and allowing it to germinate over several days. I chose barley because it requires less water to grow than other grain crops, and it contains more enzymes (one reason it is the grain to use for beer fermentation). It is thought to require less fertilizer than other grains, and as of this writing it has not yet been genetically engineered to tolerate the herbicide glyphosate (the most common brand being Roundup). However, it's quite possible that conventionally grown barley has been sprayed with glyphosate to synchronize harvesting. Fortunately for us, much of the barley that is available in our area is grown by farmers who use organic practices, even though they are not certified-organic growers.

Grain should be fed after adequate roughage intake to help slow the fermentation, limit acid production, and allow more time for the nutrients from the grain to be processed. Suggestions on how much grain is too much vary, but you can progressively reduce intake while you monitor the animal's body condition—or vice versa, slowly increase it. A full-sized goat may only need a cup or two (about 225–450 gm) at each milking; Nigerian and small-sized goats, less than half

Hedgerows and Salad Bars

Medicinal and nourishing herbs can be planted in a way that offers access to goats on an as-needed basis. It's wise to set out the plants in a long row protected by a heavy-duty fence, such as welded wire panels, with holes just big enough for the goats to extend their heads through and access the plants. Another method is to set the plants closer to a panel fence that has smaller openings, allowing the goats to only nibble the leaves that grow up against the fencing. If goats are able to extend their heads into the plant pen, then the plants must be sufficiently deep-rooted that the eager eaters don't tear them out of the ground. You can help by covering the soil with hardware cloth with openings cut for the plants. The cloth will help stabilize the soil and make it harder for the goats to rip out the plants. Ideally the animals will be able to access both sides of the hedgerow, but if not, you can harvest leaves that they can't reach and offer those separately.

Here are just a few of the herbs and plants that goats will enjoy as additions to their diet or treats when a craving strikes:

- Marshmallow (*Althea officinalis*)
- Rosemary (*Rosmarinus officinalis*)
- Peppermint (*Mentha* spp.)
- Lemon balm (*Melissa officinalis*)
- Comfrey (*Symphytum* spp.)
- Mullein (*Verbascum thapsus*)
- Parsley (*Petroselinum crispum*)
- Raspberry (*Rubus idaeus*)
- Blackberry (*Rubus* spp.)

Any other medicinal or culinary herb can be offered to goats in this fashion as well. Not only will you be allowing your goats access to their own "medicine cabinet," but you'll have a pretty herb bed to enjoy yourself.

---

that. As with every part of holistic herd management, learn to observe your own herd and make decisions based on what you see happening.

Garden Crops

Many garden crops can provide more feed and nutrition per acre than can pasture or hay. Of course they still require work on your part, along with an awareness of nutrient content and the needs of your goats, but they can certainly help with the feed bill and—even more important—supplying nutrients. Research your growing zone to determine what crops might work best in your area as a food source over a long harvest season. Some plants are easy to incorporate into an existing pasture or non-irrigated range—especially chicory, which is after all technically an invasive species. Chicory will easily reseed itself even when heavily grazed, meaning it can be a self-perpetuating crop. It's easy to find information online about chicory

Figure 5.11. Mangel fodder beets being grown for winter feeding at Pholia Farm. The beets are pulled just before the first freeze. The leaves are fed immediately; then the roots are stored in a spot that's cool but protected from freezing, and fed throughout the winter.

Christmas Trees: A Good Gift for Goats?

Goats love to eat conifers—pine and fir trees—but a commercially grown Christmas tree is likely laden with a great deal of herbicide and pesticide residue. In fact the application of synthetic chemicals to conventionally grown Christmas trees (which aren't intended to feed animals or people) is believed to be among the most intensive of any crop. In addition, the types of chemicals used are not necessarily those labeled for use on feed crops, so they're not overseen in the same manner—in other words, no testing has been done to determine whether eating treated crops is safe. If you want to enjoy the New Year by sharing your aging holiday decorations with your goats, select an organic or biodynamic tree or harvest a less groomed one from the forest. At our farm, we buy a living tree (likely also sprayed) and then plant it in the forest after the holidays. Someday it will help feed the goats—along with, we hope, helping the planet.

Figure 5.12. Barley fodder systems are an option for growing feed on farms with limited access to other feed choices.

these days, as well as to purchase seed (they even sell it on Amazon), but when I first researched forages for goats years ago, I couldn't come up with much about chicory. Finally I searched for plants meant to be grown on deer plots (which some folks plant in order to fatten up future venison). I figured goats and deer can share the same feeds, so why not try it? I ordered some chicory seed and sowed it in our seasonal pasture. It has grown well. Interestingly, the goats avoid it when it's young and tender, then eat it with relish once it begins to go to flower.

Brassicas, such as turnips (tops and roots), kale, and rape (the source of canola oil), are another category of forb that can provide wonderful nutrition, but they must be fed carefully because most can block the

absorption of iodine. It is believed that ruminants can tolerate up to 10 percent of their diet in the form of these types of feed. The leafy part can be browsed and the roots harvested and stored as a winter feed.

Other root crops, such as beets, carrots, and radishes, can be grown and stored to feed through the winter. Large, hard roots, such as the giant yellow Eckendorf and mammoth red mangel beets, will need to be chopped some before they can be easily consumed. But they will keep longer than smaller roots.

Goats also relish pumpkins and other winter squash. Put an intact pumpkin in a feeder, and the goats will easily eat their way through to the seed cavity at the center. Winter squash can also be stored into the winter season.

Figure 5.13. Dairy goats at California's Redwood Hill Farm enjoy cuttings from tree lucerne, also known as tagasaste. *Photo courtesy of Jennifer Bice*

Many other prime or rejected garden crops, or parts of them, such as lettuce that has bolted, insect-damaged leafy greens, and garden peas, can be good food for goats. If you're in doubt, check the poisonous plant list in appendix D or consult with a regional feed and forage specialist.

Fresh Fodder

A variety of grains can be sprouted and grown without soil to the seedling stage and then fed, roots and all, to livestock. Barley is one of the most popular grains for this option. Grains are soaked for a period of days, spread in shallow trays, and then watered frequently and allowed to grow. The roots form a dense mat, and at several inches tall the entire tray of seedlings is peeled out of the container and fed. The goal of a well-planned system is to have the amount of fodder you need daily maturing at a steady rate. Skeptics cite the expense of operation, the loss of dry matter (how most nutritionists compare feed value), and the lack of fiber compared with a similar portion of conventional feed. Proponents of these systems cite nutritional data and the ability to control your feed supply. Of course someone else is growing the grains, so that cost is out of your hands as well as the management of those crops. While it isn't a closed circle, it does provide options for those without other available feed options and those able to invest in building a proper system and able to support the labor and energy costs involved.

Fodder Trees and Shrubs

There is tremendous nutritional value stored in plants that sink their roots deep within the earth, bringing minerals to the surface for goats to harvest. Some trees and shrubs are grown specifically for the purpose of feeding livestock (this practice is one type of *agroforestry*). More arid parts of the world, such as Africa and Australia (inspired by over a decade of record drought from the mid-1990s on), have learned to make use of fodder trees and shrubs that require little water once established and can grow in harsher

Figure 5.14. Hybrid forage willows from Australia (*left*) grow rapidly. They can be browsed to the ground and will then regrow, or branches can be trimmed and fed as needed. I harvest them and place them in the regular hay feeders (*right*). The goats enjoy both the leaves and bits of the bark.

conditions and topography than conventional feed crops. Black locust, hybrid willows, tree lucerne, and many more offer great potential as feed sources. Tree lucerne (*Cytisus proliferus*, or tagasaste) offers the additional benefit of increasing nitrogen content in the soil, just as the more familiar lucerne, or alfalfa, does. Typically these crops are cut and taken to the goats, but they can be rotationally browsed. Pay close attention to the soil and climate needs of these plants before you invest. (I spent about $500 on seedlings of one variety and hand-watered them through a long, hot summer, only to have them perish in the water-logged clay soil of the field I had selected.)

Foraged Feed

The word *forage* can be a noun, as we have used it so far, but it is also a verb meaning "to go out, look for food, harvest it, and bring it back." Goats browsing are foraging, but when they can't go to the forage, you can bring it to them. Grapevine trimmings from an unsprayed vineyard, tree trimmings from power companies clearing away branches near power lines, and plants or trimmings that are being removed from yards are all possibilities. Of course some effort is involved, and the size of your herd may limit the impact that this option can have on your feed bill. But goats benefit immensely from having variety in their diet, so consider it for health's sake if not budget. Avoid plants that are growing near public roads and irrigation ditches, as they are likely to contain herbicide residues.

Mineral Supplements

From chapter 4, you may recall that the precise amount of most minerals and vitamins needed for optimal goat health is not known. Companies that manufacture mineral mixes for goats, often with added vitamins, do their best to determine the average need of goats spread across a large geographic area. Sometimes the formulations are based on whether the goats are fed a grass- or legume-based diet. It is impossible for most companies to formulate a mix based on regional soil deficiencies and excesses, as

Figure 5.15. In many countries farmers forage and harvest feed for their goats. On the left goat kids at Monte Azul goat dairy in Costa Rica enjoy poró (*Erythrina poeppigiana*), a nutritious and economical feed. The needs of the goats on the right are being met through feeding harvested forage at Rainbow Children Home's farm, Pokhara, Nepal. They are tended lovingly by Goma Dhaka. *Photo at right courtesy of Daniel Laney*

well as goat breed differences. Nevertheless, supplying mineral, and likely vitamin, supplementation for your herd is imperative!

Mineral mixes typically contain all of the macro minerals (including sodium chloride—table salt) and most of the nutritionally recognized micro minerals. They are designed to be fed as the sole supplemental source of both salt and other minerals. Whether this is adequate or not is something you will have to assess over time (this is covered in the Determining Mineral Requirements in Goats section on page 78).

Mineral mixes should always be available for the goat to consume as desired, called free choice. Goat minerals can be fed as loose granules or in blocks. Blocks are long-lasting and more impervious to high moisture conditions, but most goats don't like them as much as loose mixes. Goats tend to avoid contact with something on which another goat's saliva has been deposited. (You can test this out with an apple or carrot—let one goat nibble on it and then see if another goat wants it.) But in some circumstances a block may

be the best choice. Blocks designed for meat goats often have a high percentage of protein and more salt to encourage them to utilize the block. Be aware that some mineral blocks or mixes labeled for meat goats include a coccidiostat (an antimicrobial), so they aren't suitable for organic practices. Also, some blocks are labeled for sheep and goats, and don't contain enough copper for goats but are safe for sheep.

Table 5.1 shows a comparison of several popular commercial mineral mixes. You can see that there are great differences among them. Some contain minerals that others don't, some contain a great deal of salt compared with the others, and one includes probiotic bacteria. You may not have many choices when it comes to finding a locally stocked mix. The one I prefer for our dairy herd, Sweetlix Caprine Magnum-Milk, is not available locally, but we stock up when we travel through the nearest city that does carry it. I find the Purina mix to be too high in salt; it attracts more moisture, and my goats don't seem to find it as palatable. If you feel that your mix might

Table 5.1. Mineral Mix Comparison Chart for Popular Blends

	Manna Pro	Sweetlix Caprine Magnum-Milk (for legume-based diets)	Sweetlix Meat Maker (for grass-based diets)	Purina	Golden Blend
Calcium min.	16%	7.5%	14%	9%	13%
Calcium max.	19.2%	9%	16.8%	10.8%	15.6%
Phosphorus	8%	8%	8%	8%	7%
Salt min.	12%	10%	10%	40%	20%
Salt max.	14.4%	12%	12%	45%	24%
Potassium	1.5%	Not listed	Not listed	0.1%	0.9%
Magnesium	1.5%	4.5%	1.5%	1%	1%
Cobalt	Not listed	240 ppm	240 ppm	Not listed	0.01%
Copper min.	1,350 ppm	1,750 ppm	1,750 ppm	1,750 ppm	0.15%
Copper max.	Not stated	1,810 ppm	1,810 ppm	1,800 ppm	Not stated
Manganese	2750 ppm	1.25%	1.25%	—	0.3%
Zinc	5,500 ppm	1.25%	1.25%	7,500 ppm	0.4%
Selenium min.	12 ppm	50 ppm	50 ppm	25 ppm	12 mg/lb.
Selenium max.	Not stated	Not stated	Not stated	30 ppm	Not stated
Iodine	Not listed	Not listed	Not listed	Not listed	0.01%
Vitamin A	300,000 IU/lb.	300,000 IU/lb.	300,000 IU/lb.	140,000 IU/lb.	220,000 USP units/lb.
Vitamin D_3	30,000 IU/lb.	30,000 IU/lb.	30,000 IU/lb.	1,100 IU/lb.	45,000 USP units/lb.
Vitamin E	400 IU/lb.	400 IU/lb.	400 IU/lb.	750 IU/lb.	220 IU/lb.
LAB (lactic acid bacteria–probiotics)	1.5 million CFU/lb.	None	None	None	None

be lacking in an element that is likely to be deficient in your feed, such as selenium or copper, you can address that with additional supplementation. I advise talking to your vet first, or at the least consulting other experienced goat breeders in your area.

I'm a convert to the free-choice individual mineral approach, sometimes called cafeteria- or buffet-style (which I described in the Offering Individual Minerals Free-Choice section on page 79). I'm only aware of two companies in the United States offering a full line of products that fill this need. Both are listed in appendix D.

Miscellaneous Supplements

Some important little extras round out a complete nutrition and holistic health program. Sodium

Antibiotics Masquerading as Feed Adjuncts

Antibiotics are wonderful drugs that can save lives and limbs, but the extent of their use today in livestock and humans is truly a tragedy. Meat and dairy, produced on a mega scale, would likely not exist without the extensive use of low, or subtherapeutic, levels of antibiotics (sometimes administered as "rumen modifiers") intended to shield young livestock from risks of coccidiosis and to promote weight gain—up to 20 percent faster than without antibiotics.[5] Rumen modifier antibiotics (not all rumen modifiers are antibiotics) include monensin (approved for use by the FDA[6] and found in meat goat minerals and feed mixes), tylosin, virginiamycin, lasalocid, and bambermycin. Rumen modifiers also assist rumen activity and reduce the incidence of bloat when a high-grain diet is fed, and they're commonly included in the feed of adult dairy and beef cattle.

The most frequently used rumen modifiers are from a group of antimicrobials called ionophores; this group includes monensin (Rumensin) and lasalocid. These aren't related to the antibiotics used in human medicine, so residues in milk and meat don't seem to be of concern to those who make these decisions, because they are thought "unlikely to contribute to an increase in antibiotic resistance."[7]

Rumensin, while approved in meat goat feed and for lactating dairy cows, isn't approved for lactating goats. I wasn't able to uncover the logic for this, but it is likely simply because the possible side effects (which are extensive in cows) haven't been properly studied. Probably this is a lucky thing for those dairy goats. There is concern among some, however, that ionophores might provide a competitive advantage to pathogens such as *E. coli*, which is gram-negative and not affected by these drugs.[8]

---

bicarbonate (baking soda), kelp, yeast, specific vitamins, bentonite clay, and probiotics are all a part of the feeding plan at our farm. Baking soda is cheap, but the others are not inexpensive—but neither is a herd that isn't at its peak. I want to remind you that, truly, all supplements are meant to bridge a gap created by domestication and farming of livestock. Because we can't duplicate a goat's ideal natural circumstances, it is our job to assess their nutritional needs and fill the gaps. As I researched for this book, I was reminded that as agriculture gets bigger, agricultural products are produced through ever-more-complex compromises. The synthetic fertilizers, herbicides, and pesticides that crops are subjected to and the general unhealth of the land on which

crops are grown mean that the goat farmer seeking to base herd health on nutrition is likely to need more supplements than ever before. Please keep that in mind as you read on.

BAKING SODA AND BUFFERS

Sodium bicarbonate, popularly called baking soda, is an often overlooked yet critically important supplement that should be available to goats at all times.

> Sodium bicarbonate, popularly called baking soda, is an often overlooked yet critically important supplement that should be available to goats at all times.

It doesn't supply direct nutrition, but it's a vital digestive aid. Goats produces buffers in their saliva that help them maintain the ideal rumen acid balance, or pH. There are times when they cannot produce enough to keep the right balance and will seek out and consume baking soda, or another provided buffer (several are available through feed suppliers), to help with the job. As with minerals, put out very little at a time, as a way to maintain freshness (think about how baking soda can be used to absorb odors in your fridge).

KELP

If you happen to have beach access for your herd, then you might be able to browse your goats on kelp that has washed ashore. If you do, be sure you are aware of the possible pollution or unfortunately even radiation contamination that might exist where the kelp has grown. Kelp meal, on the other hand, is harvested from the cleanest parts of the sea and then dried and flaked. Goats adore kelp meal. I have read that if goats are allowed kelp free-choice, they will binge at first and then moderate their intake. I tested this at our farm, and after going through about two bags per week (at about $60 per bag at the time) for a couple of weeks, I figured my goats would choose to live on the stuff if allowed. So deciding how much kelp to feed is a matter of budget and needs. Goats that have access to a great variety of browse, feed, and an iodine source are less likely to need the multitude of iodine and micronutrients available in kelp. Goats under stress or with health issues might benefit from more kelp. I basically decided how much kelp we could afford to feed, and that is what my goats get.

YEAST

Yeast is a natural rumen modifier. While research is not clear on just how yeast in the goat's diet helps the rumen bacteria, it is clear that it does. In 100 percent of the experiments involving dairy cattle (all done by the yeast industry, I should note), rumen bacteria increased by an overall average of 45 percent. This resulted in a slight increase in rumen pH and other positive effects. In most cases fiber digestion and milk production increased as well, but interestingly, milk protein decreased a bit.

One of the better-performing brands, and the one I use, is Diamond V XP (or XPC—the *C* stands for "concentrated"), which is a dead yeast. The belief is that it's the by-products of the yeast, not the living yeast itself, that supply micronutrients to help the rumen bacteria thrive and ferment. Yeast is not inexpensive, but it is fed only in very small amounts. Diamond recommends dosing XP at the following rates per day: kids ¼ ounce (7 g); adults ½ ounce (14 g). If you're using the XPC, cut that amount in half. And remember, you can give even less and probably still see benefits. Diamond XP is available in certified organic form. Even the non-certified form is non-GMO according to the company. If this brand isn't available, there are other equally effective products available. Yeast is also found added in some mineral blends and performance supplements.

BENTONITE CLAY (SODIUM BENTONITE)

Earlier in this chapter, I mentioned the likelihood of the presence of mycotoxins, a by-product of mold growth, in grains as well as in hay. Although unseen, mold growth on grain produces many variations on the potentially damaging toxins. Unless your feed is regularly tested for mycotoxin presence, I advise offering free-choice bentonite clay powder to your goats to reduce the chances of illness. You can purchase it bagged as sodium bentonite. My goats choose to eat it periodically. Bentonite is a form of clay and is used commercially for many purposes, including clarifying wine (more often as calcium bentonite), making beauty masks for people, and manufacturing clumping cat litter—pretty useful stuff, don't you think?

PROBIOTICS

Probiotics are bacteria that provide a known health benefit in the lower digestive system, or gut. I cover probiotics later in this book as a treatment for various ailments, but they should also be considered as a feed supplement to support animal health. In addition to helping maintain optimal gut health and performance, probiotics also have some other fantastic effects. In feedlot cattle, probiotics have been shown to reduce the shedding of the deadly *E. coli* 0157:H7 by more than 50 percent.[11] In mice,

Glyphosate Remediation Products

If you're concerned about the herbicide glyphosate, aka Roundup, but not able to guarantee that all of your land and the feed you purchase is organic, you might want to offer a remediation product to your animals. There is a great deal of research, often coauthored by German scientist and researcher Dr. Monika Krüger, regarding the detrimental effects of glyphosate on the health of animals. Most of this research is dismissed or countered by opposing research by industrial farming concerns, including those that market and sell the most widely used herbicide in the world. This is quite understandable because established procedures, industry reliance, and a lot of money all hinge on the perceived safety of this product. (One is reminded of the dismissal of concerns over the use of the pesticide DDT and the efforts to discount reports and insinuations of its risks in the 1960s.)

Glyphosate is not only used on crops engineered to survive its application, but also on glyphosate-sensitive, non-GMO crops to perform helpful tasks such as killing an entire crop of wheat plants at one time so that the harvest can dry quickly and evenly, or chemically "ripening" sugarcane (its application increases the sugar content of the plant and so the yield and profits). The uses are growing. Monsanto, in recognition of the chemical's ability to kill some bacteria, has even patented it as an antibiotic.

In several papers coauthored by Dr. Krüger, glyphosate is linked to many things, including changes in the level of minerals and urea in blood serum levels of dairy cows, as well as increased levels of enzymes indicative of toxicity[9] and increased cases of botulism in dairy cows.[10] Many other peer-reviewed documents, easily found via an Internet search, implicate glyphosate as suspect in other unsavory health issues for humans (to be fair, there are as many as or more of these supporting its safety). My point here is not to attempt to prove the issue, but to let you know that if this chemical and its residues are of concern, but you're unable to purchase and use only organic products, you do have options.

What are those options? In another paper coauthored by Dr. Krüger, 380 dairy cows were evaluated as part of the research cited above into botulism. Glyphosate was neutralized in the animals' systems through the feeding of charcoal, charcoal combined with sauerkraut juice (it's a German study), or humic acid matter (as in soil humus). Advanced Biological Concepts, one of the companies I recommend as a source of individual minerals, makes a product mix that includes as its first ingredients "lactobacillus acidophilus fermentation product and reed sedge peat moss," so basically humic acid. It also includes minerals and supplements meant to support the challenges to animals' systems that consuming glyphosate and plants exposed to it is thought to cause.

raised free from any gut bacteria and then purposely infected with another deadly pathogen, *Listeria monocytogenes*, probiotic bacteria protected against death.[12] Probiotics act in two ways to help protect the animal: first through competitive exclusion, where their population helps crowd the gut wall so that there is basically no room for a pathogen, and second through direct enhancement of the gut lining's immune response.

Probiotics can be given individually, but for feed supplementation they are best mixed in with the ration. I add probiotic feed granules (they are gritty, not a fine powder) to the kelp meal once a week. Be sure to store all probiotic supplements as directed. They contain live (but inactive) bacteria that will suffer and become less viable if not stored properly. Probiotics are now frequently found in other mixes and feed supplements.

Putting It All Together

At this point, are you feeling overwhelmed by so many choices and options? Are you hoping I'll give you my formula for how many pounds of hay, grain, and other components to feed your goats on a daily basis? Even back in my dairy cow days, I couldn't wrap my mind around calculating feed rations. Now I understand my aversion to this approach. For one thing, it requires the assumption that we know the complete needs of the animal. Second, it assumes that we are willing to fully control what our animals eat. This is the way that confinement feeding on concentrated feed operations must occur, but it is the antithesis of what I teach at workshops and what I practice on my own farm. I want you to understand the primal needs of the animal, assess your options (which are likely unique for every farm), and then learn to daily evaluate your herd and how it is thriving.

I can give you this starting point: Provide fresh, pure water; a free-choice mineral blend, and free-choice baking soda at all times. Daily, or as frequently as possible, provide your goats with access to as much browse or graze as they will eat, and then throw in a bit of high-quality grass or other hay—but only as much as they will clean up quickly and still be hungry enough to go out to browse again. From this starting point, modify and adapt what you provide based on the types of feeds you have access to. And always, always watch how your goats feed and behave, and how their bodies look—they will show you if they aren't well nourished!

PART III

Managing Herd Health

CHAPTER SIX

Fundamental Skills and Knowledge for Goat Care

In most cases, when a goat's psychological, physical, and nutritional needs are being met, that goat will be trouble-free, healthy, happy, and productive. The goat enjoys a state of homeostasis, where all systems are in balance. But there are times when goats, especially those working hard to produce offspring or make milk, are vulnerable to illness or injury. Young does and bucks also experience natural times of increased stress, such as at weaning and during breeding season, and therefore increased risk of sickness.

When illness does strike, keen observation skills and quick action are the keys to preventing disaster. This might mean providing supportive therapy through remedies you can give on the farm, including nutritional, herbal, and homeopathic, or involving a veterinarian. However, there will be times when no matter what you do, the worst happens. This chapter will help you do the best you can to care for your goats and return them to a healthy, vibrant state of homeostasis.

Top Priority: Finding a Good Veterinarian

Whether you are a brand-new goat owner, an experienced goat farmer, or are just thinking about getting

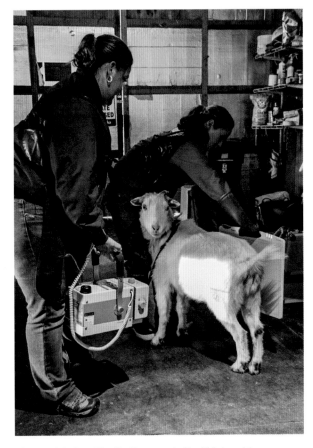

Figure 6.1. Our Pholia Farm vet, Dr. Kristen Mason, takes a radiograph (X-ray) of an ailing goat, a wether, to rule out the presence of urinary stones.

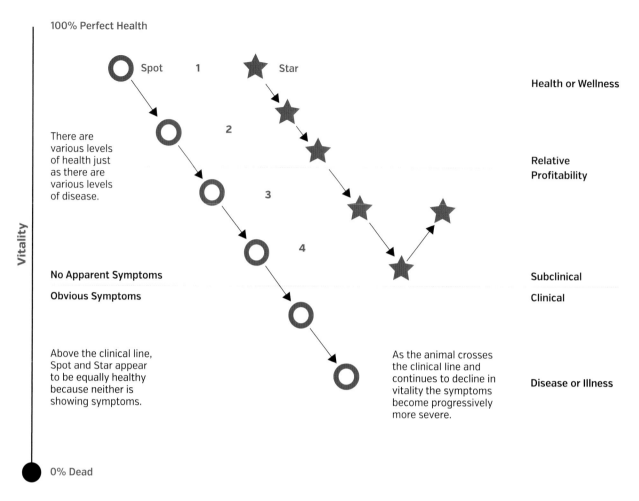

100% Perfect Health

Spot   1   Star

Health or Wellness

There are various levels of health just as there are various levels of disease.

2

Relative Profitability

3

Vitality

4

No Apparent Symptoms

Subclinical

Obvious Symptoms

Clinical

Above the clinical line, Spot and Star appear to be equally healthy because neither is showing symptoms.

As the animal crosses the clinical line and continues to decline in vitality the symptoms become progressively more severe.

Disease or Illness

0% Dead

Figure 6.2. I love how this chart helps us visualize that even when a goat looks healthy, it is actually in a fluctuating state between ideal health and subclinical illness. In this example two animals, spot and star, were both in declining health, but star recovered on her own before symptoms were observed. When symptoms are present, a rapid response is important in order to prevent a decline. *Reprinted with permission from Advanced Biological Concepts*

goats, my first health care recommendation is that you select a veterinarian and formally become a patient or client. If you don't have a vet who has you as a client, you're not likely to be able to count on help if an emergency arises. I experienced this after my husband, kids, and I moved our herd from Southern California, where my husband had completed his last duty station with the US Marine Corps, back to my family's original farm in Oregon. A young kid became very sick, and no local vet could see us for weeks. The kid died, and I learned a tough lesson. Now we are

fortunate to have developed a great relationship with a wonderful veterinarian, Dr. Kristen Mason.

Finding a conveniently located or mobile vet who is adept at goat medicine and willing to take you on as a client is not an easy quest. Finding one who also takes a holistic approach is even more challenging. The goat industry is a very small part of agriculture in many countries. Because of this, goat medicine is addressed only briefly in vet school. The approach of many vets is to treat goats as almost a cross between a cow and a sheep.

Ideally you'll find a vet who takes a holistic approach and is adept at using conventional (sometimes also called allopathic) medicine but also eager to try alternative, or *complementary*, approaches. The term for this approach is *integrative medicine*, which implies that all approaches can and should be used to help heal the sick and ailing. Complementary medicine focuses on healing and strengthening the whole body, rather than simply treating symptoms. This approach includes nutrition, physical therapy (exercise and activity), herbal remedies, chiropractic care, homeopathy, energy bodywork, acupuncture, and more. A growing number of licensed veterinarians throughout the world are embracing many of these practices, especially veterinarians who care for small animals (dogs and cats). Whatever vet you work with, let them know you're willing to explore any treatment that might help your animals—but are especially interested in those meant to heal the body overall.

Instead of making experience with goats your top criterion for selecting a vet, make your first priority an interest in and willingness to learn about goats. And preferably, seek out a vet who takes a holistic, multifaceted treatment approach. The best veterinarians are those who are not only skilled but also that treat you, the goat owner, as an equal partner in problem solving. Of course, this means that you have a huge responsibility to educate yourself so you can provide your vet with helpful, accurate information.

A Primer on Complementary Therapies

There are a good number of complementary approaches to healing, also sometimes called alternative medicine, which can be useful in treating animals. Here's a brief description of those that I've had some exposure or experience with.

Acupuncture and acupressure: Both of these therapies originated in ancient China and have been used there for centuries. They are based on stimulation of a specific point on the body that correlates with another part of the body, often internal. In acupuncture a very fine needle is used to pierce these points, stimulating the flow of healing energy along a "meridian" or pathway. Acupressure uses the same concept, but by applying finger pressure to the points.

Learning acupuncture requires significant training, but laypeople can learn some basic acupressure. This can be done on the advice of an acupuncturist to help follow up the treatment they have provided.

I've used acupuncture only one time, on a horse that had a joint problem. It didn't provide a long-term solution, but I also wasn't able to afford many treatments! The therapy is growing in acceptance.

Chiropractic: This treatment option originated in the United States in the late 1800s. It involves the manipulation of the spine, or "adjustments," in a way that is believed to improve nerve conduction throughout the body. If you read the introduction to this book, you know that my parents were both chiropractors—adjustments were a part of my rearing! I've used the services of other chiropractors several times in my adult life, with some great results. I've had chiropractic work performed on a goat only once and several times on a horse (the same one I mentioned above). As with acupuncture, budget constraints are a limiting factor for thoroughly experiencing chiropractic treatments that are meant to be ongoing.

Herbal and botanical: This type of complementary therapy is quite accessible to producers, and a great deal has been written on it by experts. I cover this approach in more depth later in this chapter beginning on page 136.

Homeopathy: This form of therapy originated in Germany in the late 1700s and early 1800s. Remedies contain extremely diluted solutions of a substance that, when taken in stronger forms, causes symptoms identical to the malady being treated. This philosophy is known as "like cures like." It's a difficult concept to commit your trust to, but because the remedies are so extremely diluted (in fact, the more diluted they are, the more expensive they

Holistic Vet Susan Beal

DVM, CVH, ONTARIO, CANADA

I first met Sue Beal at an Acres USA conference where she presented a class on parasite control in goats and sheep. I was impressed by the depth of information and holistic treatments she presented, and even more so by her frankness about the burden that producers face in selecting for hardy animals in their breeding programs—in other words, culling. This was a brave woman, both encouraging and chastising at the same time.

If any vet embodies the term *holistic practitioner*, it's got to be Sue. She graduated with her veterinary degree from the University of Guelph in Ontario, Canada, in 1987 and went on to train in acupuncture, chiropractic, and craniosacral therapy. She received her master's degree in homeopathy from the Hudson Valley School of Homeopathy in 2006. Sue also regularly incorporates herbal and energy bodywork into the treatment of many animals. She's practiced on small and large animals both in Canada and in the United States. Her work with goats began in a roundabout fashion through treating horses—many of which have a goat companion.

Much of Sue's work focuses on sharing her knowledge with producers and veterinarians alike. In 2002, as a part of the group Homeopaths Without

Figure 6.3. Dr. Susan Beal tends to all varieties of livestock. Here she takes a break with heritage hogs at Forks Farm in Orangeville, Pennsylvania. *Photo courtesy of Tracey L. Coulter*

Borders, she was the first vet to travel to Cuba to share homeopathic and other holistic principles with vets and medical doctors. She's a part of the American Holistic Veterinary Medical Association. (You can visit the AHVMA website and search for a member vet at www.ahvma.org.) She also speaks regularly at sustainable agriculture conferences in North America.

Sue has some great advice and guidance for goat farmers wanting to improve their holistic skills and practices. "Always start by approaching the whole system—the animals and the farm—when troubleshooting problems. You have to first gather information and try to choose the best resources to solve the problem. If you don't, you'll dip your toe in the river of conventional medicine and suddenly find yourself 75 miles downstream." She recommends the book *The Pet Lover's Guide to Natural Healing for Cats and Dogs* by Barbara Fougère (listed in appendix D), saying, "It's easy to translate the recommendations to goats and other livestock." In the appendix you can also find a list of searchable websites for holistic vets.

Since finding a holistic vet in your area is unfortunately often a futile search, here are Dr. Beal's tips for working with a conventional veterinarian:

- Identify what you need the vet for, then make that your priority when selecting.
- Do you need them for hands-on work such as blood draws, helping with deliveries, euthanasia?
- Do you need them for regulatory reasons such as health certificates?
- Do you need them for access to prescription drugs? (For example, the vitamin E and selenium supplement Bo-Se requires a prescription.)
- Learn to gather information on your herd health before contacting the vet to make sure you get the most out of the experience as well as facilitate the vet's efforts.
- As a holistic producer, work toward building bridges that bring conventionally focused vets toward the holistic side. Do this by not directly challenging their conventional paradigm but utilizing the resources they offer while you demonstrate holistic practices.

are—as they are thought to be more effective), they are very safe to try. I've only recently started using these remedies but have known so many people for whom they have been effective. The critical part of using homeopathy involves the very accurate reading of symptoms, because symptoms' details are the basis on which to select the remedy. For example, you can't simply select a remedy based on "headache." The headache location (right side, left side, front of forehead) and other symptoms matter greatly. This is why the trained homeopath is so valuable.

Reading the Signs

You don't need a crystal ball to tell whether your goats are healthy; you just need a set of discerning eyes and a pair of skilled hands combined with knowledge and common sense. If you observe your goats on a regular basis, the entire herd—from young to old and on all parts of your farm—you will learn almost everything you need to know about their health. Looking at and listening to your goats is not sufficient, though. As part of your inspection, use your hands to feel and palpate for anything amiss. Your sense of smell, too, can provide clues. When any of your senses detect a clue that something is not right, you can then use other tools—a thermometer, stethoscope, symptom-driven tables (such as those in part 5 of this book), even a microscope—to further troubleshoot the problem. You will be able to use your ability to read the signs and symptoms to take the diagnosis to the next level.

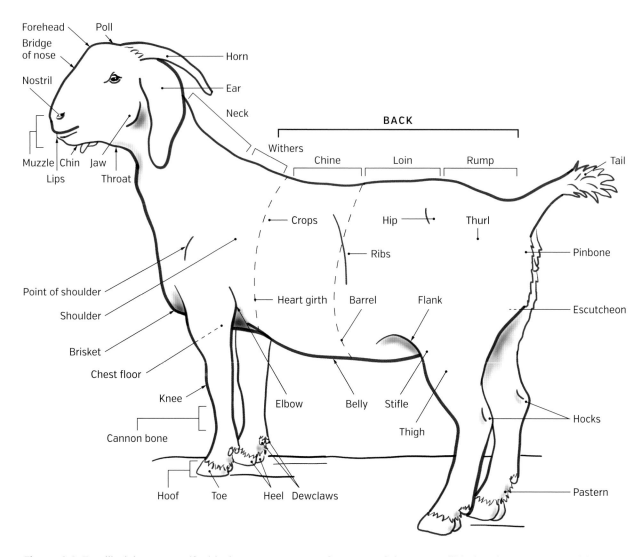

Figure 6.4. Familiarizing yourself with the correct names for parts of the goat will help when you're trouble-shooting and solving health problems on the farm, as well as describing symptoms to your veterinarian. This illustration shows a gender-neutral goat. See figures 15.1 and 16.1 for the mammary system of the doe and the reproductive system of the buck.

If inspecting your herd is new to you, it will take some practice to develop your skill. You should routinely assess demeanor; posture; the condition of the eyes, coat, and skin; body condition; feces; and vital signs of your goats.

Demeanor

I have a saying: "If a young goat isn't obnoxious, then something is wrong." Although this is a bit of an exaggeration, a healthy goat is active, eager, inquisitive, and even, perhaps, a bit annoying. A sick goat is tired, disinterested, and depressed. A healthy goat carries its tail curved over its back or upright, ears alert and erect (unless the goat has naturally floppy ears). Its gaze is intense, and it responds to sounds and movement with interest. That's a good general description, but you must develop a sense of each individual's personality

Table 6.1. Troubleshooting Demeanor Symptoms

Signs and Symptoms	Possible Cause	What to Check
Holding tail down for more than a few minutes at a stretch or ignoring usually interesting events	• Depression or illness	• Check temperature • Watch for appetite and decreased rumination time
Grinding teeth, frequent shaking of entire body, or pressing head against wall	• Pain	• Watch for appetite and decreased rumination time • Observe feces for signs of dehydration • Observe for abdominal distension, bloat • Watch for rumen movement
Lack of appetite	• Pain, rumen dysfunction, metabolic imbalance, and other conditions	• Check for rumen movement • Observe for abdominal distension, bloat • Assess age and risk factors for coccidiosis, pneumonia

Note: This table is a starting point for diagnosing problems and taking action. Consult part 5 for more detailed lists of symptoms, and be ready to consult with your vet when the situation requires it.

Desirable Demeanor Checklist

☑ Keeps tail up
☑ Responds to movement and sounds with interest
☑ Behaves normally (in keeping with its usual personality)
☑ Shows normal appetite
☑ Ruminates periodically

---

in order to determine the degree of change in the animal. For example, if the goat that's normally the bossiest in the herd suddenly hangs back, not pushing their way to the feeder or running to the milk stand to be first for grain, then you should assume that something is wrong with them. A normally docile and passive goat will be a bit more difficult to assess, but if you are aware of her personality, you'll know when it's not normal.

DEMEANOR CASE STUDY

Some life lessons are learned through tragedy. In the spring of 2015, I sold a valuable young buck kid to a home that had eagerly awaited his birth, as he was practically a prince—the offspring of our best doe and probably the greatest sire we have ever owned. He was one of only two Nigerian Dwarf bucklings that year to qualify for the American Dairy Goat Association's Sire Development Program—based on superior genetic potential. Needless to say, his price also made him quite an investment for the new owners. He left our farm in good health at two weeks of age.

About a month later, they gave me the sad news of his death from coccidiosis. This subtle and insidious killer can sneak up on any goat; in a very short period of time an animal that appears normal ends up dead. The only way to avoid a coccidiosis disaster is through extreme vigilance and by anticipating high-risk times. In the case of Jasper Blue, the owners had been occupied with a goat show (they didn't take him to the show). The next day, they were tired from being at the show. They noticed that Jasper looked sleepy that evening, but it didn't ring the alarm bells that it should have. By the following morning he was obviously suffering and by noon he was dead.

Those unhappy folks learned their lesson and have implemented a plan to check inner eyelid color of their goats to watch for anemia, one of the major symptoms of coccidiosis. (See the Inner Eyelid Color and FAMACHA section on page 126 for more about interpreting inner eyelid color.) And no doubt, demeanor changes will now sound the alarm in time to increase the odds of successful treatment.

Posture

A healthy adult goat spends most of their day eating, ruminating, and sleeping—often with a bit of frolicking and fighting thrown in. They may spend some time lying down, either to sleep or to chew their cud. When sleeping, goats usually lie with their sternum directly on the ground and their heads resting on the ground in front or curved back to one side of their bodies. Goats also lie on their sides to sleep, but only for short periods of time, usually less than 30 minutes. If they remain on one side for too long, then gas may build up in their rumen, lead to bloat, and put deadly pressure on other organs.

Goats also spend quite a bit of time standing and chewing their cuds. They should stand with the back level, and head either up or level with the back. They should stand evenly on all four feet with their front legs directly under their shoulders and their back legs also even. They will shift their weight now and then, but not stand with one back leg bent, like a horse does. For some reason, many goats seem to enjoy standing with their heads under a ledge or through a hole in the fence or other opening. While not quite the same as the metaphorical image of an ostrich with its head in the sand, it is still quite amusing to see some of the places they like to poke their heads.

POSTURE CASE STUDY

Marigold was the smallest doe in our pen of yearling does. The last daughter of one of our finest does ever, Marigold was sweet and promising. Most of the young does, born the last season, had been bred, but not Marigold. The pen, while spacious, was at maximum capacity when it came to feeder space.

One December morning, I noticed Marigold standing on her own near a play structure in the pen—in the rain. Not a good sign. Abnormal behavior, as goats abhor rain. I immediately moved her to a sheltered adjoining pen. As soon as I moved her, she cried out with the mournful, moaning cry of a goat in pain.

Her temperature was low, 99.9°F (37.7°C). While a low temperature can indicate that body systems are failing and a goat is dying, in this case she could simply have been chilled from standing in the rain. Her eyes, especially the right, appeared to be slightly bulging, her gaze intense, and she held her head slightly tipped to the left. It almost appeared as if she couldn't see well.

I immediately thought of listeriosis and polioencephalomalacia, aka goat polio. Listeriosis causes a goat to lose vision and response on one side, making it move in a circle. And goat polio, a severe vitamin B_1 (thiamine) deficiency, will cause varying degrees of blindness and will initially manifest as "stargazing," when the goat looks upward and appears somewhat dazed. The other possibility was trauma.

I covered Marigold with a warm goat coat, gave her an injection of B complex, T-Relief, echinacea and goldenseal, and probiotics and offered feed and warm water—both of which she eagerly attacked. Her temperature returned to normal quickly, and I observed her ruminating a bit later. She responded to my fingers moving in front of both sides of her head, and her ears were clear of any discharge. When I manipulated her spine, she did not object until I lowered her head; then she cried out.

Desirable Posture Checklist

- ☑ Stands with back level and feet even
- ☑ Moves evenly, holds head level
- ☑ Sleeps mostly in sternal position

Table 6.2. Troubleshooting Posture Symptoms

Signs and Symptoms	Possible Cause	What to Check
Standing with back arched	• Discomfort or fever	• Check for fever • Watch for decreased rumination time • Watch for loss of appetite • Watch for signs of pain, such as grinding teeth • Observe feces for signs of dehydration
Shifting weight frequently from one leg to another; holding up one limb, walking with limp	• Abdominal or leg/foot pain • General discomfort	• Palpate and assess abdomen for distension and discomfort • Palpate each leg, and note any swelling or reactions of discomfort • Lift each foot in turn, and inspect hooves for inflammation and wounds; note whether this causes a pain response from a particular foot or leg
Kicking at belly	• Abdominal pain	• Palpate and assess abdomen for distension and discomfort • Check for rumen movement • Watch for straining during urination (especially in bucks and wethers, because they are more prone to urinary stone formation)
Walking in a circle	• Brain inflammation, such as from listeriosis	• Check blink responses on each side of face
Holding head tipped to one side	• Inner ear problem or trauma	• Inspect ears for discharge • Gently move head and neck of animal and note any reaction of discomfort

After a few days and no improvement, I gave her a couple of days of Banamine injections—an injectable anti-inflammatory and pain control. After a week, she showed no improvement. Worried that I had overlooked something, I took her to our veterinarian. Her diagnosis: trauma to the upper spinal column. She recommended administering Banamine for a longer period of time, making sure that the doe could eat and do everything at a comfortable level, and giving it time.

You can't totally isolate goats for a long period of time. They will be lonely and, upon return to the herd, be treated as if they are new goats. So I moved Marigold's closest herdmate into the isolation pen with her and continued with the treatment. But she grew worse.

She began crying out when getting up or lying down and in anticipation of me touching her. Her demeanor seemed to shout terror. The dilemma: How long do you wait? How long do you let an animal suffer in the hope that she will heal? The longer I kept her separated, the harder her return to the herd would be. The same for her companion, who also was not relishing the confinement. Would she ever heal to the point of a life without pain? These are the times I detest—trying to make the right decision now based on what you think the future might hold. When I'm faced with a decision like this, I sometimes reflect on my own experiences, and those of my friends, with enduring pain. We are all going to experience pain some time in our lives, occasionally so intense that we move and sound as if it's insufferable, but we aren't ready to be put down! So I make the judgment for the animal based not only on the degree of pain it seems to show but also on how its situation progresses and how it's coping with it—does she eat and ruminate? If so, I give it more time.

So how did the story of Marigold end? The following spring, Marigold kidded with three lovely babies and is now a fully functioning part of the herd—with a cute tilt to her head.

Eyes, Coat, and Skin

A healthy goat's eyes should be bright and clear, free of clouding or drainage. The skin around the eyes and inner eyelid should be dark pink to rose—not pale or deep red. The coat should be smooth and glossy (unless it's naturally of a different texture). A dark-colored goat should look shiny, but a white goat may not, simply because white hair doesn't reflect light as well. The coat of a short-haired goat should lie flat. When a goat doesn't feel well, or has a fever, its coat will be fluffed up, with hairs standing on end. In the winter the coat should be full and thick. In colder climates, even goats

Desirable Eyes, Coat, and Skin Checklist

☑ Eyes clear and bright
☑ Skin of inner eyelids is dark pink to rose
☑ Coat is sleek and lies flat
☑ Skin is supple and free from blemishes

that aren't fiber producers will likely grow a thick underlayer of fine cashmere. A goat's skin should be supple and free from flaking, scabs, or hair loss.

INNER EYELID COLOR AND FAMACHA

The most common reason for a goat to be anemic is the loss of blood from the infestation of parasites coccidia and *Haemonchus contortus*, or barber pole worm, with *Haemonchus* being the primary cause. Other parasites might be present, but they don't cause anemia. A healthy goat who isn't suffering from anemia will have an inner eyelid color that is medium pink to rose. As the color gets lighter, it means that the goat is more anemic. Pure white is seen just before the animal is likely to die.

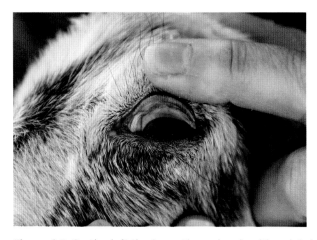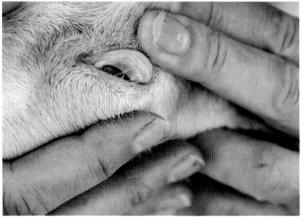

Figure 6.5. On the left the inner tissue is a healthy pink, but on the right it's pale. The doe kid on the right died shortly after this photo; a necropsy revealed the cause to be coccidiosis.

With a little training and practice, every sighted, non-color-blind goat farmer can learn to quickly and accurately find out if their goats are anemic—meaning their blood volume or quality is low. The most popular approach to this technique goes by name of FAMA-CHA, but you can also learn to do it without taking an official FAMACHA class. This is a free class offered through the University of Rhode Island; you must take the class in order to obtain the color card used for rating eyelid color. The color card is quite valuable as a tool to make sure that your appraisal is consistent each time you check the animal. See appendix D for information on how to register for a class.

You can also find many images of the card online, and you may feel comfortable using that as your starting point for understanding the range of eyelid colors you may see. The most important aspect of using this system to determine the status of your goats is to do it regularly on all animals until you are familiar with the color variations and how that correlates with the health of the animals. Do it immediately on any animal showing symptoms of anemia or coccidiosis, and do it very frequently on high-risk groups, such as kids at six to eight weeks of age or animals on wet pasture. By that age, kids have likely been exposed to enough oocysts to potentially develop an infection

Table 6.3. Troubleshooting Eye, Coat, and Skin Symptoms

Signs and Symptoms	Possible Meaning	What to Check
Coat standing on end, goat's entire body is "fluffed up"	• Fever • Pain	• Check for fever • Watch for loss of appetite and decreased rumination time
Eyes partially closed	• Pain • Exhaustion	• Check to see if goat will "wake up" to stimuli such as feed or exercise • Observe for signs of pain
Lack of blinking response on one side when stimulated	• Central nervous system problem, such as listeriosis	• Treat as for pain; consult with vet • Observe for appetite and ability to chew and swallow
Drainage from eye	• Foreign body in eye • Eye injury	• Inspect eye and inner eyelid carefully (wear gloves)
Eye cloudy or bloody	• Eye injury • Infectious conjunctivitis	• Treat for eye problems • Observe the rest of herd in case of contagious cause
Inner eyelid not bright pink to rose	• Anemia due to parasites or loss of blood from other cause	• Analyze for parasites by conducting a fecal float test • Treat as for dehydration • In case of blood loss, give nutritional supplement with iron
Skin scabby, patchy, or dry and flaky	• Nutritional deficiency • Lice or mites • Ringworm or other fungal problem	• Look for signs of pests such as lice or mites (see chapter 13 for more on skin problems) • Rule out ringworm

and are likely undergoing stress because of weaning. Don't forget that other causes of stress, including weather changes, that overlap weaning will increase the odds of an outbreak.

A SKIN CASE STUDY

At age three, a LaMancha-cross doe started having patches of scabby, itchy skin on her withers and around her tail head. The scabs were dry and flaky but became bloodied thanks to Idgy's successful attempts to scratch the irritated areas. She was so bothered by this condition, in fact, that we started calling her Itchy instead of Idgy. No other goats in the herd showed any similar symptoms, so a contagious condition was unlikely. I tried several remedies, both topical and oral. An intern at our farm, who happened to be licensed as a homeopath in Ireland, administered homeopathic treatments, but those also gave no results. By the time Idgy kidded the following January, the scabs had spread to her udder and belly. Some were so thick that they extended about 1/4 to 3/8 inch (6.4–9.5 mm) from the surface of her skin.

A month following her kidding, I began offering a selection of free-choice individual minerals, including a sulfur block. Idgy was particularly interested in the sulfur block, and within a few months her coat was back to normal. While it is possible that this recovery was coincidental, I feel strongly that most chronic, non-infectious skin and coat problems are directly related to nutritional deficiencies. Rather than "treating" the problem, it is much easier to provide the solutions and let the animals treat themselves. (Offering free-choice minerals was covered in detail in chapter 4.)

Figure 6.6. Scabs cover Idgy's skin. This photo was taken before the sulfur block was made available to her.

Body Condition

Goats are often so covered with thick hair that it is difficult to simply look at them and determine if they are too fat or too thin—unless an animal is extremely under- or overweight, that is. In addition, some goats have very large, expanded abdominal walls (called a dropped stomach; shown in figure 10.2) that to the untrained eye makes them look overdue to have babies or quite portly. To properly analyze the body condition of the average animal, you must put your hands on it.

Unlike cattle, goats do not deposit fat in muscle tissue. (This is one reason that goat meat is never as tender or succulent as beef: In the vocabulary of the meat industry, goat meat does not marble.) Goats accumulate fat mostly inside the abdominal cavity and on organs. Even a goat that feels and looks slender on the outside might reveal a tremendous amount of fat on the inside when butchered. These fat deposits are quite unhealthy and cause myriad problems with health and fertility. When you can feel and see *external* fat deposits, called pones or rounds, you know the goat is seriously overweight. Once a goat becomes overweight, it is very difficult to safely and effectively "put them on a diet" because a ruminant needs a constant input of fiber and nutrients in order to keep the rumen working well.

An underweight goat might have just used up its excess stores of energy, or it might be seriously ill. A too-thin goat is just as much at risk for developing other health problems as is an overweight goat.

Body condition can be checked and scored by feeling the animal's back, especially over the loin. The muscling along the loin, as well as the presence of a layer of fat under the skin, is judged by touching the top of the spine and feeling for its prominence, then gripping the transverse processes and feeling the layer of muscle and fat with the heel of your hand. If a goat is overweight, you can feel external fat deposits just behind its elbows and often on its sternum. If external fat deposits are palpable, there is likely also a great deal of internal fat. Keep in mind that there is some variation on what would be considered acceptable and desirable body condition scores depending

Figure 6.7. This Boer doe is recovering from an almost deadly parasite problem and still shows poor body condition.

Body Condition Checklist

- ☑ Ribs are palpable but not prominent
- ☑ Ribs are barely or just visible (dairy goat) (when the coat is short)
- ☑ Tissue behind elbows is smooth and flat to body wall
- ☑ Spine topline (spinal process) is not elevated, and muscle tissue running along the sides of the spine is not indented

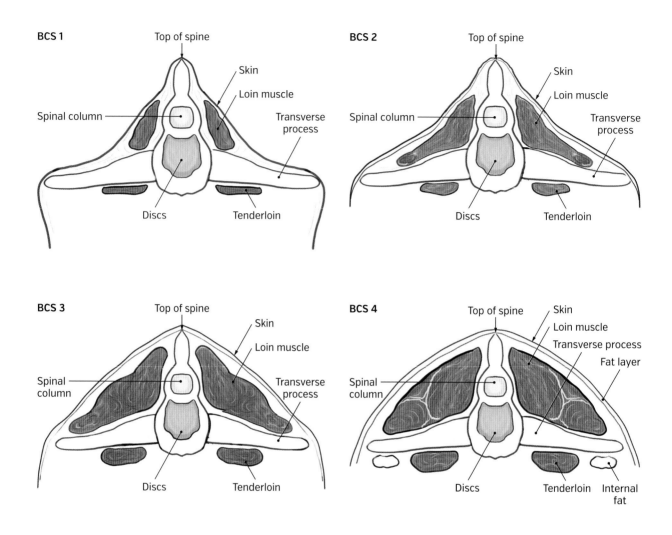

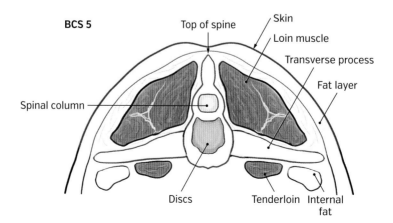

Figure 6.8. Scoring the body condition of a goat is a way to evaluate whether the animal is too thin (BCS 1) or too fat (BCS 5).

on what type of goats are being raised, their age, and job. Here are descriptions of body condition scores, which correlate with figure 6.8:

BCS 1: The spine is prominent and sharp, with no palpable layer of fat, and very little muscle.

BCS 2: The spine is prominent, but not as sharp, with a thin layer of fat and muscling.

BCS 3: The spine isn't prominent but can be felt. There is a good layer of muscling and a thin layer of fat.

BCS 4: The spine isn't prominent and is covered by a thin layer of fat. The muscling is heavy and is covered with a fairly thick layer of fat.

BCS 5: The spine can't be easily felt and is covered with a layer of fat. The muscling is heavy but concealed by fat. The barrel extends outward to accommodate internal fat deposits as well.

When a goat is in good condition, it should be easy to feel its ribs, but they shouldn't be highly visible. There should be no palpable fat behind its elbows, and it still should have adequate muscling for its job in its legs and back. An adult doe that's pregnant should have a bit more body condition as she prepares for kidding—but not too much! A buck will lose a good deal of body weight, both muscle and fat, during breeding season—even if he is not being used. So he must begin breeding season with a bit better body conditioning. Again, don't forget that the different types of goat—whether for milk, meat, or fiber—have different looks. For example, a milking dairy doe will look more slender than a meat doe nursing kids. You need to be familiar with the ideal general appearance of your breed of goat. (For an overview of how to evaluate characteristics of different types of working goats, refer back to chapter 2.)

Feces

When you are strolling among your herd doing your daily health assessment, be sure to include a good look at the fresh manure in the pen. While manure pellets are not exactly crystal balls, they are nuggets of information about your goats' health. Goats usually poop when standing but will also make a nice little pile while they are lying down sleeping. They also poop on the move (but usually stop to pee).

While urine characteristics can tell you something about the health of an animal, its feces reveal much more. A healthy goat should have well-formed pellets (sometimes affectionately called nanny berries, Milk Duds, or Raisinets) that are dark brownish green and moist enough to easily squish into your boot treads. An unhealthy goat might have runny, watery diarrhea (also called *scours*); mucousy poop; pellets that are strung together like beads;

Normal Feces Checklist

- ☑ Even, round or oval-shaped pellets
- ☑ Color is dark brown/green
- ☑ Texture is firm but moist
- ☑ Output is consistent, with defecation taking place several times each day

Abnormal Feces Characteristics

- Runny
- Soft "dog poop" appearance
- Bloody
- Mucus present
- "Rosary bead" appearance
- Small, hard, shriveled pellets
- Foul odor
- No feces or unusually small quantity of feces

Figure 6.9. Abnormal, mucus-coated feces sometimes form when a doe is pregnant, usually close to the end of pregnancy. They rarely recur and don't seem to be indicative of any lasting problem. If they continue, a fecal sample should be analyzed for parasites, and the animal should be observed for digestive symptoms.

Vital Signs Checklist

☑ Rectal body temperature is between 102 and 104°F (39–40°C)
☑ Breathing is even and deep, with respirations of:
　　☑ Kids: 20–40 per minute at rest
　　☑ Adults: 10–30 per minute at rest
☑ Rumen makes frequent rumbling sounds and undulating movements at a rate of no less than one per minute; when listening with a stethoscope, the optimal sound is a "roar," not just a gurgle or simple rumble
☑ Lungs sound clear, with no crackling sounds or wheezing apparent
☑ Heart rate (remember to count every *lub-dub* as one beat, not two!) is even, at a rate of:
　　☑ Kids 140–200 (too fast to count easily!)
　　☑ Adults 70–90 (the rate can vary greatly in some breeds)

malformed poop; no poop; or dry, hard pellets. You may notice a particular goat producing one of these telltale by-products, but also be sure to pay attention when cleaning out goat pens. It might alert you to a problem you weren't aware of.

Vital Signs

Vital signs are readings of body temperature, rate of breathing, and heartbeat—vital information for helping troubleshoot problems. No goat owner can thoroughly assess a goat's symptoms without the use of a thermometer. An inexpensive digital thermometer can be found in the infant or health care section of any grocery store or pharmacy. Checking the goat's temperature is the first thing to do when it doesn't act normally or seems to be ill. A stethoscope, while not absolutely necessary, is super handy in many cases. Having one of these will help

you listen for rumen sounds, check a goat's pulse, and listen to the lungs. Inexpensive versions can be found in any pharmacy.

The Remedy and Treatment Cabinet

"Better safe than sorry" is a wise axiom to follow in regard to goat care. A well-stocked and -maintained treatment cabinet will save both you and an ailing goat a lot of misery. When calling a veterinarian or goat mentor for advice over the phone, take stock of your supplies first. If you don't have basic tools and treatments on hand, there is little that can be done other than summon the vet for an emergency visit, take the animal to an emergency clinic if they will accept a goat, or wait until morning—which

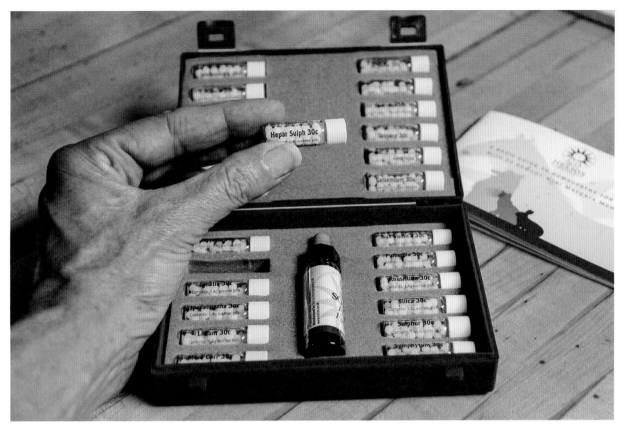

Figure 6.10. My small homeopathy kit, purchased online. Good for all the animals of Pholia Farm, even the two-legged ones.

often proves to be too late. The remedy cabinet often focuses on supportive and intermediary care, rather than cures. An otherwise healthy goat can often be supported through an illness and regain their health without major, and expensive, intervention. I'll repeat, though: No matter how passionate you are about holistic and alternative health, it's very important to have a good working relationship with a vet.

There are two lists of supplies that I recommend for all goat owners. The first is a starter kit that ideally should be gathered before a goat even sets a tiny cloven hoof on your farm. (See The Basic Remedy and Treatment Kit on page 134.) The next is a list of supplies that takes your cabinet to the advanced level and includes items needed for care of goats at kidding time, as well as their newborn young. (See

Advanced Remedy and Treatment Kit on page 135.) Both of these lists are based on a blend of conventional, herbal, and homeopathic treatments that I use at Pholia Farm. As you expand your knowledge about alternative medication and develop your own philosophies, you will no doubt add and remove items as you see fit.

Some medications and remedies, such as the antibiotic oxytetracycline and Bo-Se, don't come with a label that mentions how to use them properly in goats. This doesn't mean these medications aren't good to use or helpful, but they are in a category often referred to as off-label or extra-label, meaning you would use them under the direction of a vet, ideally, who has provided directions (usually by adding a new, or extra, label to the container), including proper withdrawal times.

The Basic Remedy and Treatment Kit

Here's my list of tools, supplies, remedies, and medications that every goat owner should keep on hand. It's also helpful to find a quick reference guide or collect information from multiple sources to help diagnose problems, verify dosages, and find answers. You should also have a calendar handy—I like a large-format wall calendar—to record treatment and other critical dates. Be sure to post your veterinarian's phone number and an emergency backup number or contact in a prominent spot in your barn and in your house. If others might be in charge, be sure to include your farm's address, as it might not be known or easily remembered by others.

TOOLS AND SUPPLIES

- Record keeping journal or calendar
- Digital rectal thermometer for human or animal use
- Drenching syringe or gun or a repurposed plastic syringe
- 3-cc syringes
- 20- and 22-gauge 1-inch needles
- Rubbing alcohol
- Hydrogen peroxide
- Goat coats of several sizes
- 4×4 gauze bandages
- Vetrap (or other stretch, self-cling wrap)
- Bandage tape
- Duct tape
- Weight tape
- Mortar and pestle for grinding tablets or herbal mixtures

REMEDIES AND MEDICATIONS

- Probiotic paste
- Injectable B vitamin complex
- Tetanus toxoid and tetanus antitoxin injectable
- Powdered electrolyte mix
- High-nutrient supplement such as Power Punch or Nutri-Drench
- Echinacea and goldenseal tincture
- Ow-Eze
- Garlic capsules, tincture, or "tea" (made ahead and stored in the refrigerator)
- Rescue Remedy
- Tea tree oil
- VetRx liquid nasal drops
- Vitamin C powder or tablets
- Topical wound spray, such as aloe vera wound gel
- Blood stop powder (I like the kind used by dog groomers rather than the feed store/livestock variety, because it's formulated with more effective ingredients. It also costs a bit more, but since it will rarely be used, it's worth the extra money.)
- Coccidiosis treatment (not preventive, but for emergency treatment), if raising kids
- Homeopathic remedies: *Arnica montana* and *Cinchona officinalis*

Antibiotics Have Their Place

There is a common misconception that the organic farmer never uses antibiotics. In reality, according to government regulations in the United States and other countries, even organic farmers are obligated to administer antibiotics in circumstances where their use is likely to prevent further suffering or save an animal's life. In the US individual animals on certified-organic farms that have been treated with antibiotics must, according to National Organic Program (NOP) standards, be subsequently removed from the herd and never returned to organic status. In Europe and Canada, however, an animal treated with antibiotics for humane and lifesaving reasons can be returned to the herd after an extra-long withdrawal time to ensure no residues remain in the meat or milk. I find this approach far more realistic, fair, and honest. In fact, it is because of the NOP regulations

Advanced Remedy and Treatment Kit

While I call this set of supplies advanced, you should add these things to your kit as soon as possible. I know how overwhelming it is to have to buy so many supplies all at once. I understand because I often find myself out of something that should be in my kit! Note that the conventional medications included in this list may not be allowed for use on organic farms; check regulations to be sure.

TOOLS AND SUPPLIES

- Sterile syringes in 10, 20, and 60 cc
- 1-cc or 0.3-cc insulin or tuberculin (TB) syringes (available for purchase online)
- Scalpel
- Bandage scissors
- Stethoscope
- Mortar and pestle for grinding tablets
- Clippers, such as the popular Oster brand, model A5
- #40 or #50 surgical blade for clippers
- Lubricating jelly
- Iodine wash
- Strong iodine solution (7% equivalent)
- Dental floss
- Balling gun
- Stomach tubes: one adult, one kid
- Small wire saw
- Red-top blood collection tubes
- 18-gauge needles
- Vials for milk samples (any small sterilized plastic bottle with a tightly sealing cap)
- Microscope
- California Mastitis Test, CMT, or other somatic cell test to monitor for mammary infections
- Fecal parasite test supplies (see chapter 7 for more information)
- Glass rectal thermometer as a backup in case of failure of digital
- Disbudding iron for cauterization of broken scur or horn
- .22 pistol or captive bolt for humane euthanasia

REMEDIES AND MEDICATIONS

- Activated charcoal (capsules or tablets)
- Calcium gluconate (IV solution)
- Calcium drench
- Sterile normal saline and sterile water
- IV fluids, such as lactated Ringer's
- Homeopathic remedies: *Bellis perennis* and Carbo Vegetabilis (or homeopathy kit)
- Optional conventional medications that I like to keep on hand (prescription from vet required to obtain these): Banamine, Bo-Se, oxytocin, Lutalyse, xylazine, and a broad-spectrum antibiotic

regarding antibiotic usage that we have never sought organic certification for Pholia Farm.

That being said, I believe that antibiotics are misused more often than not. Farmers frequently dose animals for less-than-life-threatening conditions and without proven need and without following through to prevent resistance. Antibiotics are too often given as preventives or at subtherapeutic levels, rather than as treatments. When giving antibiotics, it's paramount to follow label directions and drug withdrawal times, both for the animal's health and to ensure food safety for humans. Scrupulous record keeping is critical to protect animal health and to ensure that antibiotics are kept out of the human food supply.

Common Remedies

Now let's learn more about how to use remedies, equipment, and medications. I've included some

Herbal Medicine—
The Most Popular Medicine in the World

While researching information on wormwood, I came across a statement in an article in the *Journal of Intercultural Ethnopharmacology*[1] that I found to be a great reminder of the wisdom that Western medicine seems to have lost: "The WHO [World Health Organization] survey indicated that about 70–80% of the world's populations rely on non-conventional medicine, mainly of herbal source, for their primary healthcare." The authors of the article then emphasized the power of these traditional herbal medicines by pointing out that these medicinal plants are "rich sources for naturally occurring antioxidants" and that "have the ability to scavenge free radicals and enhance immunity and antioxidant defense of the body."

Ethnoveterinary Botanical Medicine is an excellent reference book devoted to studying the traditional herbal medicines of the world and how they can benefit animals. For example, the book lists 31 plants used around the world as natural deworming medication (anthelmintics). The importance of continued scientific study into the proper use of these plants for both humans and animals cannot be overemphasized. Herbal medications can be effective, and I recommend that you investigate them. However, because very few herbal remedies have been studied, other than anecdotally, be sure to choose a supplier that has done proper research and training. Just because herbs aren't legally considered drugs doesn't mean they aren't powerful medicines!

My favorite herbalist book is *Herbs for Pets* by Gregory Tilford and Mary Wulff. This book is easy to read, well illustrated, and quite balanced in its approach—repeatedly citing the fact that the holistic herbalist approaches the use of herbs by first looking at the underlying health and immune function of the animal. Another classic is *The Complete Herbal Handbook for Farm and Stable* by Juliette de Baïracli Levy. It has a lot of great information but some, such as the use of turpentine internally, that is dated and no longer recommended (the book was originally published in 1952). See appendix D for more information on books about herbal medicine.

medications that require a veterinarian's prescription, and those will come packaged with dosage and application instructions. It is not within my expertise to give you any specific recommendations for usage of prescription medications, but I do explain what they are often used for. It's beyond the scope of this book to cover every possible remedy and medication that a great goat manager might have and use! In appendix D I've recommended several great books that are devoted to natural and alternative treatments and Internet links to information about conventional medications.

I begin this discussion with probiotics because they are the first remedy I turn to for a range of possible problems. Then I continue with other remedies, from the ones I use most to those I use least.

PROBIOTICS

Probiotics are useful anytime a goat might not be at her peak of health, and you can never go wrong by administering them. As I described in chapter 5, I believe in them so much that I routinely feed probiotics to my goats to support good gut health.

Anything that makes a goat feel unwell is likely to have negative repercussions on its appetite, rumen function, vitamin production, and bacterial balance. Often, whatever made it feel bad is easy to fix, but the cascade of negative effects to its metabolic

Homemade Electrolytes

Some recipes for making homemade electrolyte solutions include corn syrup, others table sugar, and still others molasses. You can use whatever you have on hand. Also, varying the amounts of ingredients a bit won't harm the animal. So if you need to mix a batch up in a hurry, don't worry if your measuring isn't precise!

BASIC ELECTROLYTE SOLUTION

1 quart (1 L) warm water
2 tablespoons (30 ml) blackstrap molasses
½ teaspoon (3 g) table salt (sodium chloride)
¼ teaspoon (2 g) baking soda (sodium bicarbonate)

If you wish, you can add 1 teaspoon (5 g) probiotic powder and 1 pump Nutri-Drench, Power Punch, or other nutrient supplement to the basic ingredients. This recipe should last for months.

JAN NEILSON'S FRAGA FARM ELECTROLYTE RECIPE

1 quart (1 L) warm water
¼ cup (60) ml organic whole-milk plain yogurt
2 tablespoons (30 ml) honey
½ teaspoon (3 gm) table salt
¼ teaspoon (2 gm) baking soda

Mix it all together, and use it right away or store in your fridge until needed. This mix should be used within a week to ensure that the probiotics are still active.

system—the system that keeps blood and body chemistry in balance—can cause such adverse effects that her life might end up at risk. Supporting goats during an illness with remedies and nutritional supplements such as probiotics is a part of a multifront approach that can save their lives. For support during a health crisis, I use a probiotic paste that contains a wide variety of probiotic bacteria.

Be sure to store all probiotics in a cool or cold dry place. Heat will diminish or deactivate the bacteria, and humidity will deactivate probiotic powder.

ELECTROLYTES

You can make your own electrolyte solution or you can keep a premixed powdered version on hand for times of need. Administer electrolytes anytime you are concerned that an animal might be headed toward dehydration. (If an animal has any degree of diarrhea or is simply not interested in drinking, then dehydration is a risk.) Some goat owners use electrolytes when weaning kids from the bottle. In that case you can replace all or a portion of the milk

so that the kid still gets its "fix" from having nursed a bottle, but instead of being full from drinking milk, it's hungry and begins eating more solids. (For more about kid management, see chapter 9.)

NUTRITIONAL BOOST SUPPLEMENT

Several companies make a liquid nutritional "boost" product for goats or for sheep and goats. Two common brands are Nutri-Drench and Power Punch. These supplements usually contain several vitamins and minerals along with a sugar source, all meant to give the animal a quick boost of energy. The sugar source is primarily propylene glycol, which will bypass the rumen and be quickly available for the goat to convert to glucose, which then enters the bloodstream. This is extremely critical for some conditions such as ketosis (see the Ketosis entry on page 263 for more details). Boost products are very helpful when an animal is experiencing weakness or exhaustion, such as during a long, hard labor for a doe or if a kid is born weak or fails to thrive. A dose can be given straight or mixed in with electrolytes and probiotic powder.

Some goats don't like the taste, while others relish it. Too much can be rough on the animal's digestion, so don't overdo an otherwise good thing!

ECHINACEA AND GOLDENSEAL TINCTURE

This terrific combination of herbs is a staple around our farm, for goats and humans, too. As with probiotics, you can't go wrong administering it. I use it in conjunction with other remedies in cases (in my goats) such as coughs, runny noses, and fever. Most tinctures are quite tasty to goats, so they're usually easy to administer orally. Be aware that many tinctures sold for use by humans come packaged with glass eyedroppers. Be very careful when putting a glass eyedropper into a goat's mouth! I slip the dropper in the gap between the front teeth and molars, where it is unlikely to break.

OW-EZE

This remedy is a tincture of valerian, white willow bark, feverfew, Saint-John's-wort, and licorice root made by Molly's Herbals. Use it before painful procedures such as disbudding or castrating. You can order it online from www.mollysherbals.com.

VITAMIN B COMPLEX INJECTABLE

I use this injectable almost as frequently as I use probiotics. B vitamins, as you learned in chapter 4, are for the most part manufactured in the rumen. Whenever there is some change in how well the rumen is working, then there is also likely going to be an inadequate supply of B vitamins. Because the B vitamins are water-soluble (meaning that the body will eliminate any excess without harm), you do not have to worry about mistakenly giving a goat too

Figure 6.11. An old street sweeper brush helps goats remove their winter cashmere undercoat and also helps them satisfy the need to scratch. I tried for years to find a used street sweeping brush, to no avail. Then I met a woman who got them from the local county road maintenance crew. Once you get one, it will last indefinitely.

much. B vitamin complex comes in several strengths, so be sure to check the label. I prefer the type that has 100 mg of thiamine (B_1) per ml.

VITAMIN C

In most adult mammals, vitamin C is produced in the liver in sufficient quantities to meet all daily needs (exceptions are humans, primates, and guinea pigs). But the babies of those mammals that produce their own vitamin C, goat kids included, don't start making their own until they are ready to be weaned. Before that time they get what they need from their mother's raw milk (vitamin C is destroyed by pasteurization) or from supplementation. During times of stress, though, extra vitamin C can be helpful to all animals. You can buy powdered vitamin C (ascorbic

Goat Coats

There are times when a goat kid or adult might benefit from wearing a goat coat. It takes a lot of body energy to create heat, so any illness or condition that demands energy will mean that the animal might not have enough to spare to maintain its body temperature. By relieving that burden with a coat or other warming device such as a heat lamp or animal-safe heating mat (ridged plastic heating mats can be purchased from pet and livestock suppliers), you free the animal's body from this extra demand.

Goat coats can be purchased or made. In a pinch, you can repurpose a T-shirt, sweatshirt, or vest for the job. I've even used a large sock with half the foot cut off and two openings made for the legs for runt-sized Nigerian Dwarf kids who look like they're wearing a turtleneck sweater once they're squeezed in.

Goat coats should fit in a manner that allows the animal to move without feeling restricted. A goat should be able to easily lie down and get up with the coat on, and it should not have loose parts that goats might want to chew on.

Figure 6.12. Two healthy Nubian kids and one Sable at the Carl Sandburg historic goat farm in North Carolina sport hand-knit goat coats.

Molly Nolte

MOLLY'S HERBALS AND FIAS CO FARM, OKEMOS, MICHIGAN

For many goat farmers Molly Nolte's website www .fiascofarm.com has served as their bible of goat care wisdom. My first goat notebook is still filled with pages printed out from her site many years ago. Her herb company, Molly's Herbals, offers her carefully researched and personally formulated and tested herbal formulas, as well as advice on herbs and holistic goat health care.

Molly began working with herbs and botanicals over 28 years ago. At the time she lived in Chicago and suffered from asthma and frequent bouts of bronchitis. She was using two inhalers, one of which was a steroid. Having always believed in a more natural approach to medicine, she began studying herbs, looking for a solution to her dependence on conventional medication. It wasn't long before she developed an herbal blend tea and says that after diligently drinking her pot of tea daily for an extended period of time, "I was completely symptom free of asthma and haven't experienced bronchitis since."

In 1993 Molly left Chicago to homestead in Tennessee. Goats became a part of the farm in 1996. She didn't start out with dreams of an herbal remedy business; it came about organically. Her homestead was in need of a little extra cash, so she put her other talents to work and began designing logos and websites for other small start-up businesses wanting to sell their handcrafted products online. She used these skills, combined with her interest in helping people learn, to build her own website, hoping to "share my mistakes so others could avoid making them in their goat keeping." As she shared what she had learned, including the use of herbs to boost health in her herd, people began to ask her where they could buy these helpful herbal formulas. "At the time, I was designing websites with shopping carts, and knew how to set this kind of thing up, so I thought what the heck, I'll offer my products for sale."

Figure 6.13. Molly Nolte is the founder of Molly's Herbals and Fias Co Farm. Her work on herbal remedies for goats and her informative website have made her a goat mentor for many people across the world. *Photo courtesy of Paul Nolte*

Molly recommends you always have a bottle of tea tree oil and lavender essential oil on hand. Tea tree oil (or TT) has many uses, including as an antiseptic, anesthetic (mild), antibacterial, antimicrobial, disinfectant, fungicide, and germicide. Molly says, "Use it straight from the bottle and apply to insect bites, cuts, sores, anything fungal (including ringworm), or any little bump that you don't know what it is." She adds, "Be careful when buying tea tree oil; it's available commercially in a range of grades, but it's important to use the best quality oil you can find." Look at the percentage of the compounds cineole and terpinen-4-ol. For cineole, the lower the number the better. Tree oils with high cineole content are thought to be of poor quality and more likely to cause skin irritation. For terpinen-4-ol, the higher the number the better. Terpinen-4-ol appears responsible for most of the antimicrobial activity of tea tree oil.

Lavender has a calming and mood-lifting effect. Lavender essential oil may work to calm a nervous or excited animal. Molly advises, "In depressed or aggressive animals, lavender can be used as aromatherapy to lift spirits and adjust attitudes. An open bottle of the oil can be waved under the animal's nostrils, or a few drops are put on a piece of cardboard that is placed near the animal's bedding."

I asked her if there is any one particular story that she could share regarding how herbs have helped goats. She said, "Actually there are so many, I don't even know where to start! I love to help people and animals, and it makes me so happy to know that I've not only helped with quality of life but have actually helped save lives." You can learn more at her website www.fiascofarm.com.

powder) or you can grind vitamin C tablets and mix either with a bit of water to administer orally.

BO-SE

A vitamin E and selenium supplement in injectable form, Bo-Se is used to treat and prevent white muscle disease and boost fertility. (You may come across a product called Mu-Se, which is similar but higher in selenium and meant for cows.) Bo-Se is a prescription-only medication and must be purchased from a veterinarian or from a supplier using a vet's prescription.

GARLIC

While some goats enjoy noshing on whole garlic cloves, you are more likely to be able to administer garlic to an ailing goat if it is in capsule, tea, or tincture form. Garlic has a natural antimicrobial, antifungal, antiviral, and immune-system-boosting effect that can be useful in many compromised health situations.

ARNICA OR T-RELIEF (PREVIOUSLY SOLD AS TRAUMEEL)

A homeopathic remedy for pain, trauma, bruising, and inflammation, arnica, also called *Arnica montana*, comes as tablets or drops. T-Relief contains arnica as well as other "natural medicines" and comes in tablet or ointment form. I keep both on hand. It's fairly easy to get the tablets into the goat's mouth, and goats don't seem to find the taste unpleasant.

LINIMENT–TOPICAL

Usually made from natural ingredients that stimulate circulation, liniment is useful for strains, sprains, and udder inflammation.

VEGETABLE OIL OR BLOAT TREATMENT

Vegetable oil can be used to treat frothy bloat. Bloat treatments, such as Therabloat, act in a similar way to break up foam in the rumen. Mineral oil (made from petroleum) found in pharmacies and sold for use as an intestinal lubricant can be used, but there is concern that since it is flavorless, the goat might resist swallowing it and inhale some instead. A little peppermint oil can be mixed with either type of oil to increase its palatability.

VETRX

Put a drop of this in each nostril when the goat is stuffy. It helps open the airways and clear breathing. Think of it as Vicks VapoRub for goats.

TEA TREE OIL

A natural antimicrobial, antifungal, and germicide that can be used on wounds and skin lesions.

MOLASSES

While I avoid using any sweetened feeds on a regular basis, having some tasty, iron-rich blackstrap molasses on hand can be helpful when you need to administer

Table 6.4. Remedy and Treatment Chart

	Fever	Loss of Appetite	Lack of Rumination	Diarrhea	Rumen Distension/Bloat	Grinding Teeth	Conjunctivitis	Pale Inner Eyelid	Stargazing	Taking Antibiotics	High Somatic Cell Count	Dehydration	Limb Injury
Probiotics	Yes	Yes	Yes	Yes		Yes	Yes		Yes	Yes	Yes	Yes	
Electrolytes		Yes[a]		Yes								Yes	
Nutritional supplement products		Yes		Yes				Yes[b]	Yes				
Echinacea and goldenseal tinct.	Yes		Yes	Yes			Yes			Yes	Yes		
Vit. B complex injectable		Yes	Yes						Yes	Yes			
Vit. C oral	Yes			Yes						Yes	Yes		
Vit. A							Yes						
Garlic oral	Yes			Yes						Yes	Yes		
Arnica or T-Relief						Yes							Yes
Liniment–topical											Yes[c]		Yes[d]
Veg. oil drench					Yes								
Bloat treatment					Yes								
Molasses or maple syrup								Yes[e]					
Fecal test; then address parasite issues				Yes				Yes					
Rumen transfaunation		Yes[f]											
IV or SQ fluids				Yes[g]								Yes	

Note: These supportive treatments can be used in combination with proper medical advice and diagnosis, not as simple cures.

[a] If not drinking

[b] With iron

[c] Udder massage

[d] On limb

[e] Blackstrap mixed in warm water

[f] If condition persists more than a day

[g] If dehydrated

something less tasty. Depending on the medication or remedy, you can mix it with the molasses and give it by spoon or tongue depressor (or wooden craft stick), or mix it as a liquid and administer by drenching syringe.

IV FLUIDS

Your vet can sell you a bag of IV fluids that you can use intravenously, subcutaneously (under the skin), or by tube feeding. If a goat isn't eating well or taking in fluids, administering fluids can help keep it alive while it recovers from the initial illness.

Basic Medical Skills

After assembling that long list of supplies, you might feel as if you are becoming an amateur veterinary technician, and in a sense you are! You are your goat's best bet for quick relief and a long healthy life. If you embrace this responsibility, you will quickly become an accomplished technician. Now that you have your supplies, let's go over the skills you will need to become comfortable with when working with goats. Please don't forget: One of your best medical skills is to know when to call in a pro!

Taking Body Temperature

As I hinted earlier, a thermometer is one of the most important things to have on hand to provide good care for your goats. Now that you have one, how do you use it? Goats' temperatures are taken rectally. The thermometer tip must be lubricated for both the goat's comfort and ease of use. While you can use lubricant made specifically for this purpose, I find a quick spit into the palm of my hand and then on to the thermometer tip will do. While some goats may not mind having such an intrusion, most will need to be restrained before the offending instrument can be inserted. Try clipping the goat's collar to a nearby fence with a short lead and using your body to stabilize the goat's body against the fence.

> Don't forget: One of your best medical skills is to know when to call in a pro!

Turn on the thermometer just before inserting it. For adult goats, use your free hand to lift the tail, then insert the tip of the thermometer into the anus (that's the hole on top for a doe), about ½ inch (1.25 cm) for adult goats and ¼ inch (0.6 cm) for babies. You can hold a kid under your non-dominant arm with its rear end facing forward. Then use your dominant hand to insert the thermometer into the rectum.

It should take a digital instrument only a few seconds to register the reading. If you are not sure if the results are accurate, take the temperature of another goat. A cold morning may mean that everyone's body temperature is a bit low, and vice versa on a very hot day. You'll notice that in the Advanced Remedy and Treatment Kit section on page 135, I recommend having a backup sturdy glass rectal thermometer; if you don't have that, then at least have a second digital, just in case your main one doesn't seem to be working right, or you want a "second opinion." If you use a glass thermometer, don't forget to "shake it down" first and verify that the colored liquid is down toward the bottom, or the reading will be wrong.

Figure 6.14. Taking a kid's temperature.

Oral Dosing

Oral dosing can be liquids or capsules. For liquids, you'll use a dosing syringe when giving the dose; for capsules, a balling gun. You'll need to restrain the goat when you give an oral dose, and the method I described above for holding a goat while taking its temperature can work. Let's go over a few more ways to keep your goat still, too.

For an adult goat, take hold of the collar and straddle the goat's neck with your legs. Use your knees to hold the animal steady. If the remedy isn't very tasty, or this particular goat has an aversion to oral invasion, back it into a corner or—better yet—have an assistant steady it either by holding its head or straddling its neck and then stabilizing its head. Once the situation is stable, release the goat's collar and use your non-dominant hand to steady

the goat's chin and hold the head slightly tilted up, as shown in figure 6.15. Insert the dosing syringe or balling gun at the side of the goat's mouth, in the gap between the front and back teeth. If the dosing syringe is a drenching gun (this has a curved end), insert it into the mouth through the side, then angle it back over the top of the tongue. When you're giving capsules with a balling gun, it's helpful to keep the balling gun in the goat's mouth for a few seconds after you eject the capsules. This stimulates the goat's swallowing reflex and usually short-circuits its attempt to spit out the capsules. If you cover its nostrils, you will also inspire the goat to breathe in through its mouth—and as a part of that response it will swallow first.

For oral dosing, tuck newborn kids under your arm, with their heads facing front. Administer the

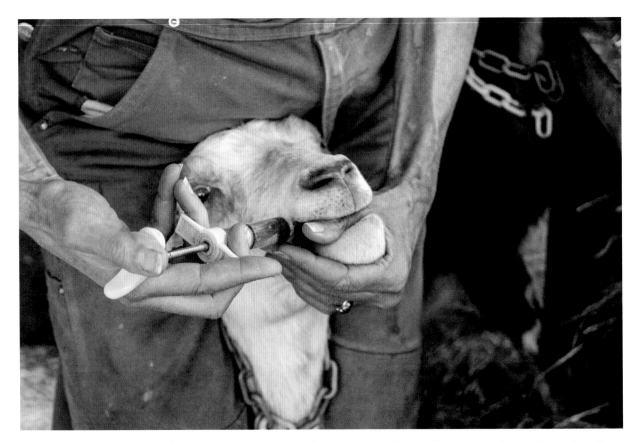

Figure 6.15. When giving oral medication with a drenching syringe, I often hold the goat's jaw with one hand, using my thumb inserted in the mouth to hold the animal still.

remedy by cupping the kid under the chin with your non-dominant hand and using your thumb to slightly open the mouth; insert the dose in the gap between the front and back teeth. Then use your dominant hand to insert the dosing syringe on the opposite side of the mouth.

Bucks are usually more difficult to restrain because of their size and the fact that they are usually focused on only one thing. . . . Plan on securing the buck to a post or fence, and then grip his beard firmly, steady his muzzle, and dose the remedy. Also, keep in mind that working with the less-than-fairer sex of the species always leaves you with a telltale odor. I have a separate pair of "buck overalls" that I leave hanging outside the door.

Injections

Goats, being ruminants, cannot readily assimilate some substances through their digestive systems, so oral dosing isn't always effective. The best way to administer many medications and remedies is by injecting them under the skin (such injections are called subcutaneous, or sub-Q) or into the muscle (intramuscular or IM injection). Becoming comfortable with giving shots is important.

Inevitably, injections cause discomfort. There's the pain due to the piercing of skin and muscle, which can be proportional to the size of the needle used. Also, some remedies are irritating to the body tissues, and that's a secondary source of discomfort. Single-use needles will ensure sharpness for improved comfort and will prevent the spread of blood-borne disease.

Many medications and remedies that are labeled for IM injection can actually be given sub-Q instead. A sub-Q injection of a substance often causes a goat less discomfort than an IM injection would. Be sure to get your vet's advice if you deviate from what the label recommends, though.

With any piercing of the skin there is the risk of introducing a contaminant inside the animal, so always use careful procedure when giving shots. Be sure that injectable remedies are not outdated, have been stored properly, and have not become

Figure 6.16. To give a subcutaneous injection, hold the goat with your legs, pinch a bit of skin, pull it outward, and then insert the needle at the base of the lifted skin.

contaminated. Most are best stored in a cool, dark place. Make sure labels are easy to read, so that you don't accidentally give the wrong remedy. You must also make sure that you are careful when drawing up and giving shots so that you don't accidentally dose yourself with a bit of the medication!

In the livestock industry, it's not uncommon to use a single syringe and needle to administer medication to multiple animals, especially with sub-Q injections. This is acceptable only if you are certain that none of the animals being dosed are carriers for a disease, such as caprine arthritis encephalitis (CAE), that can be passed to an uninfected animal via a needle stick. It's a personal choice whether or not to reuse needles. It does improve efficiency and offers a cost savings. But keep in mind that in addition to the risk of possible disease transmission, needle reuse can be painful for the animals. Disposable needles come out of the package extremely sharp but become progressively more dull with each penetration of the skin. My advice is that it's worth the small price to change needles for each animal.

Needles come in a variety of lengths and thicknesses. Thickness is expressed as gauge size—the smaller the gauge number, the fatter the needle. For sub-Q injections you can use a short needle of ¾ to 1 inch (2–2.5 cm). For IM shots, you will need a 1- to 1½-inch-long (2.5–3.8 cm) needle. You must choose a needle of sufficient thickness to allow the solution to be drawn into the syringe. Some liquids are quite thick and viscous and will not pass through the bore in a tiny needle. Remember that if a solution is hard to draw into a needle, it is also going to come out of the needle in a thin, pressurized stream, making it more painful to the goat. Keep a variety of needles in supply. Tiny insulin syringes that hold only 1 cc or less (and come with the needles attached) work well for dosing amounts less than ½ cc.

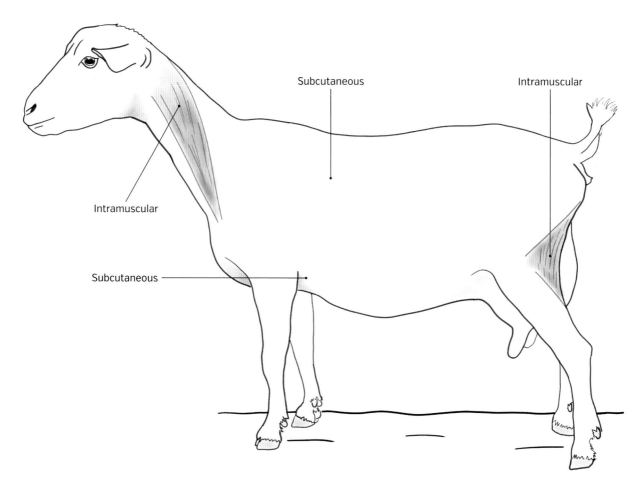

Figure 6.17. The most common sites for giving injections on goats are on the neck, in a back leg (*not in meat goats*), and behind the elbow. In meat goats, it's common to give all injections subcutaneously to protect the muscle from damage.

Injectable remedies and medications are packaged with a rubber stopper that is left in place. Each time the stopper is pierced by a needle to withdraw some of the liquid, there is a risk of contaminating the remaining liquid. The rubber stopper will eventually become worn from repeated piercing. It will end up with a hole so large that the contents are no longer sealed. You'll know the seal has failed when you try to inject air into the bottle before pulling out the medication. You won't feel any resistance because there is no back pressure left in the bottle. At this point you should discard the medication, which has likely oxidized, and replace it with a new bottle. Some people like to leave a needle in the stopper, without a syringe attached. This does preserve the stopper, but the open end of the syringe is a clear route for air to enter the bottle. Remember that air, unless filtered and sterilized, is teeming with bacteria, yeasts, molds, and viruses, so leaving the contents exposed to the air is worse than wearing out the rubber stopper. You can insert a short needle and then attach a new, sterile syringe to the top, to plug the needle.

Before drawing up the solution, use a clean paper towel or cotton ball wetted with rubbing alcohol to rub the surface of the stopper. This cleans and mostly sanitizes the surface. Attach a new needle to the end of a new syringe, making sure to not touch the tip of either. Pull back on the plunger and draw up air into the syringe at the same volume as the remedy you will be using. For example, if you are giving 3 cc of B vitamins, first draw up 3 cc of air. (Yes, this air will likely have a certain number of contaminants. That's one reason why expiration dates appear on bottles that are meant to be used over time—not just a single dose.) Insert the needle through the stopper and into the bottle. Inject the air into the bottle. This will help you to draw the same amount of liquid out of the bottle by increasing the pressure in the bottle. Then hold the bottle in your non-dominant hand, hold the syringe with your dominant hand, invert the bottle, and withdraw the dose. Pull the needle out of the bottle and clear the syringe of any air bubbles by holding the syringe with the needle upward, tapping the syringe, and then ejecting the air bubble that will rise to the top. Carefully replace the needle cover.

Don't carry syringes and needles in your mouth or pockets, only in your hand or a caddy! While gripping a syringe between your teeth while you strategically maneuver your patient may make you feel like a cross between Crocodile Dundee and Dr. Dolittle, the practice is not only dangerous but unsanitary. By the same token, a needle and syringe placed in your pant pocket can quickly become a sharp surprise when you bend over to take your boots off, as well as a possible source of infection for you!

For sub-Q injections, secure the animal or straddle its neck and grip it with your legs, facing the back of the goat. Select the area you want to give the shot. You can clean the site with an alcohol swab or alcohol-dampened cotton ball if you like, but it isn't necessary unless the animal is very filthy. With your non-dominant hand, grip a good tweak of loose skin behind the shoulder or elbow of the goat and raise it to make a "tent" of skin. Remove the needle cover and quickly, without hesitation, insert the needle into the base of the tent. The needle angle should be parallel to the animal's body so that it only enters the space inside the tent. Inject the solution, and pull the needle out. Don't massage the area after giving a sub-Q injection; it's meant to sit undisturbed and be absorbed slowly. Also, massaging the site might make the fluid leak back out. Don't be surprised when you are first learning to give sub-Qs if you send the needle clear through the tent of skin and inject the dose right out the other side. It's happened to all of us.

For IM injections, secure the animal by clipping its collar to a post or fence, and then use your body to immobilize it against the fence. Most IM injections are given in the animal's thigh or large neck muscle. (If the animal is to be used for meat, avoid the thigh—the needle stick could cause formation of scar tissue in a valuable cut of meat.) Gently grip the area to be injected to keep the muscle tissue still and away from nearby bone. You can clean the site with an alcohol swab or alcohol-dampened cotton ball first if you wish. Don't get the site too wet with the alcohol, as it might cause the injection site to bleed—and sting!

Keep in mind that injections in the muscle are much more likely to be irritating than sub-Q injections, so be prepared for a stronger pain reaction. Remove the cover from the needle with whichever hand you have free, then quickly and without hesitation plunge the needle into the center of the muscle. This may or may not be the full length of the needle, depending on the needle length and the size of the muscle. Pull back gently on the plunger. If blood appears in the syringe, that means the needle tip has entered a blood vessel. In that case, ease the needle out a fraction of an inch, and then pull back on the plunger again. If no more blood appears, you can inject the solution. Keep in mind that the larger the quantity of solution you are injecting, the slower you will want to inject it. Flooding tissue with solution too rapidly can cause tissue damage and pain. Once you've finished injecting the dose, withdraw the needle and gently massage the area.

When you are first learning to give shots, I recommend practicing by giving B vitamin or sterile normal saline injections. Extra B vitamins will not harm goats, and you will be grateful that you've gained a comfort level with the procedure when you're suddenly called on to give an injection in a stressful situation. The old technique of practicing first on an orange, as I did in nursing school back in the 1980s, can help you overcome the initial fear of working with a needle.

Treatment Suggestions

Even while you are troubleshooting a health problem, you can begin to treat the symptoms. In most situations, treating the symptoms will help the animal to improve, even if you haven't figured out the underlying cause. But remember, signs and symptoms can indicate a serious disorder that must be further investigated. Use the treatment suggestions here to help the animal while you and your vet consider other health issues.

Fever

Place the animal in an isolation pen with easy access to feed and water. Provide a variety of feed choices, including fresh leaves and greens. If the weather is cold, provide warm water for drinking. Give probiotics,

vitamin C, garlic tea or the equivalent, and echinacea and goldenseal (or an immune support tincture that contains the same ingredients). If the weather is cold, cover the animal with a goat coat or provide a place for it to warm itself up. Check its temperature again in four to six hours. If there's no improvement within 6 to 12 hours, the fever is higher than 105.5°F (41°C), or the animal is showing other signs of distress and pain, then it is advisable to consult a vet and consider starting conventional treatment with antibiotics.

Loss of Appetite

Place the animal in an isolation pen with easy access to feed and water. Provide a variety of feed choices, especially fresh edible tree leaves and greens. Willow leaves can be helpful if you suspect rumen upset. If the weather is cold, provide warm water for drinking. Administer injectable vitamin B complex as well as nutritional boost supplement, and drench with electrolytes if the animal won't drink.

Decreased or Absent Rumination

This usually follows or is concurrent with loss of appetite, so treat for that as well. Give probiotics and injectable vitamin B complex every six hours at least. If rumination doesn't resume within a day, rumen transfaunation is a good idea (see the Rumen Transfaunation section on page 171 for instructions).

Signs of Dehydration, Including Diarrhea

Offer warm water with or without electrolytes. If the animal won't drink, use a drenching syringe and include electrolytes. If dehydration is severe, consult your vet and give subcutaneous or IV fluids.

Abdominal Distension

If the distension is on the left side, administer the appropriate bloat remedy and walk the goat around if possible (uphill can be helpful). If it's lying down, roll it over a couple of times and elevate its head. See the Bloat entry on page 250 for more details.

If the distension is on the right side, colic or other blockage is possible. Consult with your vet as soon as possible.

If the distension is in the lower belly, it may be urolithiasis (see the Urolithiasis entry on page 323). Other diseases may also cause distension during their end stages.

Limb or Joint Strain or Pain

Isolate the animal or remove physical stress so that it can rest as needed. Give arnica or T-Relief, or another herbal or homeopathic pain remedy. Rub the affected joint or limb with liniment. Consider giving Ow-Eze or aspirin or another NSAID if the pain is severe. Consult with your vet to rule out more severe injury such as a fracture.

Depression Without Loss of Appetite or Fever

Give Rescue Remedy, offer fresh tree leaves, and continue to troubleshoot to determine the underlying cause.

General Discomfort or Pain

While troubleshooting the underlying cause of this symptom, give arnica and/or Ow-Eze. If the pain is severe give an NSAID such as aspirin or Banamine. Isolate the animal to limit its stress.

Eye Discharge or Changes

Inspect the eye for foreign matter or trauma. Clean it with saline solution. Give vitamin A supplement or a food source of vitamin A. If the animal is sensitive to light, try to ensure that it has access to shade.

Wounds

Clean the wound with a mild soap and warm water solution and inspect it for foreign matter. Give probiotics and garlic for three to four days. Tea tree oil or an herbal salve can be applied to the wound. Dress the wound if it's more than superficial or to protect it from manure and other contaminants. Isolate the goat if necessary to keep the wound clean. If the wound is deep, check the animal's tetanus vaccination status, and administer tetanus antitoxin if necessary. Watch for redness, swelling, drainage, and fever.

Maintenance Practices

Vaccines are substances that are designed to provide immunity to certain diseases and conditions. Dewormers are treatments given to reduce the

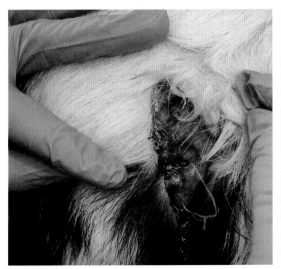
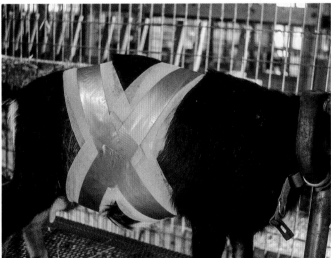

Figure 6.18. This wound (*left*), which penetrates into the abdominal muscle, is the result of a goat's being pushed into a small protrusion of heavy steel wire from a fence. After being cleaned with hydrogen peroxide, the wound was bandaged, and the goat was given probiotics and garlic (no antibiotics). One dressing change and one week later (*right*), the wound had healed nicely. Duct tape makes a great, nearly goat-proof wound binding!

number of parasites inside an animal. Use of vaccines and dewormers is a regular management practice on many farms, but scheduled, routine use is not mandatory for a healthy herd. Indeed, the routine use of dewormers has caused widespread resistance among many parasites to these previously effective treatments.

Vaccinations

Vaccines are medications classed as biologics (which also includes antitoxins, toxoids, and antibodies). In the United States the most common vaccine for goats, and the only one currently labeled for goats, is called CD&T. This injection is meant to protect the goat from tetanus and enterotoxemia —commonly called overeating disease. Unfortunately goats vaccinated for enterotoxemia do not retain adequate levels of protective antibodies for more than a couple of months.[2] I do not use it in our herd. When some veterinarians learn this, I get everything from a patronizing eye roll to out-and-out accusations of negligence. Yet in my decade-plus of managing a dairy goat herd and many years prior to that with pet goats, I have never had a goat experience the diseases that CD&T is meant to prevent. Of course it could happen tomorrow.

Just to be clear, I do keep on hand two injectables for tetanus: toxoid, which is a vaccine, and tetanus antitoxin, which is a treatment. I administer the antitoxin when a wound occurs or prior to a procedure such as tattooing. I administer the tetanus toxoid, the vaccine, when extended protection is needed, such as after castration. I encourage all goat owners to read up on the subject and make their own decisions. The same choice is not right for everyone!

There are diseases for which I would happily vaccinate my goats, if a vaccine were ever developed. High on my list are Q fever and Johne's disease, for which there is no perfect prevention or testing. So however you feel about vaccines, keep your mind open, and understand the risks as best as you can.

Internal Parasite Control

Hands down, parasite problems are the biggest hurdle facing goat farmers, especially when working toward full organic management practices. Dewormers, also called anthelmintics, vermifuges, or vermicides, are often used to reduce the number of parasites inside an animal. For better or worse, most internal parasites have developed or are developing resistance to all synthetic deworming medications. I say "for better" because the silver lining of this news is that scientists are much more motivated to research some of the many holistic options and new, natural approaches.

Parasites are life-forms that must spend a part of their life cycle on or in another living thing (one that—while unwilling and unaware—is still called a host). The host provides something that the parasite needs—such as nutrients, a safe environment, and even transportation. Most animals are able to host a limited number of parasites without suffering catastrophic results. It is only when the number of parasites exceeds what the host can accommodate that major problems result. The numbers of parasites increase for several reasons: emotional and physical stress (including from other concurrent disease), poor genetics, low nutritional status, or exposure to extreme concentrations of parasites.

Animals that can successfully deal with parasites without our medical intervention are called either *resistant* or *resilient*. A resistant animal is healthy enough and has genetics that are strong enough to limit the number of parasites that can exist inside its body—especially by limiting the parasites' ability to reproduce. A resilient animal is one that might host a good-sized population of parasites but doesn't suffer poor health because of it. When you observe resistant and resilient animals, they will both look equally healthy. Fecal samples from the two animals, however, would show different results—the resilient animal's sample is likely to show fairly high counts of parasite eggs. This doesn't mean it should be dosed with deworming medication, though!

For a long time, common wisdom, along with the terrific advertising ability of companies producing

Table 6.5. Common Vaccines in the United States

Vaccine	Labeled for Goats	Disease Prevented	Protocol*
CD&T (*Clostridium perfringens* type C and D and tetanus toxoid)	Yes	Enterotoxemia (overeating disease) and tetanus	2 mL sub-Q. Repeat in 21 to 28 days and once annually or as advised by your vet. See Enterotoxemia entry on page 258 for more on immunity and vaccination. Usually given to does during 4th month of pregnancy to provide protection of kids until they are 1 month old. Then vaccinate kids at 1 month of age and again 1 month later.
Tetanus toxoid	Yes	Tetanus	Use when CD&T is not given. 2 IM doses of 0.5 mL each approximately 30 days apart, then once annually. Usually given to does during 4th month of pregnancy to provide protection of kids until they are 1 month old. Then vaccinate kids at 1 month of age and again 1 month later.
Ovine Ecthyma	Yes	Soremouth (orf)	Because this is usually a live vaccine, use it only on farms where the disease is already present. On such farms, develop a plan with your vet to protect animals based on need and risk. (For example, on extensively managed farms where kids are unlikely to be able to be bottle-fed during an outbreak, vaccination will help decrease mortalities.)
Mannheimia Haemolytica–Pasteurella Multocida Bacterin	Yes	Pasteurellosis (pneumonia)	Inject 2 ml sub-Q. Administer a 2nd dose at 2 to 4 weeks. Animals vaccinated at less than 3 months of age should be revaccinated at weaning or at 4 to 6 months of age.
Corynebacterium Pseudotuberculosis Bacterin	Non-lactating goats only	Caseous lymphadenitis (CL or cheesy gland)	This vaccine is not approved for lactating goats. For non-lactating goats, give 1 ml sub-Q in one side of neck. Repeat in 14 days with 1 ml SQ in opposite side of neck. May cause injection site swelling.
Rabies	No	Rabies	Requires a veterinarian's approval. Only advised in high-risk areas.

Note: Consult with a veterinarian and dairy goat specialist in your region before you begin any vaccination program.

* According to manufacturer's directions

synthetic chemical dewormers (commonly simply called chemical dewormers), dictated that a responsible herd manager would follow a program that involved automatically dosing the entire herd with deworming medication. Originally, a single brand was used, and then people were encouraged to rotate brands and types. After years of this onslaught, parasites, in turn, have adapted, developing resistance to the dewormers. Current recommendations suggest their continued use, but without rotation (which is now known to speed resistance), until resistance is proven by fecal testing or suspected because of symptoms and signs present in the herd. In addition, farmers are encouraged to use other methods to identify animals with a parasite problem and then treat only these goats. Still resistance is growing. This, and the increasing popularity of organic farming, is motivating researchers to take more interest in holistic alternatives.

Dealing with parasites without the use of synthetic dewormers requires a change in the conventional paradigm of utilizing medications to fight the problem. Instead, the focus shifts to improving nutrition and immune function, including plants that provide anthelmintic benefits in the herd's diet, and choosing genetics that are either resistant or resilient to parasites. Luckily, as you have read already, this is great for every aspect of the goat's health, not just worm worries! Let's go over the hopeful and proven alternatives to the conventional approach: environment and management; botanicals (herbal treatments); copper oxide wire particles (COWP); tannins; and genetics.

ENVIRONMENT AND MANAGEMENT

As you learned in chapter 1, goats are naturally adapted to somewhat dry, mountainous land with plenty of shrubs and tall forage for food. This is the type of climate and habitat they have thrived in over the ages. They also spent their days roaming and wandering, never staying long in the same place. Goats on farms rarely live in anything that comes close to such ideal conditions. Every step away from that ideal brings with it an increased likelihood of problems with parasites. When given the best conditions and feed choices, goats will have far fewer problems with parasites, and you will have far less nursing to do. So the first step in parasite control is meeting the habitat and dining needs of your goats. The key factors in management of your goats to minimize parasite problems are:

- Monitor height of plants grazed: no shorter than 4 inches (10 cm), ideally taller than 6 inches (15 cm)
- Graze only when grasses and plants are dry
- Always avoid marshy, boggy, wet soils
- Provide access to browse—trees, shrubs, tall forbs
- Practice rotational grazing or interspecies grazing if possible

Don't Forget the Immune System!

There is no doubt that a healthy, hardy goat is more capable of resisting or recovering from parasite problems than one in poor health. Achieving this healthy status is accomplished by following the key management practices I've described and ensuring that your herd has all of its nutritional needs met (these topics were covered in detail in part 2).

I want to restate that without optimal nutrition, your goats *will not* have properly functioning immune systems and therefore *will* likely have parasite problems. Therefore, as you read about the holistic options for dealing with internal parasites, always remember to go back and look at what gaps might exist in the animals' overall health, nutrition, and genetics.

Herbal Dewormer Reality Check

I had been treating my goats with herbal dewormers for years, and they had had no parasite problems. One day, however, it dawned on me that maybe the lack of problems was not thanks to the dewormer, but rather stemmed from the conditions of our farm and the health of our herd. I stopped using the dewormer, and lo and behold, still no parasite problems. The takeaway is that, as with synthetic dewormers, it's smart to figure whether your herd really needs them before you spend your money and waste your time. Try other natural management methods, build up your herd's nutritional status, and do all of the other things that nature insists goats need before treating them. If you do decide to try an herbal program, I recommend Molly's Herbals (see appendix D for contact information).

BOTANICALS

Across the world and the centuries, people have always looked to plants as medicine. In most parts of the world, plants are still the main source of remedies and treatments for both animals and humans. Many medicines used by conventional practitioners are derived from plants, and yet many conventional practitioners are convinced that there is no validity to herbal approaches to deworming. Scientific research into botanicals is growing, with many herbs and plants that have long been used by indigenous peoples showing great promise. Unfortunately, I couldn't find any commercially available botanical deworming product on the market at this time that has been the successful subject of academic-based research. Yet there is great interest and support of their success.

Oregano essential oil has proven effective as a parasite control in poultry. Regano is an Italian brand of oregano oil in a proprietary blend available from an online supplier for use in poultry. One research project concluded that Regano proved effective in reducing coccidia populations in goats and sheep. In the tests, the oil was given at the rate of 2 g/100 pounds (45 kg) of body weight on a regular basis—so not as a treatment, but as a preventive.[3] This product is a bit hard to find and is quite expensive. The smallest size I have been able to find is a 1-gallon (4 L) container, which costs about $200. At the top of the list for other botanicals that have some proven success is garlic, presumably for its support of the immune system. I know several organic goat farmers who regularly feed garlic cloves to their goats. Most will learn to relish the cloves or even capsules straight from your hand or from the feeder. The flavor does taint milk, so if you are a milk producer this is a bit of a taste issue. One cheesemaker I know solves the problem by making flavored cheese the next day! CEG Remedy, which is a blend of cayenne,

Figure 6.19. I learned to make garlic tea for treating mastitis from an organic goat farmer. Organic production brings unique challenges to the goat farmer, including alternatives to the ubiquitous use of antibiotics.

echinacea, and garlic, can also be used to support immune function during times when coccidiosis risk goes up or if other internal parasites are a problem. Garlic tea can be stored in the fridge (it gets quite pungent, so make sure you have a jar that seals well!). Fertrell makes an easy to use organic garlic oil tincture (see appendix D).

Other plants and herbs that are thought to provide natural anthelmintic help are black walnut (toxic to horses), wormwood (must be used with some caution as it can cause liver, kidney, and nervous system damage if used frequently), pumpkin seeds, and yucca. There are many others used throughout the world. I recommend investigating usage and dosages by talking to a practicing holistic herbalist, listening to anecdotal advice from other producers you trust, and following instructions in herbal care books. For my own herd, as I explain in Herbal Dewormer Reality Check on page 153, I no longer use any type of dewormer, herbal or otherwise.

As I mentioned earlier the mechanism by which some botanicals help reduce worm populations may involve support of the goat's overall immune system and direct digestive system health. Some plants can reduce parasite populations, and there is scientific support for this. In most cases these plants are also natural feeds for the goats. In that case, they are considered food that's also medicine. A real health food! An example of one of these plants is my favorite forb, chicory. Several studies have documented its ability to reduce parasite loads, especially of barber pole worm. The way that chicory (and perhaps other natural herbal dewormers) works might be due to its high nutritional benefits, including protein that bypasses the rumen to help repair digestive tract tissues (refer back to chapter 4 for more details on digestive tract function) and the immune-boosting trace minerals the plant contains. The part I love best about this research is that it demonstrates the goat's natural wisdom about what is good to eat.

DIATOMACEOUS EARTH

Food-grade diatomaceous earth, DE (made from the finely powdered fossilized remains of tiny creatures

called diatoms), is FDA-approved for use in animal feeds, and many people believe it has a direct effect on internal parasites in the digestive system (not on lung worms or liver flukes). According to organic regulations, it can't be fed in amounts greater than 2 percent of the animal's daily feed. Other sources recommend no more than 1 percent. Fertrell Herbal Capsules contain DE, cayenne, and garlic and are marketed as a "digestive aid." Some producers offer DE as a component of a free-choice mineral and supplement buffet, but Dr. Susan Beal (profiled in chapter 6) says that it must be given intermittently, not continuously.

To date studies have not shown DE alone to affect parasite egg counts when given internally. It can be helpful when directly applied to the goat's living environment (usually by spreading a fine layer on a newly cleaned pen or paddock), however, and helps reduce parasites in the soil and external parasites such as lice.

COPPER AND COPPER OXIDE WIRE PARTICLES

Many studies show that ingestion of copper oxide wire particles (COWP) reduces parasite loads. The tiny copper wires embed themselves in the stomach wall lining of the goat, but their precise mechanism

Figure 6.20. Copper oxide wire particles from Santa Cruz Animal Health.

of action is unknown. However, when it comes to parasite resistance and resilience, overall animal health is of the utmost importance. The goat with an optimal immune system and healthy lining in its digestive system has the best chance of avoiding problems with parasites.

Copper has a complex role in goat health (as discussed in chapter 4), and optimal dosage amounts are uncertain. Depending on the mineral content of soils where your feed is grown, and on the mineral supplements available to your animals, your goats may or may not be getting enough copper. Even if they are, COWP can be given with less fear of copper overdose than potent copper sulfate, because copper oxide isn't absorbed well into the bloodstream. Copper oxide will increase liver levels of copper, however, so it should be used carefully, especially if you live in one of the few parts of the country with adequate to high soil copper levels.

The particles of copper oxide wire are administered inside a gelatin capsule, or bolus. The size needed for goats is much smaller than that more readily available for cattle. You can purchase them prefilled from Santa Cruz Animal Health (in the United States) in 4 g and 2 g boluses, for adults and kids respectively. If you're using copper boluses for parasite control, up to two 2 or 4 g boluses can be given during each parasite season. Be sure to work with a holistic vet if you are dealing with a parasite problem that isn't remedied by the above dosage and receive guidelines should that dosage need to be exceeded. Laini Fondiller of Lazy Lady goat dairy in Vermont said for her pasture-fed goats parasites are "always an issue but never a problem." She doses them twice a year with COWP boluses.

TANNINS

Tannins are natural substances found in many plants. You might have had a close encounter with tannins if you've ever tried eating an unripe persimmon and felt the astringent, drying effects in your mouth. Tannins exist in two forms in plants. The most common form, condensed tannins or CT, has proven itself very effective against internal parasites, including

Figure 6.21. While out browsing, the Pholia Farm herd eagerly devours the bark from a ponderosa pine tree, a source of condensed tannins. Fortunately these mature trees don't suffer any ill effects from the eager goat nibbling.

coccidia and the dreaded barber pole worm. The other type of tannin, hydrolyzable tannin (HT), is believed to be detrimental to animal health, although goats may be less affected by all tannins than sheep or cattle thanks to their larger salivary glands and the production of protective proteins in their saliva that bind with tannins.

CT are thought to help reduce parasite loads by a twofold mechanism. First, CT are known to affect the way that rumen microbes process protein. They reduce the breakdown of protein in the rumen, meaning more protein is available for processing in the abomasum and the small intestine. The small intestine and stomach can then use the protein to make repairs and keep the intestinal lining strong. The extra protein also directly supports the animal's immune system, which helps the animal be either resistant or resilient to the parasites.[4] The other way that CT might help are by directly affecting the

Table 6.6. Forage Plants and Browse with High to Mid-Levels of Condensed Tannin

Common Name	Latin Name	Type of Plant	Where Found	CT Level*
Bird's-foot trefoil	*Lotus corniculatus*	Temperate legume	Europe, Mediterranean, United States	48 g/kg of DM†
Big trefoil	*Lotus pedunculatus*	Temperate legume	——	77 g/kg of DM
Sulla	*Hedysarum coronarium*	Temperate legume	Mediterranean	51–84 g/kg of DM
Sericea lespedeza	*Lespedeza cuneata*	Tropical legume	——	46 g/kg of DM
Chicory	*Cichorium intybus*	Herb	Worldwide	3.1 g/kg of DM
Mulga	*Acacia aneura*	Tree	Australia	>50 g/kg of DM
Yellow box	*Eucalyptus melliodora*	Tree	Australia	>50 g/kg of DM
Oak	*Quercus* spp.	Tree	Worldwide	Not available
Maple	*Acer* sp.	Tree	Worldwide	Not available
Birch	*Betula* sp.	Tree	Worldwide	Not available
Willow	*Salix caprea*	Tree	Worldwide	Not available
Pine	*Pinus* sp.	Tree	Worldwide	Bark 11–13% of DM

Source: Byeng Min and Steve Hart, "Tannins for Suppression of Internal Parasites"; Antonella Cannas, "Tannins: Fascinating but Dangerous Molecules," Cornell University Department of Animal Science website.

* 10 g/kg of dry matter is considered high.

† Dry matter

parasite larva by combining with, and therefore making unavailable, nutrients that the larva would have used to thrive.

The CT content of many tannin-creating plants isn't known, but studies are under way for a few species that can be commercially grown or marketed. Currently the most studied CT source is the legume forage plant *Sericea lespedeza*, which can be grown in many areas and also turned into a pelletized feed supplement that supplies ample tannins. Pine bark is another proven source of CTs. Because pink bark can be a by-product of the timber industry, it should be possible to make it broadly available as a feed supplement. According to Dr. Byeng Ryel Min of Tuskegee University, "Goats on a pine bark (PB) powder diet (30% PB powder mixed with TMR diet) had 74 percent lower fecal egg counts and 5 percent better animal weight gain compared to control diets

during three-month trials."[5] As you can see in figure 6.21, goats find pine bark quite tasty.

Table 6.6 lists other forage plants from various parts of the world that are known to be high in CTs. The list is in no way exhaustive, and there are no doubt many wild plants that goats will select that are also high in these beneficial tannins. Fortunately, when goats are offered an ample variety of forages, they have the opportunity to select what they need when they need it.

GENETICS

In an online article called "Stop Selecting Sissy Sheep (and Goats)!", Dr. Gareth Bath points out that "nature is hard and dispassionate" (unlike most of us goat owners) and those that cannot cope with parasites perish. The advent of synthetic chemical dewormers, however, has allowed us to breed animals that aren't naturally resistant or resilient to parasites. It's speculated that the positive characteristic of being able to resist parasite infestation or survive well with it is quite heritable, at about 25 to 30 percent. With decades now of generations that are lacking this resistance, the progressive, holistic breeder has to take the place of that harsh reality that nature would have provided if we are to create new herds of healthy, naturally parasite-problem-free animals.

Before you can decide who stays and who goes, however, you first must make sure that you are providing all of the other things that nature would have, as I've emphasized more than once so far! So do the future of goats a big favor—when a goat that is given the best environment, feed, and nutrition still suffers from internal parasites, cull the animal. You don't have to resort to slaughtering if you don't want to, although some will rightly point out that these animals will simply continue to spread generations of resistant parasites in the environment. The goat could become a pet, a brush goat, or a pensioner. If you follow this guideline, before long you won't be spending any more time, energy, and money on battling worms.

Hoof Trimming

In the wild, goats naturally wear down their hooves by climbing and moving about on dry, rocky ground. Even on the best naturally managed farm, however, hoof trimming must be a regular part of care. People often ask me how often hooves should be trimmed. I can't give a definite schedule. The best advice I can give is, "Whenever they need it." Each animal has different rates of hoof growth and patterns of wear. In the wet winter months hoof growth might slow down a bit, but so does activity on hard, rough ground that would otherwise wear off some of the hoof. In the spring and summer hoof growth might speed up a bit, but activity should also. Much depends on the

> With decades of generations that are lacking parasite resistance, the progressive, holistic breeder has to take the place of that harsh reality that nature would have provided if we are to create new herds of healthy, naturally parasite-problem-free animals.

type of terrain your goats walk on. And the day-to-day weather of your area dictates not only the softness of the ground, but also whether or not the goats will want to be out walking around.

When trimming hooves it's important to visualize the structure of the foot as you work. The hoof of the goat is made up of two toes (designed to help the goat find footing on steep, treacherous hillsides and rocks). The toes are covered with a tough keratinous growth—the wall of the hoof. The bottom of each toe, the sole, isn't protected by this hard hoof wall. The sole is soft, but not sensitive. Just beneath the pliable sole, though, the living, internal part of the hoof is very sensitive. Structures called laminae attach the hoof wall to this internal tissue. Each hoof also contains a phalangeal bone (analogous to our finger bones). (See figure 13.6.) The goal in trimming is to get rid of the portions of the hoof that would normally have worn off on their own, but to avoid cuts that would result in the goat's

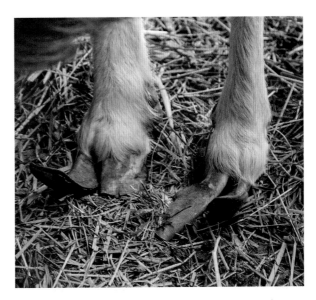

Figure 6.22. Here's an example of goat hooves that are severely in need of attention.

experiencing pain when it walks, and also to avoid cutting away so much hoof that you cause bleeding. If that happens, let the hoof bleed a bit to flush out bacteria, then clean and dress the hoof as in the Wounds section on page 149. Verify that the animal is up to date on its tetanus protection or administer a dose of tetanus antitoxin.

Where possible a dedicated area for hoof trimming will help make the task more efficient. A stand or raised area for the goat to stand on lessens the need to stoop over, which is easier on your back. Hoof trimming tools can consist of a pair of hoof nippers (which look much like pruning shears) or a flat, handheld file (a drywall file works well). A file works

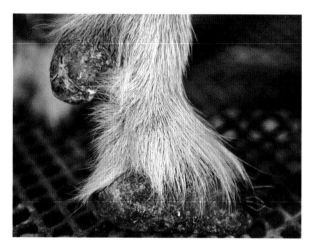

Figure 6.23. This back hoof doesn't look too bad in side view.

Figure 6.24. However, when the hoof is lifted, it's clear it needs cleaning and trimming. You can use one blade of the hoof trimmers to clean out debris from the sole. This removes pebbles, which could dull the blades, and makes it easier to see what you're doing while trimming.

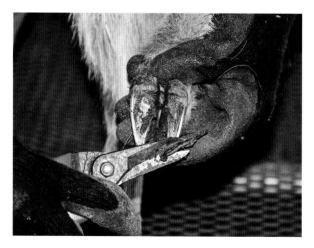

Figure 6.25. After trimming the hoof wall evenly all the way around, nip off the overgrowth of the soft heel.

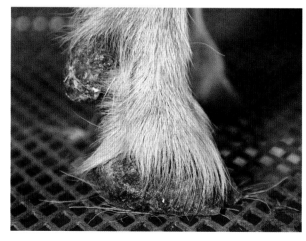

Figure 6.26. Set the hoof back down and view it from the side again. Note the more upright angle of the pastern. This looks much better!

well only if a goat's hooves have been well maintained, so that there is little hoof wall overgrowth. When you first start trimming hooves, it is likely to feel very awkward until you get the hang of it. I tell people it's a bit like taking a pair of scissors and trying to cut paper using your non-dominant hand.

Biosecurity and Zoonoses

These two complicated words might seem like they would apply only to large-scale industrial farms, but they are important for all goat owners to know and understand. Biosecurity is the protection of your farm and animals from exposure to diseases and parasites from other farms and animals. Zoonoses (*zo-A-no-sees*) are diseases that can be passed from animals to people and are a very real concern for those who work with animals and consume their products. I cover specific zoonotic diseases in greater detail in part 5.

Many parts of the world have problems with highly contagious diseases that can spread throughout an animal population, causing suffering and death—not to mention a huge emotional and financial burden to the farmer. Many of these diseases are reportable to government agencies such as the USDA. Biosecurity focuses on preventive steps that aim to eliminate or limit the chances of such disasters. Don't be shy about enforcing policies such as having to walk through a sanitizing foot bath or wash hands before handling kids. Let your visitors know you are happy to extend the same courtesy when you visit someone else's farm. See appendix D for more information on biosecurity and zoonoses.

Zoonotic diseases can be passed from animals to humans through direct contact and airborne contact and via milk and meat. Most diseases suffered by goats are not a threat to people, but a few are and a few others are inconclusively linked. Problems such as undulant fever (called brucellosis when it manifests in the animal) and tuberculosis have all but disappeared in many states, in both goats and humans but remain serious health issues in many parts of the world. (Both are reportable diseases in the United States.) Other diseases, such as Q fever, are becoming more common. In different parts of the world, the concerns will vary greatly. When you look at long lists of zoonotic diseases, please keep in mind that very few will be a major concern for the well-chosen and well-cared-for goat herd. Don't let

Figure 6.27. A foot bath greets all those who enter this goat dairy, Mountain Lodge Farm, in Washington State. It serves as a biosecurity measure to prevent bacteria from other farms being brought onto the premises.

Biosecurity Checklist

- ☑ Stay current on biosecurity threats in your area.
- ☑ Ask all visitors if they have been in areas with biosecurity threats.
- ☑ Enforce policies to prevent threats. These may include (1) ensuring that any footwear that has been worn on any other farm where goats, sheep, or cows are kept has been cleaned and sanitized, and (2) following a closed-herd policy of not allowing your goats to have direct contact with goats from other herds.

the length of these lists and the descriptions of the diseases prevent you from getting goats, but do let them inspire you to continuously improve your herd management practices. Remember, some diseases are "reportable"—when detected, they must be reported to your state vet and to the USDA vet.

Table 6.7. Most Common Zoonotic Diseases of Goats in North America

Disease	Method of Transmission	Symptoms in Humans	Notes
Brucellosis or undulant fever [*Brucella* spp.]	• Aerosol • Direct contact with urine, raw milk, or semen • Contact with fetal membranes (during and after delivery or cesarean)	• Fever, chills, sweating, loss of appetite, constipation, headache, abortion, inflammation of testicles	• Highly infectious; contact with only a few *Brucella* organisms can result in infection • When present in herd, this disease must be reported to state or federal animal health agency
Caseous lymphadenitis [CL]	• Direct contact with pus from cyst • Oral contact with feces (fecal/oral)	• Abscesses of lymph nodes, painful skin sores	
Contagious ecthyma [orf, sore mouth]	• Direct contact with animal • Direct contact with objects that are contaminated with the ecthyma organism	• Skin sores, especially on the hands	• Virus may remain viable in scabs for months • Disease is usually self-limiting, remission in 2 to 4 weeks
Q fever [*Coxiella burnetii*]	• Aerosol • Consumption of raw milk	• Fever, chills, loss of appetite [anorexia], abortion, liver disease	• Caution *especially* for those that are pregnant or immuno-suppressed, or have a heart valve disease or replacement
Yersinia pseudotuber-culosis	• Oral contact with feces (fecal/oral)	• Acute abdominal pain, fever, vomiting, diarrhea • Arthritis, iritis, nephritis • Septicemia in immuno-compromised individuals	

CHAPTER SEVEN

Advanced Skills, Procedures, and Management

There are many health care skills that can enhance your experience and augment your success as a goat farmer, ranging from drawing blood and giving goats pain medication to rumen transfaunation (which sounds more complicated than it is in practice). Some of these are fairly advanced and may not be something every goat owner will want to attempt, but understanding what's involved in these health care tasks will help you better care for your goats. It's taken me years to feel comfortable with all of the skills described in this chapter, and no doubt I'll be learning more over the years! Don't feel daunted by the challenge: Becoming an accomplished "goat health technician" takes time and practice.

Fecal Testing

Fecal testing is not a chemical test; it's observation using a microscope to check for the presence of parasite eggs in goat feces. Although this is an advanced skill, it's fairly simple to master, and doing it yourself will save the cost of testing. It's also very convenient to be able to do this test on your farm, versus taking a sample to the vet's office. Sometimes knowing whether parasites are present will determine how you care for the goat, and ascertaining

this information quickly can help prevent suffering. Fecal testing isn't without its negatives. For one thing, the rate that eggs and oocysts are shed in manure isn't always proportional to the number of parasites inside an animal. However, when considered in conjunction with other observations, it is useful data.

It takes time and a bit of practice to get good at reading fecal float results (so called because the eggs float in the test solution). The eggs of various species of parasites often look similar to one another, and in the beginning it's difficult even to distinguish eggs from plant cells and air bubbles. It's helpful to have an experienced mentor coach you as you do your first tests, but even if you don't have a mentor, you can learn. When I first started doing these tests, I couldn't tell if I just wasn't seeing the eggs or if there really weren't any there! See appendix D for a link to an online video course in how and why to do goat and sheep fecal egg counts. The video has great descriptions that are helpful in learning to identify eggs. There is also a link to a helpful illustration of goat parasite eggs and oocysts.

You can collect fecal samples either just as they are produced or in piles on the ground. As long as they appear fresh (not dried out) and aren't mixed in with debris, you can use them. You might want to collect individual samples or a bit of poo from several sources. In the beginning, I recommend testing individual animals until you feel comfortable that the count is low. A mixed sample will in essence dilute the count of each individual.

As to how many eggs or oocysts are acceptable, even a healthy goat will always have a few eggs present in its feces, but probably no more than a total of 10 per slide. This is a subjective number and suggestion, so please consult with your vet and/or other resources if you have concerns and questions about how to interpret the test.

Equipment

The biggest roadblock to doing fecal tests at home is the need for a good microscope. I recommend one with a 10x eyepiece and 10x lens. A second lens, 40x, is also helpful. Be sure to get one with a movable stage—the place you put the slide. Try hunting around on sites such as craigslist and eBay for used scopes. If you buy new, consider a snazzy digital model that allows you to view the slide on a built-in screen or hooked up to a computer—no need to squint through an eyepiece!

There are a few different variations on equipment for a fecal test. You can assemble your own set of supplies utilizing regular glass slides and coverslips (see Make Your Own Test Kit below), or McMaster slides or egg count chambers. (See appendix D for sources of supplies.) Using these slides, the test is called the modified McMaster test; eggs are much easier to count this way. The initial investment is a bit more, but if you are starting from scratch to assemble a kit, I think it's worth it. Figures 7.1 through 7.4 show a fecal test being performed using the McMaster method.

Another option is to buy a box of "fecalyzers" or fecal float devices (see appendix D). These small plastic disposable devices come in a box of 50 for about $30. The fecalyzer serves as the single container for the entire process, eliminating the need to purchase test tubes, a test tube holder, and beakers. I'm not a big fan of them, simply because I don't like the disposable approach.

Fecal Float Solution

Fecal float solution, often called simply fecal solution, can be purchased by the gallon for $10 to $15. You can also make your own at home quite inexpensively. All

Make Your Own Test Kit

The supplies needed for a standard kit and a McMaster kit are similar, but not identical.

STANDARD KIT

Disposable gloves
Wooden craft sticks (Popsicle sticks)
Small scale that can measure in grams
2 (50 ml) glass beakers
20 ml test tube
Test tube holder
Small fine-mesh sieve or tea strainer
Glass slides and coverslips
Fecal float solution

MCMASTER KIT

Disposable gloves
Wooden craft sticks (Popsicle sticks)
Small scale that can measure in grams
30 ml syringe
Small bowl
1 cc syringe (insulin type) or eyedropper
Small fine-mesh sieve or tea strainer
McMaster slides
Fecal float solution

you need is a bag of Epsom salts (magnesium sulfate), which is available at most grocery stores and pharmacies in the first-aid section. Here's how to make the solution:

Step 1. Fill a clean glass quart (liter) jar halfway with hot water.

Step 2. Add about ¼ cup (75 g) Epsom salts; cover, and shake until dissolved.

Step 3. Continue adding more Epsom salts, ¼ cup or less at a time, and shaking after each addition, until no more salt will dissolve in the water.

Step 4. Let the solution sit at room temperature overnight.

Step 5. The next morning, if no salt crystals are visible at the bottom of the jar, add about 1 additional tablespoon (20 g) of Epsom salts, and shake.

Step 6. Just before use, observe the solution for crystals. If any are visible, don't shake the solution before using it for a float test, because you want the crystals to remain at the bottom of the jar. Crystals in the solution on a slide could be mistaken for parasite eggs.

How to Do a Fecal Float Test

The basic procedure for doing a fecal float is to collect some feces, mix a sample of feces into solution, put a drop of solution on a glass slide, and then look at the slide under the microscope and count eggs.

USING MCMASTER SLIDES

Step 1. Don a disposable glove, and collect some feces. Pull off the glove so that it turns inside out, with the feces trapped inside. Smash up the pellets.

Step 2. Place a beaker on the scale, and tare it to zero. Use a craft stick to weigh out 2 g of smashed feces. Stir in 28 ml of float solution, mixing well with the craft stick.

Step 3. Strain the mixed solution through the sieve or tea strainer into a second beaker. Use the craft stick to press as much liquid through as possible.

Figure 7.1. Fecal float test step 1.

Figure 7.2. Fecal float test step 2.

Figure 7.3. Fecal float test step 3.

Figure 7.4. Fecal float test step 4 using McMaster chamber.

Step 4. Use a disposable syringe to draw up 1 ml of liquid. Fill the slide with the liquid and put a coverslip in place.

Step 5. Place the slide on the microscope. Slide the 10x lens into place, and adjust so that the green lines on the slide or any air bubbles under the glass are in focus. Then find the corner of the grid and slowly move the bed of the microscope, counting eggs as you go. Use the grid lines to guide your progress, making sure you look at and count the eggs in each square of the grid. If needed, use the 40x eyepiece to get a closer look for identification purposes.

USING GLASS SLIDES

Step 1. Don a disposable glove, and collect some feces. Pull off the glove so that it turns inside out, with the feces trapped inside. Smash up the pellets.

Step 2. Place a beaker on the scale, and tare it to zero. Use a craft stick to weigh out 3 g of smashed feces. Stir in 28 ml of float solution, mixing well with the craft stick. Let set for two minutes.

Step 3. Strain the mixed solution through the sieve or tea strainer into a second beaker. Use the craft stick to press as much liquid through as possible. Let set for two minutes.

Step 4. Stir, and pour slowly into the test tube setting in the test tube holder, filling it to the brim. If

bubbles are immediately present, add a bit more of the manure mixture to cause the bubbles to overflow out of the test tube.

Step 5. Carefully lower a coverslip onto the test tube (more solution will probably dribble over the edges). Wait 20 to 30 minutes.

Step 6. Carefully and slowly lift the coverslip straight up and off the test tube. Place it on the center of a clean slide by lowering one edge so that it touches the slide and then allowing the rest of the coverslip to lower onto the slide.

Step 7. Place the slide on the microscope. Slide the 10x lens into place, and adjust so that the air bubbles under the glass are in focus. Then find the corner of the coverslip and slowly move the bed of the microscope, counting the eggs as you go, moving up, over a bit, then down, on the slide. You'll know you're finished when you reach the opposite bottom corner of the coverslip from where you started.

Drawing Blood

If you've never drawn blood from a goat, or any other animal for that matter, the first time can be nerve racking. I have to confess, for years I avoided learning how to do it. But once I did, I was amazed at how easy it is. It's best to practice on an adult goat of standard size, if you can. Nigerian Dwarfs, Pygmy, and young goats have shorter necks and smaller blood vessels, making the process much more difficult for beginners.

Blood is drawn from a goat most commonly from the jugular vein in the neck. You must use a fairly large-gauge needle to allow red blood cells to easily enter the needle. I prefer an 18 gauge, but some people use the next size smaller, a 20 gauge. You can draw the blood into a syringe and then inject it into the proper blood vial. Or you can collect it using a special needle holder into which the vacuum tube is inserted once the needle is in the blood vessel. I prefer using a syringe rather than a vacuum tube because I had poor luck attaching the vacuum tube to the needle holder without displacing the needle's position in the jugular.

Lab Lingo

When you're having blood or feces tested for disease, you'll learn that there are different types of laboratory tests. Usually your vet or other authority (see the resources in appendix D) can suggest the most useful test for your situation. Here are some of the most common options (there are variations within some of these options!):

ELISA: The most common test type that you'll encounter is the ELISA test. The acronym stands for "enzyme-linked immunosorbent assay." This basically means that the test searches for antibodies in a blood or milk sample. Antibodies indicate that an animal has been exposed to the pathogen being screened for. The presence of the virus is also conclusive of disease. ELISA testing is quick and affordable.

AGID: This is a fairly common test. The acronym is short for "agar gel immunodiffusion." This test also searches for antibodies (like the ELISA test) in blood and is more accurate and specific than ELISA. False positive tests are rare, but false negatives occur when the animal has very low levels of the viral antibody being screened for, which is common in cases of early infection.

PCR: Short for "polymerase chain reaction," this DNA test is able to analyze all types of samples and detect genetic material that indicates infection. Samples must be refrigerated, and the test is more expensive. For some diseases it can be helpful, especially if it's used after either ELISA or AGID to confirm results.

Culture: Samples of blood, manure, soil, and tissue can be cultured to determine the presence of microorganisms. A culture literally attempts to grow the organism being tested for. This type of test takes the longest—many weeks, in fact—but is seen as the most conclusive.

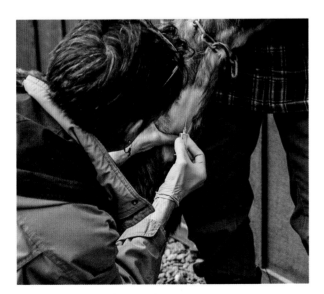

Figure 7.5. Drawing blood for a pregnancy test. You can stand or kneel, whatever is more comfortable. I find that it's easiest when I'm at eye level with my target site.

There are two jugular veins, one on each side of a goat's neck (see figure 12.1). It doesn't matter which vein you draw the blood from; choose the side that feels more comfortable to work from, depending on your handedness. Until you are quite good at drawing blood, it's best to use a clipper to shave off some neck hair before you get started (I still do!); that way it's easier to see the jugular vein. Locate the jugular by having a helper hold the goat's head and tip its chin up a bit. Some people recommend turning the goat's head to point away from the person drawing the blood, but I find this stretches out the vein, making it harder to find.

Run the fingers of your non-dominant hand across the front of the goat's neck. You will feel its trachea, with its ridged texture, and a bit to the left and behind the trachea, you might feel the esophagus, which is smooth. The jugular vein runs in a groove between the trachea and esophagus and the musculature of the neck. (See figure 12.1.) (The carotid artery is located deeper, behind muscle tissue that runs parallel to the jugular vein.) The jugular veins are smooth and soft, but a bit resilient, and move easily side to side. The veins will easily roll away from any pressure. When you think you're in the right area, apply a bit of pressure to what you think is the jugular vein. You're trying to stop the blood flow and make the vein bulge up. When you see this, you know you're in the right spot. Remember, the vein returns blood from the head down to the heart, so you will apply pressure lower on the neck than the spot where you will be inserting the needle. If you are having trouble finding the jugular vein, take a break and take the goat for a short walk. This will raise its blood pressure, helping fill the vein.

Once you've found the right spot, you're ready to get started.

Step 1. Clean the skin thoroughly with an alcohol wipe or paper towel soaked with a bit of rubbing alcohol. Rub vigorously to clean the spot.

Step 2. Hold your needle and syringe (or tube holder and needle) in your dominant hand. With your non-dominant hand's thumb, block the flow of the vein again (below the spot where you will draw the blood). Once the vein pops up, hold the syringe so that the needle is at about a 30-degree angle with the neck with the bevel side up. That places the sharp point toward the neck and the opening in the needle toward the blood flow.

Step 3. With a decisive but gentle push, insert the needle through the skin and in about 1/8 to 1/4 inch (3–6 mm). You should feel a slight resistance as the needle goes through the wall of the vein (you might not notice this sensation on your first few blood draws). If the vein slips out of the way (and they often do), leave the needle in, angle it toward the vein, and push about 1/16 inch (1.6 mm) farther. Be very careful not to go too deep or you could go right through the vein, creating a hole through both sides of it. A hematoma, a swollen area filled with blood, could result. While that's likely not a big problem, you'll feel bad, and the goat won't appreciate it, either. If a hematoma develops, use warm packs on the area and give the animal arnica. Watch that it doesn't get any larger.

Step 4. Once the needle is in the vein, there may be enough pressure from the blood to push the syringe plunger out a bit. If you don't see any flash of red in the syringe, try drawing back a bit on the plunger—keeping your thumb still on the vein. If you get blood, quickly draw up the amount needed.

Step 5. Take your thumb off the vein and remove the needle. Hold a cotton ball or small piece of bandage gauze over the puncture site for a moment.

If no blood flows into the vein when you pull back on the plunger, you have two options. If you can still see the vein and the needle appears to have moved the vein to the side, angle the tip of the needle toward the vein and see if you can penetrate it. If you can't see the vein at all, it's probably better to start over. Use a new needle because even one penetration of the skin will dull the tip. And, reusing a needle that will be inserted directly into a blood vessel increases the odds of introducing a contaminant into the animal's bloodstream.

Calculating Medication Dosages Accurately

Math has never been my strong suit. Even so, getting comfortable with the arithmetic required to properly measure medication doses was a must for when I became a nurse—and it has turned out to be very handy in my life as a goat farmer, too. The hesitation some veterinarians might feel when prescribing more than a single dose of drugs with potentially dangerous side effects is in great part due to concerns over producers' understanding of pharmacological math.

When you are administering a prescription or extra-label medicine, you can usually refer to what your vet has recommended regarding the amount to give according to the size of the goat—and that

Figure 7.6. A weight tape designed for use with goats is a fairly accurate way to estimate a goat's weight. The tape is placed around the animal's heart girth (the narrowest part just behind the withers and elbows), and the weight in pounds or kilograms is read from the tape. I've found this method to be accurate within 5 pounds (2.3 kg) on goats of all sizes and breeds.

information will appear handily on the bottle of said medication. When you want to administer over-the-counter medications and treatments, though, or a prescribed medication for the whole herd, you will need to calculate for yourself what the appropriate dosage is for an individual goat. Let's go through an example in detail to clarify the process.

Liquid medications and vitamin solutions come in different concentrations. The amount of active ingredient is usually quantified by how many milligrams or units are present in a certain number of milliliters. For tablets or capsules, the active ingredient listing will be milligrams or grams per tablet or capsule. Powders will be in milligrams or grams per scoop or other measurement.

The amount of a medication or vitamin that the animal needs will usually be listed as a certain number of mg or units per kilogram (mg/kg or lb.) of body weight. If you are lucky enough to live in a sensible country when it comes to measurements, you'll already be using the metric system. If you are from the United States, Myanmar, or Liberia you will have to do even more math and convert the goat's weight from pounds to kilograms (kg).

Let's use the example of treating Winston, a goat who is suspected of having polioencephalomalacia (also called stargazing or goat polio), a softening of brain tissue that can be treated with vitamin B_1 (thiamine). The goal is to give Winston 10 mg/kg (5 mg/lb.) of vitamin B_1 every 6 hours for 24 hours. According to the weight tape, Winston weighs 85 pounds. (Note that weight tapes are fairly accurate, but if the dosage range is small, finding a more precise way to weigh the animal might be advisable.)

Step 1. Convert pounds to kilograms. One pound equals 0.45359237 kg. You can round this to 0.45 kg for most dosage situations. To do the math:

$$\_\_\_ \text{ lbs.} \times 0.45 = \_\_\_ \text{ kg}$$

Winston Example:

$$85 \times 0.45 = 38.25 \text{ kg}$$

Step 2. Now you're ready to figure out how many mg or units of the solution you need to give the animal (not how many milliliters—that's next). Check the bottle again and do this math:

$$\_\_\_ \text{ kg} \times \text{ the amount of mg/kg recommended}$$
$$= \text{ the total number of mg needed}$$

Or

$$\_\_\_ \text{ kg} \times \text{ the number of units recommended per kg}$$
$$= \text{ the total number of units needed}$$

Winston Example: According to the dosage recommendation, we want to give 10 mg of thiamine for every kg of body weight, so

$$38.25 \times 10 \text{ mg} = 382.5 \text{ mg of thiamine}$$

Step 3. Next, you need to find out how many mg or units are in each ml of solution or mg/g. Most solutions will tell you how many mg per ml; if this is the case, skip to step 4. Otherwise, take the number of mg or U and divide it by the number of ml stated. For example:

$$10 \text{ mg} \div 100 \text{ ml} = 10 \div 100 = 0.1 \text{ mg/ml}$$
$$2 \text{ U} \div 10 \text{ ml} = 2 \div 10 = 0.2 \text{ U/ml}$$
$$2 \text{ mg} \div 5\text{g} = 2 \div 5 = 0.4 \text{ mg/g}$$

Winston Example: Our bottle of B vitamin complex says that each ml contains 100 mg of thiamine, so we'll skip to step 4.

Step 4. Finally, you're ready to calculate how many ml of the solution to draw into your syringe or measure out. Go back to your calculation for number of mg or units per kg of body weight and note how much the animal needs. Take that amount and divide it by the number of mg per ml.

$$\_\_\_ \text{ mg needed} \div \text{ number of mg per ml}$$
$$= \text{ total number of ml needed}$$

Winston Example:

$$382.5 \text{ (mg needed) divided by}$$
$$100 \text{ (number of mg per ml)} = 3.83 \text{ ml}$$

Pain Management

Livestock respond very differently to pain than humans do. As prey animals they cannot afford the luxury of moping about. If they have just narrowly escaped death (whether in reality or in their own minds) they rapidly recover their ability to act normally. This doesn't mean that they don't have physiological reactions almost the same as we do. Studies on calves being dehorned looked at their heart rate and the production of stress hormones when disbudded without pain control or local anesthetic, with pain control only, with local anesthetic only, and with both. The results were conclusive: Calves experienced the least pain when both a local numbing was performed and a medication was given for pain relief after the local numbing wore off. You will hear many people say, regarding disbudding of goat kids, that the animal is yelling more about being restrained (highly traumatic for a prey animal) than the actual burning of the skull. Whether that is true is not the relevant consideration, though. If your goal is to promote good health, you need to make procedures like these as low-stress as possible.

In some parts of the world—the United Kingdom, for example—pain control is required when performing procedures on farm animals such as castration and disbudding or dehorning—procedures routinely done in the United States without pain relief. (But cats, dogs, and horses always receive pain control for medical procedures.) With humane certification and animal welfare issues becoming more prominent public concerns, along with the proven benefits of pain control for the animal, it is a good time to change our practices and make sure that animals receive adequate pain control during procedures, for injuries, and during other traumas. For moderate or temporary pain, you can try an herbal or homeopathic remedy first. If that doesn't work, you can try any of the medications below; all of them are allowed (with specific withdrawal times and veterinary oversight) in organic production.

Fortunately the skilled producer who works with a licensed veterinarian can access several options for pain control for the animal. Some training is required, but the result is that procedures can become less stressful for all. Pain management can be approached from three fronts: pain medicine, or analgesia; local anesthetic; and sedation. Let's look at these three types of pain control. (For a refresher on symptoms of pain and alternative treatments for mild pain, refer back to chapter 7.)

Non-Steroidal Anti-Inflammatory Drugs

The most common analgesics used for animals are non-steroidal anti-inflammatory drugs or NSAIDs (ibuprofen is a common NSAID used for pain relief by humans). In addition to the pain relief benefits, their anti-inflammatory effects are also helpful in complex cases where both pain and inflammation are involved, such as enterotoxemia, coccidiosis, and laminitis (founder).

Banamine: The generic name for this NSAID commonly used with goats is flunixin; there are other brand names, too. This is an injectable drug very effective in providing pain relief to the animal. It does not come labeled for use in goats but can be "extra-labeled" (a term used when a medication is prescribed that isn't labeled for use in goats) and used according to recommendations. In organic production flunixin withdrawal time must be two times the usual recommended.

Phenylbutazone: If you are a horse person, you have probably used this NSAID, which is also simply referred to as bute. This drug is currently not allowed for use in food-producing animals, not even for off-label (extra-label) use due to concerns over severe and fatal reactions in some people. One study cites its effectiveness in goats, but milk withholding times and other critical information remain unknown. Use in elderly goats or pet goats might be advised by your vet, as long as there is no risk they might be headed to the butcher for use as meat.

Aspirin: This readily available NSAID is good for pain control and also acceptable to US organic standards. It's not labeled for goats, so if you use it on a licensed farm, where inspectors will need access to your goat medications, it must include an extra-label provided by your vet. If you use aspirin the dosage must be quite high, because when given orally it is degraded in great part by the rumen. Dosages for goats are just over one 325 mg pill per 10 pounds (4.5 kg) of goat. So a 100-pound (45 kg) goat would need 10 aspirin. If you have a different strength of aspirin, calculate it based on 100 mg/kg of body weight. You can give this dose twice a day with a milk withdrawal time of only two days. I recommend grinding up the tablets with a mortar and pestle and mixing with something tasty, like agave nectar and water, and then dosing with a drenching gun, although I've had plenty of goats that will happily eat the whole tablets right from my hand.

Local Anesthetic

If administered properly, an injection of lidocaine around the site of a procedure, such as castration, dehorning, or disbudding, will deaden the nerves and make the procedure itself pain-free (the animal may still experience stress related to confinement or handling). Lidocaine (there are several brand names for this drug) can be acquired from a vet or with a prescription, but administering it properly will require training by your vet. You will need a very small needle; I recommend getting 1 cc insulin syringes, which come with a tiny 21- to 31-gauge needle. If you are already comfortable with injections, this will make learning the proper site, and technique, much easier—on you and your vet!

Local anesthetic wears off fairly quickly. To be sure that an analgesic starts providing pain relief before the local stops having an effect, administer the proper dose of analgesic medicine *before* starting the procedure. Remember, even though the kid doesn't outwardly respond in the same way a human subjected to severe burns would, the kid's body is still feeling that pain and experiencing stress.

Sedation

Now we traipse into tricky terminology and some controversy. Sedatives are substances that cause relaxation and relieve anxiety. Anesthesia is loss of feeling through the use of drugs. You can use local anesthetics, as described above, or general anesthesia to cause sedation. Sedation can range from light, when the subject is vaguely aware and some pain is felt, to deep or heavy, in which the patient is unaware and doesn't feel pain. Sedation is usually administered by a vet, but it is legal for a vet in most jurisdictions to prescribe a particular sedation drug for a producer's use. Because of the risks, however, this is rarely done.

For more than a decade I have been administering moderate to deep sedation to goats, using the drug xylazine, under the oversight (not direct observation) of the several veterinarians I've had over the years. I use sedation when I perform disbudding on our goat kids and for some procedures on adult goats. Rightfully so, my choice is met with great concern and caution by those who know the risks of this type of drug use. Done improperly, sedation can kill the patient. To be able to use this type of pain management, you must have enough medical background to be able to carefully and accurately measure quantities of a dangerous drug, monitor the respiratory signs of the patient, and know what to do if things aren't going well. (Xylazine is allowed under organic standards for "emergency situations," if it's given with veterinary oversight and followed by twice the normal withdrawal time.)

On our farm we disbud the kids before they are a week old, as long as the buds are palpable by then. At the same time I also tattoo them. Judging by the reactions I've observed, even from goats under sedation, I believe that the experience of being tattooed in the ears is even more painful than the burning from the disbudding iron. (Tattooing in the tail area seems to be less painful.) Based on what I've observed, I am very happy to have my goats unaware during this type of treatment.

If you can afford to take your kids to the vet, or have the vet come to your farm, the vet will administer a mixture of drugs that will put the kid in a state of medium to deep sedation. The mixture is also designed to sedate for a period of time and then stimulate the animal to wake. When a sedative alone is given, as I do, the kid will sleep for several hours. But for a farm our size, with about 100 kids per

Timing Things Right to Limit Stress

I want to reemphasize the importance of limiting stress to reduce the potential for health problems. For example, it's best to avoid overlapping procedures with other stressful events in the goat's life. Castration shouldn't occur at the same time as weaning; switching a kid from nursing to bottle feeding shouldn't occur at the same time as disbudding; and a pregnant goat should not be introduced into a new herd when she's close to giving birth. Keeping stress in mind when managing your goat herd will support the goal of keeping the herd healthy.

year—born over a period of three months—using moderate sedation for disbudding procedures is quite manageable. If you're working with adults, you should wake the animals using the proper reversal drug. Or you can prop them up (as shown in figure 7.11) to avoid the danger of dying from bloat. Goats need to belch every half hour or so in order to allow excess rumen gas to escape; when they're lying on their sides, they can't burp.

> In cases of extended rumen dysfunction, no matter the initial problem, the rumen problems alone can lead to death.

Rumen Transfaunation

Transfaunation refers to the transplant of fauna—live creatures. In the case of goats, transfaunation refers to moving living microbes from one or more goats' rumens to that of another goat. In extreme cases, veterinarians can perform a full rumen transplant from a less valuable, "expendable" goat to a more valuable goat. But in less extreme cases, or before a case becomes extreme, you can harvest rumen contents from healthy goats and administer them to the unhealthy one. This procedure can be lifesaving. The rumen, with its microbial population, is truly the foundation of health for a ruminant. When the rumen isn't functioning well, the entire animal suffers. In cases of extended rumen dysfunction, no matter the initial problem, the rumen problems alone can lead to death.

My first experience with this procedure was early in my goat managing experience. I had let a pregnant doe labor far too long—I hadn't understood the signals that indicated a developing emergency. The mother did deliver, and the kid was saved, but the mom was in a bad way. Over several days she continued to deteriorate. I had read about "cud transplants," and I decided to try it. The doe rebounded quite quickly.

Figure 7.7. By rinsing fresh cud from a healthy goat's mouth and administering it to an ailing goat, you can help restart good rumen function.

We also gave her other supportive measures, and she made a full recovery.

Goats that have been on antibiotic treatment or have suffered some other illness that has caused a loss of weight or a failure to recover from weight loss will also benefit from rumen transfaunation. By directly administering active, healthy rumen microbes—bacteria, protozoans, fungi, and viruses—to a sick goat, you can help "jump-start" an ailing rumen.

This procedure is easily performed without any prior experience or hands-on training. The tools needed are simple: a small bucket, a 30 ml or larger drenching gun or syringe, and about ¼ cup (60 ml) warm water. Put the water in the bucket, draw up a syringeful, and take the bucket and syringe into your goat yard. Look for an unsuspecting animal chewing her cud. (Although either gender of goat will suffice, I suggest choosing a doe simply because she is likely to be easier to work with and not leave you smelling quite so musky!) Walk over, quickly straddle her neck, set the bucket on the ground directly below her head, grasp her chin with one hand, and with the other hand insert the drenching syringe into her mouth and squirt in the water. The goat will not want to swallow it, and as the water runs out of her mouth, with luck it will fall into the bucket. Even if the goat has swallowed her cud, the water should contain a good amount of green, partially rechewed rumen contents.

Repeat this process, using fresh water each time, on a few more goats or until the water in the bucket is a nice bright green. Then go immediately to the sick goat and administer this fluid to it, making sure that it swallows it all—you'll have to hold its head with the chin tipped up. It won't enjoy the process, but it won't harm it. PS: I highly recommend wearing latex or nitrile gloves for this procedure, as the pungent odor of rumen contents is hard to wash from your skin.

Repeat these steps daily until the sick goat shows improvement; namely, regularly chewing its cud. There is never any harm in doing this procedure, other than a bit of a personal insult to all the goats involved. So never hesitate to try this if a goat has stopped ruminating for more than a day, has been on antibiotic therapy, or has recently suffered from some illness that has led to weight loss.

Scur or Horn Control

When you've chosen to manage your goats without their horns (I'll talk more about disbudding and the pros and cons of horns in chapter 9), they might experience some unwanted horn-like growth called a scur. Bucks are particularly prone to scur growth. Or you may have a horned goat but find that the horns raise safety concerns for other goats or for the animal itself (for example, if your fencing situation makes having horns a liability). If the scur is simply a cosmetic problem, then it isn't really a problem at all. But often the disfigured horn growth will develop in a manner that can harm the goat—pressing into the skull or the animal's eye. Having bucks share a pen is often the best treatment for scurs. The bucks' frequent jousting will usually keep scurs worn down. When this doesn't take care of the problem, then the scur must be removed. It will grow back, unfortunately, but can be removed again.

Adult goats with full horns can also have a portion cut off in order to limit the risks associated with full horns, should those risks exist in their environment, or if you're bringing a horned goat into a hornless herd (figure 3.18 shows a goat whose horns are managed in this fashion). In both cases the horn growth will continue, and the process will have to be repeated every year or so.

If you don't want the goat to bleed, you can attempt to cut the scur or horn above the level where the blood supply is present. Feel the horn, and try to locate the point at which it's cooler than at the base. If you want to try this, keep the goat in a shady area for a while beforehand, so that the horn will not be warm from the heat of the sun. Cut above the level of the heat.

I use the IM drug xylazine as a sedative when removing a scur. You can also perform scur removal on a goat that is awake and restrained, but with more stress for you and for the animal. If you aren't using a strong drug, at least administer aspirin, a

Figure 7.8. After the buck was sedated, I cut the scur with a coping saw at a convenient spot. A wire saw would make it easier to cut lower on the scur if that were desired, but more bleeding would have resulted.

Figure 7.9. There will be some bleeding from the cut horn.

Figure 7.10. I use a hot disbudding iron to cauterize the wound.

Figure 7.11. When you're giving moderate to deep sedation to an adult goat, it's critical not to leave the animal lying on its side for more than 30 minutes. Otherwise, gas buildup in the rumen can potentially put pressure on the heart and lungs, even leading to death. This sedated animal is propped up on his sternum between two straw bales with his head elevated on a pillow of straw.

homeopathic remedy such as arnica or hypericum, or an herbal one such as Ow-Eze. Be sure to have plenty of help to hold the animal so that you don't accidentally cut into the skin. If the goat isn't up to date with its tetanus vaccination, administer tetanus antitoxin that day. I don't dose with antibiotics after this procedure, but I do give the animal probiotics for a couple of days after the procedure.

Euthanasia

I strongly believe that every goat owner should have a plan for implementing euthanasia for a suffering animal. If you have a small suburban or urban herd and live near a 24-hour emergency veterinary clinic, then you can rely on the clinic if the need for euthanasia arises. But if you have a larger herd (which often implies that you have fewer funds per goat) or you do not have quick access to a vet, then you owe it to your animals to have prepared the means to quickly end suffering. I realized fairly quickly that managing a goat herd meant that I would have to take the place of nature. The goal is to try to mimic what nature would have offered the animals in terms of feed and to try to mimic what might happen to a goat naturally at the end of life. In the wild, an ailing goat doesn't suffer for long. A predator takes care of that. We have to become that solution for our goats.

When deciding if euthanasia is the right choice for a particular animal, start by asking yourself several questions about the animal's condition. Then make a decision based on your own philosophies and resources. Ask yourself the following three questions:

1. What are the realistic odds for an improvement that leads to a high-quality life?
2. Since goats have the instincts of a prey animal and are very stressed when they feel vulnerable, can you manage the period of recovery, or attempted recovery, in a way that will alleviate their fear?
3. Are the costs realistic for your budget given the possible outcomes? As you consider this, remember that you must manage your money wisely so

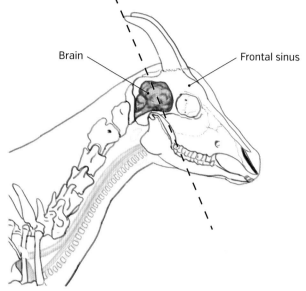

Figure 7.12. This is the proper angle for a gunshot or captive bolt to cause instant death or stunning for a goat. The mature goat's brain is located well back in its skull, compared with younger goats and other livestock. In addition, the front of the goat's skull is protected by a thick sinus cavity. A shot from the back and angled toward the jaw is the most certain to hit the brain and cause instant death.

you can address the future needs of your entire herd. Spending an extraordinary amount on one animal could compromise the health of many animals if it leaves you in a financial pinch.

A veterinarian can administer a powerful and effective drug or combination of drugs that will painlessly end the life of a suffering animal. As a farmer, you don't have access to this method, but there are a few other humane means of euthanasia. They are more difficult to enact, because they seem more traumatic from our side of the situation. The choices are to end the animal's life by using a captive bolt, a gun, or, if the animal is already unconscious, a very sharp knife. A captive bolt is a tool that instantly causes the animal to lose consciousness; in many cases it does end its life. Its use must be followed with the severing of the carotid artery if the animal is only rendered unconscious so that it will bleed out, causing death. I choose to use a gun. A small- to medium-caliber handgun, such as a .22- or .38-caliber pistol, is extremely effective for euthanizing a goat when done properly. Performing euthanasia in this manner is tough. I find it a very difficult and emotionally traumatic experience to shoot an animal, but for me it has often been the right thing to do. You will have to think through the options carefully and make your own choice.

If taking the life of an animal is new for you, you may be surprised to witness the body continuing to move after the animal is dead. This is simply electrical impulses in the muscles trying to communicate with a command center that has stopped functioning. In addition there will be some blood from the exit wound and mouth as well as steam or smoke from the mouth when a gun is used. This is normal. You can confirm death by using your stethoscope to listen for a heartbeat or by pushing gently in on the surface of the eye with your finger. If the animal is dead, there will be no movement of the eyelid or eye.

Field Necropsy and Liver Sampling

Amateur necropsy. It sounds like the title of a really bad TV reality show, but it is a procedure that I find very valuable in terms of my own learning about goat anatomy and physiology and as a way to verify or disprove my suspicions about the possible cause of death of an animal. A field necropsy is the examination of a dead goat's body on the farm, rather than in a lab setting. Of course this procedure is traditionally done by a qualified vet or technician, but when I first decided to try it on my own I asked myself, "What was there to lose?" It was an opportunity to learn something, and necropsy or no, the final outcome would be the same—I would bury the animal. So I decided to try it. In most cases, I have learned something that has been helpful for the future of my herd. I may not figure out what caused the goat to become so ill in the first place, but I do learn things that will allow me to support my herd's health in the future. It's a great idea to take photos of any questionable things you see and then consult with your veterinarian using these images. They aren't as good as a professional necropsy, of course, but might help a great deal.

Some of the most common causes of death are fairly easy to spot during the field necropsy of a goat. At the least, the procedure should allow you to rule out some causes and then decide if you should investigate other possibilities. Coccidiosis is readily apparent: You see blood in a section of the intestine and small white bumps, or nodules, where the parasite has eroded the intestinal wall (see figure 10.5). Colic or twisting of the intestine usually appears as a section of dying or dead intestine that is dark red to almost black, but without clotted blood or nodules inside. You may find a blockage in the upper digestive tract—anything from gravel to baling twine. You can take a sample of rumen contents to test the pH. Check the lungs for signs of pneumonia, such as lung sections (lobes) that are reddish-purple instead of light pink. Internal CL causes abscesses in the lymph nodes (even if you aren't sure where the nodes are, you should be able to spot the abscesses).

In addition to learning about the anatomy and physiology of the goat and looking for the cause of death or illness, a field necropsy provides the opportunity to collect a liver sample and have it analyzed. As I mentioned earlier in the book, many important minerals are stored in the liver; an analysis is a great way to get a snapshot of your whole herd's nutritional health. This is especially true if the goat you are sampling was otherwise healthy—it died from sudden complications as opposed to having been sick for an extended period of time. A small piece of liver can be dissected, placed in a ziplock bag, and frozen. Then you or your vet can ship it to a lab where it will be analyzed for mineral levels.

A necropsy must be done very soon after the animal dies. Even if you place the animal in a cold place, the heat retained in its organs will begin the process of breaking down the cells, or autolysis. This will cause changes that will confuse your search for a cause of death.

If you've ever butchered an animal, or even cut up a few whole chickens, that's pretty good preparation for trying a field necropsy. If you haven't, I recommend first participating as an assistant when someone else is dismantling an animal carcass. It's a profound and informative process, but a little daunting to the uninitiated. A field necropsy is also called a "gross" necropsy, not because it is gory, but because it takes a big-picture look—as in the term *gross income*. To conduct a field necropsy, you will need the following:

- A very sharp knife or box cutter that's easy to hold. A blade length of about 5 inches (13 cm) is plenty long.
- String is helpful for tying off sections of intestines, bowels, and upper digestive organs to help keep things tidier.
- Large pruning shears are essential if you will be cutting through the ribs.
- A table or surface that that will position the animal at a comfortable working height, or supplies to cushion your knees so you can kneel on the ground comfortably.
- A large bucket to place removed organs in.
- A tarp or other item under the animal to catch body fluids so that they don't soak into the soil, which could attract pests or predators.
- If you will be burying the animal later, have a cart ready.
- Waterproof gloves. Wear a double set if they are thin.
- Excellent lighting.
- Reading glasses if you need them, and/or a magnifying glass.
- Camera.

If you simply want to collect a liver sample from the body of a dead animal, here's what to do:

Step 1. Don a pair of gloves (or double-glove).
Step 2. Lay the body on its left side.
Step 3. Locate the rib cage along the right flank. Using the knife or box cutter, make a 4- to 8-inch (10–20

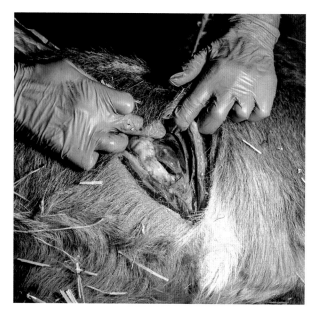

Figure 7.13. The liver is easily accessed from the right side of a dead goat by making an incision just along the line of the last rib.

cm) incision, parallel to the last rib, about halfway between the spine and the belly. You will cut through several layers—skin, fascia, and muscle.
Step 4. Use your fingers to pull back the layers. It should be easy to expose the liver and likely the bile duct.
Step 5. Reach in and pull a section of the liver (a lobe) out. Portion off two chunks, each about golf or tennis ball size.
Step 6. Place each sample in an individually labeled bag. Freeze and keep one (in case something happens during shipping that makes the other one useless), and freeze or keep the other sample cold and take it to your vet for him or her to analyze or send to a lab. In some cases you might be able to send the sample directly to the lab yourself. If so, follow their shipping instructions carefully.

A necropsy is not an easy procedure, even when you have done it several times. In the beginning you may not even know what you're looking at and whether it is normal or abnormal. But with experience, you will eventually learn what normal organs look like and be able to spot when things aren't as they should be. It's a good idea to think ahead of time about how you will deal with the remains. A healthy goat—one that died from injury, for example, or one that died suddenly and hasn't been medicated with any drugs that would require a withholding time—can be converted into meat. But if you don't know the cause of death or you have medicated the animal, butchering for meat is not an option. (Refer back to chapter 3 for details on managing mortalities.)

Here is the procedure for a full necropsy:

Step 1. Position the body of the animal with its right side down.

Step 2. Starting at the flank, carefully use your knife to cut through the skin working down toward the mammary. You must be careful to not puncture the rumen or intestines. It is likely that there will be bloating of the rumen simply from the continuing work of the rumen microbes even after the animal's death.

Step 3. Open a flap of skin from the mammary to the sternum, then up to the spine (following the rib cage) and back, peeling the skin back as you go—use your knife to help separate the skin from the muscles.

Step 4. Next, carefully cut along the same lines, but going through the layer of muscle and connective tissue. Be careful to not puncture the organs just underneath. As you open up the abdominal wall, note if any fluid has accumulated in the abdominal cavity (ascites)—often a sign of liver failure. Note the character of the liquid.

Step 5. Without disturbing any organs, take a careful look at everything you can see. Note if organs are positioned where they should be (refer to figures 10.1, 11.1, and 12.1 for a guide). Note the color of the organs. Note whether any organ, such as the stomach or a section of the intestines, is distended or empty. Look for lesions: abscesses, tumors, and trauma.

Step 6. After inspecting the left side, carefully roll the animal over and make additional cuts to reveal the abdominal cavity on the right side and inspect the organs. A liver sample can be taken now. As you cut the sample out, observe the tissue for signs of fat, bleeding, parasites, or lesions (abnormal areas in the tissue).

Step 7. Starting with the digestive system, closely inspect each organ in turn. Feel them, noting texture and color. Inspect the numerous lymph nodes that are located along the small intestines. Open each hollow organ, and carefully inspect the contents and the lining. If you wish, you can collect samples of digestive tract contents for fecal float. Also look for visible adult parasites. (Prepare yourself: When you open the rumen there will be some pretty powerful methane gas and a very unpleasant odor.)

Step 8. Cut the kidneys and any enlarged lymph nodes in half, and examine them for lesions.

Step 9. Use your knife to puncture the diaphragm. There should be a faint rush of air drawn into the cavity, unless the rib cage had been punctured earlier.

Step 10. Next, cut the rib cage open using sharp pruning shears. Cut along the spine and up the sternum, and then open the diaphragm so that you can observe the chest organs, lungs, and heart.

Step 11. Cut sections of the lungs to observe the tissue inside the lobes—note any differences from the lower to the upper lobes. Observe the heart for abnormalities.

Here's a partial list of what you might be able to suspect based on the necropsy observations:

- Heart is round instead of heart shaped—cardiac deformity
- Stomachs contain feed but no contents in intestines—blockage
- Kid too young to ruminate, but rumen has foul-smelling contents—overfeeding and milk overflow to rumen
- Blood in the chest cavity—ruptured aneurysm
- Pregnant doe with kids in uterus and fatty liver—pregnancy toxemia
- Rumen contents are milky in color, and the pH is below 5.5—acidosis

- A clog of feed in the esophagus—choking
- Black or bright-red section of intestines without signs of coccidiosis—colic
- Bright-red section of intestines with small white nodules—coccidiosis
- Visible *Haemonchus contortus* worms in aboma-sum—parasites leading to severe anemia
- Liver shows damage and bleeding—liver flukes
- Lung lobes are firm and bright red (not light pink)—pneumonia

If the animal was healthy, the weather is cool, and your investigation was relatively quick (less than an hour), you can harvest the meat if that feels appropriate (even when I necropsy a healthy goat, sometimes I'm too emotionally involved to make this more pragmatic choice). If the animal is to be buried or composted, include all of the removed organs. The fact that the body cavity has been opened will speed the breakdown of the carcass. I typically put the innards into a large muck bucket as I work. I then slide the tarp and carcass into the bucket of our tractor or into a cart and put this in the muck bucket, too. Once the animal is buried, in either the soil or the compost heap, I rinse the tarp and bucket well with water. Then I often ponder my mixed feelings of sadness and awe, hopefully along with a feeling of satisfaction at having learned something useful.

Photo (right) courtesy of Susanna Klassen

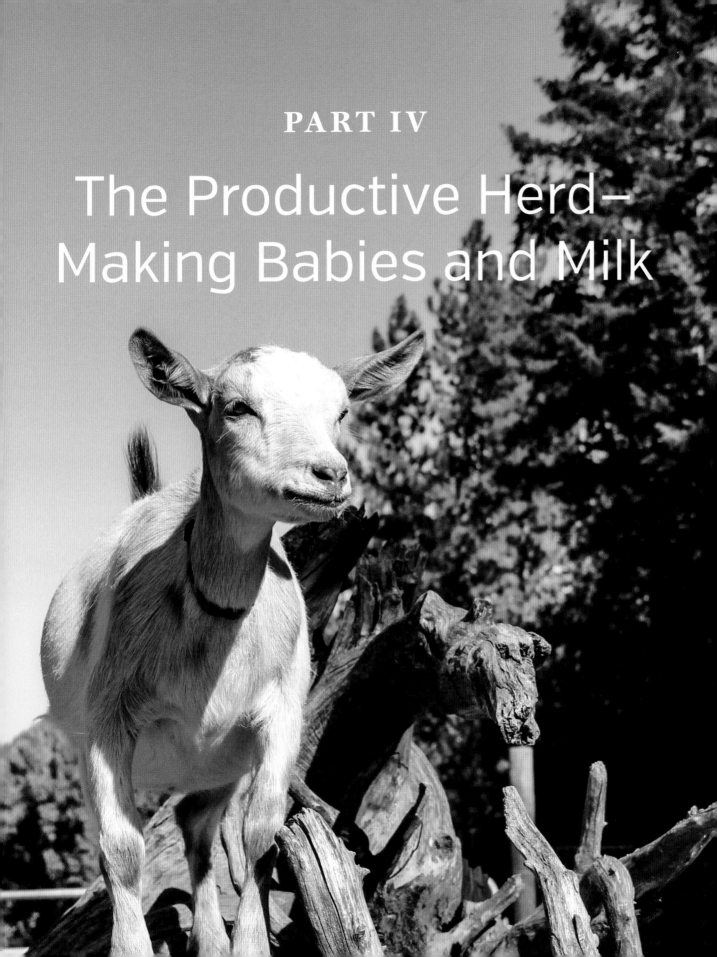

PART IV

The Productive Herd–
Making Babies and Milk

CHAPTER EIGHT

Breeding, Pregnancy, and Delivery

A well-planned and -executed pregnancy and delivery is a joyful, miraculous process. It results in a healthy mother and lively healthy offspring. I've "delivered" close to 700 kids over the years. In most cases delivering means simply standing by, drying off a nose or two, and oohing and ahhing over how cute the new babies are. But every year brings one or two difficult births that challenge my skills and make me doubt my abilities. Amazingly, in all but one of those cases, things turned out pretty well.

The delivery process must not be approached casually, or disaster can result. Even when you are an experienced goat farmer, things can go wrong, ending with the loss of a valuable and beloved doe and/or her offspring. This chapter offers some deep insight and understanding of the reproductive process of the goat, lots of troubleshooting and tips for successful kiddings, and options for raising healthy kids.

Deciding When to Breed

As with many aspects of goat care, the first task in goat breeding is planning. The maturity of the goat when she is bred and how old she will be when she gives birth are the primary considerations in deciding when to breed. Goat breeders hold strong opinions as to what is right and wrong in regard to some of these choices. Another consideration is the optimal time of year for kidding. Timing kidding in order to maximize meat and milk supply is important from a financial standpoint. But from a health standpoint, weather conditions during pregnancy are critical. Pregnancy when the weather is exceptionally hot, cold, and/or wet can cause pregnant does a lot of stress.

Age and Size

A doe should ideally give birth for the first time when she is one to two years of age. Many breeders are adamant that one year is ideal, but others say two is best. Both options have their pros and cons. I usually prefer to have my dairy does kid by age one as long as they are growing well and healthy. If I were raising meat or fiber goats, I would make this my goal as well. Remember, though, that breeding rarely goes 100 percent according to plan: A few goats simply won't get pregnant; others might be too small to breed at age one (see table 8.1). Or you may simply choose not to breed some of your does in order to control the number of kids born that year if you're concerned about having enough time and help to rear them or in finding homes for those you aren't going to keep.

I prefer to have goats kid at age one for several reasons. The first is related to economics. Waiting to age two to give birth means feeding a non-producing animal for an additional year before finding out whether she is a good mother and milk producer

Table 8.1. Weight Goals for Does When Kidding as Yearlings

Breed Size Category	Sample Breeds*	Weight at 12 Weeks Weaning[†]	Weight at 7 Months[†]
Large dairy	Saanen, Nubian, Alpine	45 lbs. (20 kg)	70 lbs. (32 kg)
Light dairy	Toggenburg, Oberhasli, LaMancha	40 lbs. (18 kg)	60 lbs. (27 kg)
Miniature dairy	Mini Nubian, Mini LaMancha	25 lbs. (11 kg)	40–45 lbs. (18–20 kg)
Nigerian Dwarf	Nigerian Dwarf	20 lbs. (9 kg)	30 lbs. (14 kg)
Large meat	Boer	50 lbs. (23 kg)	80 lbs. (36 kg)
Light meat	Spanish, Kiko	40 lbs. (18 kg)	60 lbs. (27 kg)

* Whether a particular breed (especially dairy breeds) would be considered large or light varies greatly because of individual breeder trends. Breeds cited here are examples only.

[†] Weight data based on Smith and Sherman, *Goat Medicine*; Merkel and Gipson, *Meat Goat Production Handbook*; and, for miniatures and Nigerians, author's personal experience.

(which is important when raising meat and fiber goats, too). Second, you can discover sooner whether there is value in the genetics of the parents—for example, if a doe kids at one year of age and shows great promise, you can more confidently breed her and that same buck together in the future. And finally, it can be difficult to prevent does from getting chubby if their first kidding isn't until age two. That may not sound like a serious problem, but an overweight doe can have fertility and kidding difficulties. Aged does may experience trouble, too. Yearling does will not give as much milk as a first-freshening two-year-old will, but typically yearlings don't bear as many kids, either, so their milk supply should be plenty to feed their kids.

Time of Year

Most goat breeds that originated in the United States and Northern Europe have seasonal fertility cycles. They are known as seasonal or short-day breeders; the majority of goats being farmed today fall within this category. Breeds with origins closer to the equator tend to be less seasonal or not seasonal at all. Seasonal breeders come into heat and are ready to breed in the fall. This natural pattern is governed by the number of daylight hours. As the days grow shorter in the fall, the goat's brain sends signals that

Figure 8.1. Saanen goats have a white coat, but this handsome Saanen buck owned by Fran Brown of Central Point, Oregon, is mainly beige because he has stained his face and neck with his own urine, an interesting behavior common to all bucks. In fact, there's a saying that a white buck is only white for a short part of his life.

regulate hormone production as it relates to fertility. The buck is affected by the change in day length, too; his scent and eagerness to breed are more pronounced at this time of the year.

Commercial operations of all sizes often breed goats during the off season to ensure a steady supply

Using Lights for Off-Season Breeding

Producers seeking to supply milk year-round and meat breeders who wish to produce meat kids for slaughter at specific holidays use artificial lighting to help bring does into heat in the off season. Research at Cornell University[1] documents the following practices:

During the winter, expose both bucks and does to 20 hours of light for 60 consecutive days (overlapping natural daylight hours). Correct position and intensity of the light is important. Cornell suggests 40- to 60-watt fluorescent lighting hung 9 feet (2.7 m) above the ground with 1 foot (30 cm) of bulb for every 10.5 square feet (1 sq m) of floor space. If there are windows, it's recommended that the goats not be able to see the true night sky during the hours of false daylight. After the 60 days, return the lighting to normal for 45 days, and then put the buck in with the does or in an adjoining pen. Plan on one mature buck for every 15 does that need to be serviced. Does typically don't cycle more than one time after this type of approach.

of milk or meat goat kids for a seasonal market demand. Because the animals don't naturally come into heat at this time of year, the producers use artificial lighting to extend daylight length and influence the heat cycles of the does.

When you're deciding when to breed your goats, consider not only when they would be due to kid, but what they might have to experience during pregnancy and just after delivery. For example, does bred in late winter will be kidding during the hottest part of the year—a time when there may be heat stress for the pregnant animal and when flies are more numerous, which is problematic when buck kids are being surgically castrated. Or in areas where winters are harsh, if you breed in the early fall, resulting in late-winter or early-spring kidding, the cold conditions may be tough for you and the animals during delivery, and newborns may suffer frostbite.

It's important to time kidding so that kid-rearing occurs when there are adequate hands on board to provide proper care. For example, if your farm is also a produce farm, then kidding should be timed so that it doesn't overlap with gardening demands. Always remember to consider your own availability for the expected delivery time, too. When making breeding plans, I have occasionally forgotten to look at my calendar to see what's going on five months down the line. When I remember after the fact, I realize that I am actually scheduled to be out of town on the due dates. Of course no due date is guaranteed, but it is important to do whatever you can do to not be caught off guard.

Selecting the Right Match

The ideal choice of buck will help you improve herd health overall by improving on a quality or qualities that the doe lacks and your herd needs (including vigor and parasite resistance). This is why a serious breeder often owns several bucks—a diverse range of genetic choices. On medium- to large-scale goat farms, a single buck or perhaps a few bucks are put in a pen of does with little thought to genetic improvement—there just isn't the time for record keeping and management of a more complicated breeding program. In order to make the ideal choice of buck, you'll need access to the genetic data that's been collected about him and his predicted ability to transmit important qualities. Genetic data is more commonly available for dairy goats than for meat or fiber goats.

Most breeders try to not select a buck that's too closely related to the does, while others commonly cross uncles with nieces and grandfathers with granddaughters, even occasionally father with daughter. In

What About Artificial Insemination?

Artificial insemination (AI) sounds like the perfect answer for those who don't want to or can't own a buck. AI for goats involves the use of frozen semen and special tools to impregnate the doe. In the world of the dairy cow, AI is so popular that you can actually peruse a catalog for the ideal match for your cows; a professional technician will perform the service. But in the goat world it is typically the producer who owns the equipment, sources the semen, and does the insemination. AI can be a great option for a breeder whose herd is large enough to justify the cost of the equipment, the time it takes to master the technique, and the cost of the semen.

High-quality semen, from bucks that offer proven benefits to a herd, can be difficult to source and acquire. At large goat shows and conferences, it's not unusual to see breeders pulling carts and dollies with nitrogen tanks aboard. (At -321°F/-196°C, liquid nitrogen provides the degree of coldness required to preserve sperm.) Semen swaps and sales take place in quiet corners and backrooms with great care being given to protecting the integrity of the frozen genetics.

Figure 8.2. Richard Purcella fills Pholia Farm's liquid nitrogen tank on a regular schedule as a part of his mobile business.

Figure 8.3. Taking an AI class is a great way to become familiar with the tools and techniques of using artificial insemination. At this class, Dr. Charles Estill demonstrates technique to Christina Lidner (holding the goat) and Fran Brown looking on.

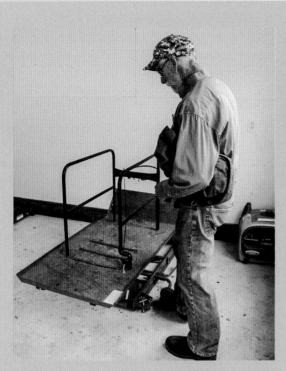

Figure 8.4. Richard Johnson of Lookout Point Ranch shows how this cleverly designed goat AI stand, a repurposed motorcycle lift, can go up and down. It also tilts forward for a more efficient AI position. Richard made this stand by adding sides and a neck hold to a motorcycle lift.

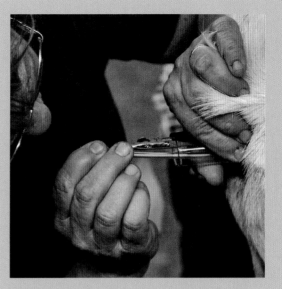

Figure 8.5. Breeding a goat by AI involves an up-close-and-personal look to find the cervix.

The success rate of AI depends on the skills of the technician and the quality of the semen. The collection and freezing of semen is a precise and highly skilled business. In addition not all semen collected, even by a superior practitioner, will be of equal quality. On average, success rates for AI are about 50 percent, which is considerably lower than live cover. If you plan on using AI as your sole method of breeding does, proficiency at the task and allowing for time to retry the insemination are important.

A liquid nitrogen tank ranges in price from a couple to several hundred dollars. Used tanks are more reasonably priced. The tank must be kept supplied with liquid nitrogen or the semen will perish. Most tanks will hold enough nitrogen to keep the contents preserved for two to three months. You can usually have the tank filled by a professional who drives from farm to farm refilling semen tanks. To locate such a person, talk to other goat, cattle, or horse breeders; a veterinarian; or even a dermatologist—who might have a nitrogen tank on hand for the removal of warts by freezing.

Although I've kept a nitrogen tank for storing semen on the farm for years, I have successfully artificially inseminated only a handful of does. Part of my failure comes from being a bit lazy and unsure of the technique. But I am glad that we've had semen from many of our great bucks collected and stored in the tank so that their genetics are available for the future.

If you are interested in learning more about AI, I highly advise taking a class or two to practice the techniques and learn about the different types of equipment available. Supplies can be purchased through several companies (see appendix D for contact information). It's wise to seek opinions from other breeders doing AI about which tools they think work best; preferences vary.

some species, this is called inbreeding, but when tight crosses are purposefully done in the hope of "locking in" certain traits it's known as linebreeding. This approach is a way to establish a more homogeneous (look-alike) appearance in a herd, but it also brings the potential of locking in recessive traits that aren't outwardly visible but may become problems for generations down the road. In dairy cattle, for example, inbreeding and linebreeding are proven to correlate with reduced fertility and diminished lifetime production. Some pedigree software that you can purchase (see appendix D for specifics) includes the capacity to run a logarithm for an inbreeding coefficient, which can help you determine if the past generations in your herd are getting a bit too close for comfort. Most breeders choose to introduce new genetics primarily by bringing in a new buck or through AI. You can search for inbreeding information on dairy goats for any goat registered with the American Dairy Goat Association at adgagenetics.org. The longer a breed has been recognized by ADGA, the more extensive the data. For example, my chosen breed—the Nigerian Dwarf—was first recognized by ADGA in 2005. Data for the breed only extends back that far.

If your breed is small or if you don't own or want to own a buck, then your choices will be much more limited. Finding another breeder willing to provide buck service can be quite a challenge. Start searching for breeders in your area long before the season when you want to breed your does. Make contact, and try to establish a rapport. If you locate a farm that is willing to either house and breed your does or allow you to bring over does in heat, be sure they are as picky about your herd's health as you should be about theirs. You may need to provide proof of health and recent testing results for diseases such as CAE, Johne's, CL, and Q fever.

Preparing the Buck and Doe for Breeding

Think of breeding season as a marathon, especially for the buck. In order to run it successfully,

conditioning and preparation must occur. Let's go over the factors that affect the fertility of bucks and does, how to feed for a successful season, and how to recognize the doe's heat cycles. Fertility issues are also covered in chapter 16.

Buck Fertility

A buck's fertility—his ability to produce and deliver high-quality semen—depends on many things. The time of year, the buck's age, scrotum size, nutritional status, and health all contribute to his fertility. A buck may show adequate libido—the desire and ability to mount and deliver semen—but if his sperm is not superior in quality and quantity, the number of pregnancies will be low.

Young bucks born in the early spring are usually ready for their first work in time for the fall and early-winter breeding season. Even baby bucks will often act amorous—flirting with, chasing, and even mounting other kids. But until testosterone is produced in enough supply, a goat's penis will not extend. It is held inside the prepuce by tissue that's present at birth. Only with the production of testosterone does the separation of the connective tissue occur, allowing the penis to extend. The buck's age also influences the quality of the semen. At the start of puberty, many of the sperm produced are poorly shaped. As the buck matures, the sperm quality usually improves. In addition, in the spring, semen from the young buck is more likely to contain dead sperm than in the fall. This is important to remember if you are planning your breeding season with a young buck as part of a year-round breeding plan.

Breeds that are seasonal—such as most European and American goats—experience the greatest testosterone production in the fall, when the daylight hours shorten. Bucks from breeds that breed year-round might experience full puberty prior to fall, though, and they should be watched for signs of being ready to breed so that accidental pregnancies of very young does doesn't occur. (Fortunately, young does usually don't experience their first heat cycle until long after the young buck is capable of the act.) The fall effect is experienced by bucks of

Feeding for Future Fiber

If you're raising mohair-producing Angora goats, increased nutrition for does during breeding, pregnancy, kidding, and beyond is an important part of increasing your overall profits. Although the extra nutrients will often make the dams' hair fibers grow thicker, leading to a decrease in fiber quality, the nutritional boost will pay off in four ways:

1. The better the dam's nutrition at breeding, the more eggs are likely to be released from the ovaries, thus increasing the number of fetuses, which subsequently leads to a greater number of fleeces of the highest value.
2. The nutrition that a fetus receives determines the number of the hair follicles in its coat capable of growing mohair.[2] The better the nutrition, the denser the coat.
3. The better the dam's vigor during pregnancy, the less likely she is to abort.
4. The better the dam's nutrition during and after kidding, the more likely she will be able to feed and nourish her litter.

all ages. Shortening of day length influences doe hormones as well, and they respond by coming into heat. The doe's heat cycle also stimulates testosterone production in the buck.

The size of the buck's scrotum is thought to represent his potential for producing large quantities of semen. The testicles should be fairly even in size and have the texture of a firm muscle when palpated (not something most bucks are keen on, so have the animal well restrained before the exam is performed).

The nutritional status of the buck affects the onset of puberty as well as the sperm quality.

Mineral deficiencies can cause sperm deformities, and overall poor nutrition affects quality and quantity of semen. Bucks are often given a dose of vitamin E and selenium (Bo-Se is a well-known brand) about six weeks before the onset of breeding season as a way to assist with semen quality. In the past, when I have not done this, semen quality was diminished. I learned this by going to a lot of trouble to have semen from the bucks collected, only to find out that all the sperm had curly tails and weren't good enough to be worth freezing.

Various disorders can lead to infertility in the buck. Pizzle rot, also called posthitis (see the Posthitis entry on page 323) makes it impossible for the buck to extend his penis. Arthritis, hoof problems, and leg problems can also reduce a buck's effectiveness.

Doe Fertility

Does that are born in the spring will almost certainly reach puberty before or by the fall of their first year. A doe kid from a less seasonal breed may come into her first heat by 12 weeks of age.

In some cases a doe is physically incapable of being successfully impregnated. For example, older does can experience various cancers that prevent pregnancy.

Intersex goats (a term that applies to both pseudohermaphrodites and freemartins) are uncommon but not rare (see the Intersex entry on page 316). These animals cannot breed successfully. Pseudohermaphrodite goats appear to have female genitals but are likely to begin acting like bucks when they reach puberty, which makes them easy to identify. Freemartins are females with partial male genes—true hermaphrodites. Freemartins can occur in cattle and goats when a male and female fetus are both present in the uterus. Births of freemartins are unusual, but when they do occur, it is difficult to diagnose other than by genetic testing. The vagina may be short, which can be determined by using a speculum designed for artificial insemination to measure the length (a visual inspection of the cervix can be done at the same time). A short vagina isn't conclusive proof of the condition, though. The freemartin doe will have external features that appear normal (unlike the intersex goat), although her teats and vulva may be a bit smaller than average. She may also show some masculine behavior.

Caprine herpes is a cause of infertility in does in the United States, Australia, and New Zealand.[3] When present, this sexually transmitted disease causes inflammation of the vulva and vagina (vulvovaginitis), short cycling of heat cycles, and infertility. See the Caprine Herpes Virus entry on page 321.

Feeding for Breeding

Both bucks and does should be in peak health and body weight before entering breeding season. The buck in particular will likely spend several months focused only on the ladies—even if he never breeds. During this time, he will likely lose weight and be under stress.

The term *flushing* is used to describe the purposeful nutritional buildup of a doe, usually by adding energy feeds such as grain but also any feed that improves her body condition, before breeding and during the first month of pregnancy. The goal is to increase the number of eggs released by the ovaries and to help ensure adequate nutrition to support the developing fetuses when they are at their most vulnerable. Not all breeders practice flushing.

During breeding season, the hardworking buck might also need extra nutrition. It can be difficult to get him to consume as much food as he should. Pay close attention to his body condition and appetite. If his condition becomes too poor, he will not be able to service as many does, and he may be more likely to succumb to illness. Bucks are so focused on their breeding instinct that once the season starts, it's close to impossible to get an underconditioned buck to eat enough to improve his physique.

Understanding the Heat Cycle

To improve the odds of having a successful kidding season, we have to start by understanding and reviewing the doe's heat cycle, or estrus. Does in breeding season undergo a heat cycle on average every 21 days. The cycle will last one day to several.

Ovulation, the release of fertile eggs from the ovaries, always happens toward the end of the cycle.

Does exhibit signs of estrus in several ways. Some become quite vocal and may position themselves as near to the buck pen as possible, calling and looking longingly in the direction of the boys. Most will show a small to moderate amount of clear mucus discharge from the vagina. The vulva itself might appear a bit swollen and more pink (in a light-skinned goat). It's often easier to notice these two symptoms on does that are being milked or otherwise inspected daily. Tail flagging—the flicking and wagging of the tail, often in conjunction with calling out—is one of the most common signs of estrus. Flagging is believed to help release and spread pheromones that help attract the buck. When the doe nears a buck, she is likely to squat and pee; nature may have designed this response to reduce the likelihood of urine rinsing the semen from the vaginal canal if the doe pees *after* being bred. Bucks will often place their muzzles in the doe's urine stream, and then raise their upper lip in the *flehmen response*—helping them interpret the doe's readiness to mate. Does in heat may mount each other and even make sounds like a buck. In this case both may be in heat, or only one. (Pregnant does that are close to delivery time will also sometimes "act bucky," mounting does and even making "buck talk.")

If you don't own a buck and will be using a stud service or AI (see What About Artificial Insemination? on page 184), keep a buck rag on hand to help stimulate the does and to identify when they are in heat (ovulating). A buck rag is simply a cloth that has been thoroughly rubbed on the buck's scented areas, especially his face and flanks, and then stored in an airtight jar. When the jar is opened in front of a doe's nose, the scent can help bring her into heat. If she is already in heat, she will respond to it with tail flagging, squatting and peeing, or even buck talk.

The presence of the opposite sex influences the fertility of both bucks and does during breeding season, for seasonal breeders, and throughout the year for year-round breeders. For does that have not been

Making Their Mark—Using a Breeding Harness

A product known as either a breeding or a marking harness is useful for some herds. This handy contraption, consisting of a harness and a block of colored crayon, is worn by the male goat. When he mounts a doe, a telltale mark is left on her back. It's also possible to use this tool to mark does that are in heat—without breeding them. To do so, you would strap the harness onto an infertile buck (one that has had a vasectomy or is sterile for other reasons) or an intersex goat (who acts like a buck but cannot breed). Or you can alter the harness by attaching a heavy piece of material to block the buck's penis from entering the doe. This last option is a bit risky but can work.

near a buck, the introduction of a buck into the pen or an adjoining pen (sometimes called a *teaser buck*) at the beginning of breeding season will bring most into heat within a week or so. Sometimes does will be more attracted and responsive to a particular buck, or a new buck replacing the familiar one.

A group of does can be synchronized to come into heat within a few days of one another. This is desirable when you're scheduling artificial insemination or the short-term lease of a buck, or trying to plan for the spring kidding and your own availability. Synchronization can be done using either the above method of suddenly introducing a buck or by the use of hormone treatments. These treatments aren't acceptable for organic certification in the United States, Canada, or the EU. They're given to the doe by inserting a drug-releasing implant referred to as a CIDR (short for controlled internal drug release) into her vagina and then following a regimen that includes an injection of another hormone, prostaglandin (often given as Lutalyse), to cause ovulation.

A phenomenon known as short cycling is common in goats. A doe will appear to be in heat but won't necessarily be receptive to the buck. Then about six days later she'll come into heat again. This can occur several times and usually happens during the first heat cycle of the season or with does that have been on extended lactations and not bred in more than a year. The second cycle is almost certainly the one in which ovulation actually occurs. If a doe continues to short cycle, you can't tell during which cycle she might actually be ovulating, so it's a good idea to have a buck cover her each time. In the United States, Australia, and New Zealand, short cycling might also be a symptom of an infection by the caprine herpes virus (see the Caprine Herpes Virus entry on page 321).

Understanding Mating

To help ensure that all of your preparations to this point haven't been in vain, it's important to understand a few more facts about the breeding process of goats. The number of does a buck can successfully impregnate depends on several factors. The most important, after health and nutrition, are the age of the buck and the system of breeding. A young buck in his first year of work may only be able to breed a dozen, or even fewer, does in a month. A two-year-old buck can breed about one per day, and a mature buck in his prime perhaps 45 to 50 per month. As an example of how the system of breeding affects fertility, imagine 10 does put in with the young buck. If they all come into heat in a week, he is unlikely to be able to settle more than a couple because he can't sustain his semen supply. The buck will mount the does and ejaculate, but his sperm count may be extremely low. Don't confuse libido with fertility.

Jan and Larry Neilson

FRAGA FARM, SWEET HOME, OREGON

Fraga Farm is Oregon's first organic goat cheese dairy, originally located in the wonderfully named town of Sweet Home. In 2004, when I was doing research and planning for the goat dairy I wanted to build on my family's land in Oregon, Fraga Farm founder Jan Neilson was one of the few people who took me under her wing. I visited her farm and picked her brain with dozens of questions. Even though she and her husband, Larry, have recently sold their brand and most of the goats, Jan is still one of my most favorite people and a role model for holistic goat care.

The couple opened the dairy and creamery in 2000, just ahead of the farmstead cheese craze, and became certified organic in 2002. Jan was, until recently, also a licensed massage therapist—although her practice was mostly limited to the massage of her livestock guardian dogs and goats after the opening of the dairy. The decision to go organic was an easy one for Jan. "I always loved taking care of them naturally. I don't want to use antibiotics or pesticides or worry about chemical residues in my food." Because she was already managing her 20 to 40 Nubian dairy goats in this manner, she says that the transition was also relatively easy, even though the inspections and documentation requirements were rigorous and time consuming.

Certification brought other challenges, including finding sources of organic feed to supplement the farm's 2 acres (0.8 ha) of lush pasture on the banks of the South Santiam River. To help meet the requirements of time on pasture, Jan and Larry also maintained an 48-acre (19 ha) farm of mixed pasture and browse nearby where they kept their dry does, bucks, nursing mothers, and young stock. (They raise kids with their dams and wean them at two months of age.) There's another 5 acres (2 ha) next door to the main farm that the milking does could access. Jan also planted what she calls "salad bars" for her

Figure 8.6. Jan Neilson with a few of her dairy goats at Fraga Farm, Sweet Home, Oregon. *Photo courtesy of Tami Parr*

goats. Similar to the hedgerows I mention in chapter 5, Jan designed her salad bars with fencing so that the goats could browse the fringes, but not destroy the plants. She included medicinal plants and herbs such as marshmallow (*Althea officinalis*), rosemary (*Rosmarinus officinalis*), mint (*Mentha piperita*), and lemon balm (*Melissa officinalis*). The first iteration was lovingly tended in the center of the pasture and protected by electric fencing, only to have the goats figure out how to break in and quickly decimate the little garden. Jan and Larry planted the second version along the pathway to the milking parlor, protected by welded wire panels. The goats nibbled plants daily as their appetite and needs dictated. Jan also regularly picked and hand-fed raspberry (*Rubus idaeus*) and blackberry (*Rubus armeniacus*) leaves and Douglas fir (*Pseudotsuga menziesii*) branches, which the goats would eat if they needed or desired.

Jan said that despite the fact that their area receives an average of 54 inches (137 cm) of rain each year, parasites and health problems weren't an issue. No doubt her other management practices, which included weekly feeding of whole garlic (obtained from a nearby organic farm), Bragg organic apple cider vinegar, and kelp meal had a lot to do with the goats' resilient good health. Jan said she believed the garlic, with its sulfur-containing molecules, was responsible for the lack of parasite problems. The vinegar, she noted, was especially helpful if a doe's somatic cell count was elevated. Also, the kids in her herd never had coccidiosis problems, and Jan said the key to that was keeping the herd population low; she averaged 40 to 50 kids per year.

In 2013 Jan and Larry passed the Fraga torch to Steven Monahan and Elisabeth Bueschen-Monahan, who continue to make cheese and raise the herd (now Alpines and Nubians) organically, but at their own farm in Gales Creek, Oregon. Jan kept a few goats at their Sweet Home farm, just to enjoy and pamper. To learn more about the current Fraga Farm operation, visit their website at www.fragafarm.com.

Matings can take place in a casual group or on a selective basis. Putting the buck in with the does is often referred to as pen breeding. Taking the doe to the buck when she's in heat is sometimes called leash breeding (as both are often held by a leash during the act). Individual pen breeding, where a doe spends the day with a single buck, is also an option. We have special enclosures just for this system that we call "date pens." Your approach should depend on both your preferences and goals and how many bucks are available for the number of does to be serviced.

When you see a doe acting as if she's in heat, and she appears ready to be bred, you can take her to a buck to verify her readiness. Some does are less amicable to the union than others. This doesn't necessarily mean they aren't in heat, but it could. Others seem to take a shine to another buck that they have interacted with through a fence and not be interested in the union you have selected. This can require a bit of deception on your part if you are determined to have her mate with the selected buck. The doe can be allowed to flirt through the fence with her dream date, while the preferred candidate is stealthily introduced from behind.

Most bucks spend a few moments to minutes wooing the doe by snorting, sniffing, and flickering their tongues in and out, and pawing at the doe. Other bucks are ready without any foreplay, at least for round one. Either way, allow the union to take place two times during the leash breeding. A successful delivery of semen usually results in the doe arching her back in response—an orgasm. A young buck, after his first breeding, might respond so enthusiastically that he falls off the doe onto his back. Rather amusing. Observe whether the doe urinates after the buck has bred her. If so, allow another union and walk the doe for about five minutes afterward to try to prevent her from washing the semen away by urinating. (The doe's bladder empties into the vagina, as shown in figure 16.1.)

If pen or group breeding is your method of choice, remember that you may find yourself, about 150 days later, with a lot of baby goats to manage all at once. To track the possible due dates for pen-bred does, write down the date that the buck was placed in the pen and the date that he left it. Then forecast the range of due dates ahead on your calendar.

Some breeders will place a "cleanup buck" in with the doe herd toward the end of the breeding season. This is to ensure that any does that haven't yet been impregnated will be bred. If you want to track parentage of the offspring accurately, be sure to let at least a week elapse from the last breeding by another buck before introducing the cleanup buck.

Verifying Pregnancy

Sometimes it is easy to tell if a goat is bred; other times they hide it quite well. The simplest but least reliable way to tell involves observing for changes in behavior and physical changes. The end of heat cycles occurs in most does; rarely, some continue to show signs of heat while bred (they will usually not be receptive to a buck, however). A doe in milk often shows a drop in production about two to three months into her pregnancy. Many does also exhibit a progressive swelling and fullness to the labia, the tissue around the opening to the vagina and the urethra. On rare occasions a doe will experience a false pregnancy that includes the development of a large amount of fluid in the uterus—making her look pregnant. When this happens, it's called hydrometra and ends when the fluid is expelled, sometimes called a cloudburst.

For more conclusive proof a blood or milk hormone test can be run 30 days after breeding. A urine test can be performed on the farm 60 days after mating. A blood sample can be mailed, without need for refrigeration or ice, to a lab that conducts the test. (Instructions for taking a blood sample are in chapter 7.) Results are fast and very reliable. A urine test, called P-TEST, is a bit less reliable at 96 percent. Some people prefer ultrasound as a means to both verify pregnancy and perhaps get a baby head count. Ultrasound can be done by a farmer who has access to the right equipment or at a vet's office by the vet or a technician. Goats can be a bit tricky to examine via ultrasound, though, because of the sheer size of the rumen and the movement of the rumen contents. The last time I had an ultrasound done on a couple

Figure 8.7. These pregnant Boer does at Sospiro Goat Ranch in North Carolina are in good shape, still able to make the trek from their shelters to the pastures and back, thanks to constant access to exercise and good feed.

of pregnant does, the vet estimated they would bear two or three kids apiece. Each doe delivered five! See appendix D for information on blood and milk pregnancy testing.

Caring for the Pregnant Doe

Note on a calendar the date each of your does is bred. Also note the due date for each, 145 to 150 days from breeding; you may need the following year's calendar for this. (Appendix B comprises a gestation chart with breeding and estimated due dates.) Then note various dates when tasks should be done to ensure good health of the doe and the growing kids. Here are some of the likely tasks you will want to include:

Table 8.2. Pregnancy-Related Problems

Symptom	Possible Causes	Remedy
Bright-red bloody discharge from the vulva	• Early in pregnancy: miscarriage • Anytime during pregnancy: bladder infection	• Observe the doe for other signs of distress such as grinding her teeth, fever, lack of rumination, and restlessness. If she seems fine, note the date of the discharge and continue to observe. If a complete miscarriage occurred, she should experience a heat cycle within a month's time. • If discharge continues for more than a couple of days without other signs of distress, consider urine sample for bladder infection.
Dark-brown or bloody mucus discharge from the vulva as the doe's due date approaches	• Partial detachment of the placenta • Death of one or more fetuses	• If the doe appears ready to deliver but does not go into labor, veterinary intervention will likely be needed. • If the doe proceeds with labor, it is important to be on hand to ensure that all fetuses are delivered.
Listlessness and progressive weakness a couple of weeks or less before the delivery date	• Pregnancy toxemia (aka ketosis; see the Ketosis entry on page 263 for more)	• Administer 60 ml propylene glycol orally or a high-energy supplement with propylene glycol as the main ingredient 2–3 times per day. If the doe does not improve call a veterinarian. Urine can be checked with ketone strips to verify the diagnosis.
Bulging or protrusion of a section of the rectum, usually when the doe is lying down or coughing	• Partial rectal prolapse	• Observe closely, and be prepared for a full rectal prolapse during kidding.
Bulging or protrusion of a section of the vagina when the doe is lying down, standing, or coughing	• Partial vaginal prolapse	• If the protrusion is only when lying down or coughing, observe and monitor. If the protrusion is when standing, see the Vaginal Prolapse entry on page 318 for how to build a harness; be prepared to observe during delivery and prevent a full vaginal or a uterine prolapse.

- If you vaccinate your herd for CD&T, make a notation 39 days from the due date.
- If you live in a selenium-deficient area, note that a Bo-Se injection should be given 30 days from the due date.
- Mark a reminder four to six weeks before kidding to trim the doe's hooves. (After that, she may become too uncomfortable to tolerate this procedure.)
- About a week before kidding, some breeders like to clip or shave the udder and tail area to help keep it cleaner during and after delivery.
- Make a note to reintroduce grain one to two weeks before delivery if you're going to feed grain after kidding. Grain is not usually needed during pregnancy unless a doe is underweight.

The pregnant doe should be housed with her peer group and maintain the same activity level that the group is accustomed to. If possible, new, previously unknown goats should not be introduced during this time, because natural herd behavior often includes a good deal of fighting as the herd establishes the

Figure 8.8. A partial vaginal prolapse that came and went as the doe lay down and got up.

new goat's status in the group. A tussle may include knocking a pregnant doe with enough force to cause miscarriage or the death of one or more of the fetuses. If goats must be moved to a new group during pregnancy, it is best to move several at one time. This way they can maintain some semblance of peer group stability. I find that housing the new group in an adjoining pen for several days and then opening the fence between the two pens allows for the most peaceful transition.

Pregnant goats can be transported without too much concern up until a couple of weeks before delivery, as long as they are healthy and have had uneventful pregnancies. They should be transported in a crate or box in which they can lean and brace themselves on either side and so reduce the risk of falling.

Throughout the pregnancy, observe the doe at least once daily for signs of distress, such as those described in chapter 6. Remember that pregnancy is

a time of increased stress, and an otherwise healthy goat is at greater risk for all illnesses. In addition there are problems that are specific to or more likely during pregnancy; these are described in table 8.2.

Dealing with Abortion

An abortion, or miscarriage, can occur at any time during the pregnancy. In healthy herds free from infectious disease, abortions are quite rare.

Early abortions (those occurring in the first month or so) are easy to miss, because few signs are present. There are several possible causes: poor nutrition, including copper and iodine trace mineral deficiency; selenium toxicity (rare in goats); and overall deficiencies in energy and protein. *Toxoplasma gondii*, the parasite associated with cat feces, can cause early and late abortions. Deformations of any internal organs in the fetus can also cause the doe to miscarry.

Early abortions are of less concern than those occurring toward the end of the pregnancy, because late abortions can be a warning sign of an infectious organism present in the entire herd. Infectious causes of late abortion include chlamydiosis, Q fever, brucellosis, campylobacteriosis, leptospirosis (rare in goats), listeriosis, salmonellosis, and toxoplasmosis. When many miscarriages occur in the same herd, or herds of farms neighboring each other, over a short period of time, it is referred to as an abortion storm and should be taken extremely seriously.

Unless you can confidently rule out infection—for example, in the case of physical trauma and/or extreme emotional stress—treat *every* late miscarriage as infectious. The fetus, the placenta, and a blood sample of the doe should be sent to the nearest livestock laboratory for analysis. **Warning:** You must protect yourself by wearing gloves and preferably a mask when collecting these items. Dispose of the gloves and mask afterward. The area where the doe delivered the aborted fetus must be sanitized. Remove the contaminated bedding, and if possible, burn it. Segregate the entire area until the test results come back.

There are other, less dire causes of late abortion; the most common are general malnutrition, trauma, and stress to the doe. Severe vitamin A deficiency, selenium deficiency, and plant toxicity are also linked to abortion. Mummified aborted fetuses have been linked to copper deficiency in the mothers. Certain medications and dewormers are also linked to late abortions. (For more on causes of abortion, see chapter 16.)

Delivery Time

Watching a healthy goat give birth without the need for assistance is magical. But if you are not prepared and a complication occurs, the event can turn into a heartbreaking disaster. Intervening and providing proper assistance to a goat in delivery distress are not skills that can be learned through reading alone. They must be practiced over time. My goal in this section of this chapter is to advance your learning and, most important, to help you understand when to get help.

The Delivery "Room"

Does should always deliver in a clean, safe, stress-free place. This might be out in a large pasture or paddock or in a snug barn. They can deliver in the pen with the herd or in segregated pens. If the main pen is clean and has ample room and the doe can segregate herself, as many of them prefer to do, this might be the best option if the delivery cannot be observed. Often other, more experienced mothers, or even an accepted livestock guardian dog, will help clean and stimulate the kids of an inexperienced or overwhelmed mother.

If you wish to easily observe the doe and be able to intervene without the distraction of other animals nearby, then individual pens are the way to go. I use a webcam or video baby monitor to keep an eye on the expectant mom. For the doe to be comfortable and clean, the pen should be bedded deeply enough to be clean and absorb the liquids of birthing, but not so deep that the kids will have difficulty standing and walking. The pens should provide enough space so that the babies aren't crushed during delivery. Remove hazards such as water buckets into which a kid might be delivered and drowned or be chilled. The pen should be positioned so that the doe can still see the main herd (otherwise she may feel stressed) and be easily observable by the human in charge.

In all but the most severe of winters, does can give birth without the need for artificial heat as long as the pen is out of the wind and is dry. Heat lamps and lights might make us feel more comfortable while we are on baby watch, but they are rarely needed for the delivery of healthy kids.

Inducing a Doe's Labor

There are times when a doe's labor might need to be instigated at the producer's desire rather than nature's. An example is when a doe is struggling with ketosis, a relatively common pregnancy-related illness. You may also want to induce labor in does at risk for complications, to ensure you will be there to help during delivery in necessary. Or you may need to separate kids from a doe immediately after delivery to prevent the transmission of disease, such as CAE, to the kids.

Several drugs can induce labor; the most common is the hormone prostaglandin. A popular brand is Lutalyse. You can ask your veterinarian to recommend a dosage. Smith and Sherman (authors of *Goat Medicine*) suggest that 7.5 to 10 mg (be sure to check the strength for the right dose in ml) administered intramuscularly in the morning will result in labor the following day, about 29 to 36 hours after giving the drug.[4] If you give the drug too late in the day, delivery could occur during the following night.

Essential Delivery Room Supply List

Before the first kidding date arrives, be sure to gather your supplies. Here's a list of what I like to have handy for every birth.

Towels—I use terry-cloth bath towels we get at a thrift store

7% iodine and a dip cup (I use an old pill bottle or a flip-top plastic milk sample vial)

Scissors, any type

Gloves, disposable latex or nitrile and long plastic OB gloves

Betadine scrub or other wash solution

High-energy supplement (Power Punch, Nutri-Drench)

Calcium drench

Dental floss

Lubricant jelly or powder (that rehydrates as you use it)

Signs of Approaching Labor

Does that are pregnant for the first time usually start developing mammary tissue six weeks or so before the due date. Colostrum production also begins before kidding. Some does exhibit minimal udder filling until they give birth; others show increasing filling of the udder as their due date approaches and become downright engorged the day they kid.

Anytime from several weeks out, the pelvic ligaments will start to soften in response to hormonal signals. (See figure 8.10.) Frequent checking will reveal that the ligaments repeatedly soften and then become more firm. Often the way the doe is standing, lying, or moving will have an effect on how the ligaments feel.

As labor nears, the doe might stand away from other goats or not appear interested in feed. She should continue to ruminate regularly, though, and not show any signs of pain or distress until labor actually begins. If she is standoffish or appears to be in pain and is not close to her due date, then something might be amiss.

The posture of the goat should appear normal, other than looking increasingly pregnant, until the day of delivery. At this time, ligaments will have softened to the point that her rump angle begins to change. Study the right side of the abdomen or

Signs of Approaching and Stage One Labor

- Isolating herself from the herd
- Not eating
- Vocalization—short, nickering sounds
- Licking—the wall, your hand, your clothing
- Pawing the ground—nesting
- Posture changes—pelvic angle steep
- Pelvic, spinal ligaments soft

palpate the belly in front of the udder; you may see or feel movement of the kids.

The cervical plug, sometimes called the mucous plug, is a thick, white glob of mucus that is expelled anytime from 30 days before to the day of delivery.

Signs as Labor Begins

Impending labor should progress through a fairly predictable pattern. Learning to spot the signs and understand the pattern requires repeated opportunities to observe pregnant does and a keen attention to details. Each doe will show the signs of normal

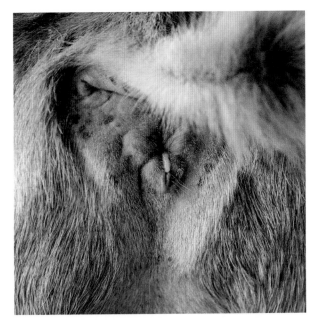

Figure 8.9. The mucous plug, which helps protect the uterus from infection by sealing off the cervix, is passed before labor begins.

progression a bit differently, and you can't rely strictly on any single sign—they need to be considered as a whole. The first stage of labor—when just contractions (not pushing) are occurring, the babies are moving into position, and the cervix is dilating—can take many hours. As long as the doe continues to periodically ruminate, she shows no other sign of troubles, and the labor signs described here continue to progress, there is no need to worry. Simply keep track accurately of timing of when the signs begin and how they progress, or you could lose an opportunity to provide vital assistance.

During the first stage of labor, the pelvic ligaments are usually so soft that they might be impossible to locate with your fingers. Some even say, "The ligaments have disappeared," or "The ligaments are gone," which of course they aren't, but it feels as if they are. As full labor approaches, the tail head becomes very loose and easy to wiggle. While it's not 100 percent accurate, checking the ligaments is

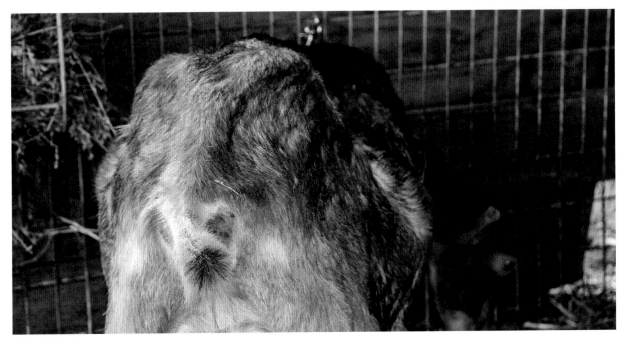

Figure 8.10. When delivery is imminent, the ligaments that support the back of the pelvis will soften to allow the pelvis to flex and open. A depression will appear along either side of the spine. The area will be so soft that if you place fingers on either side of the spine, just ahead of the tail, you may be able to push into the depressions under the tail head so deeply that your fingers will touch (with the goat's hide in between).

Figure 8.11. As active pushing labor approaches, a doe will usually begin to alternately lie down and get up again. While lying down, she will extend first one leg far out to the side and finally two legs.

something that many goat farmers use as a pretty reliable guide to determine when labor is imminent.

Impending delivery brings a great change in the doe's behavior. Usually she will stand by herself and be quite vocal. Many of us simply say, "She is talking." If she is quite close to pushing, she is likely to paw the ground and appear to make a nest. If you hold your hand to her muzzle she may lick your hand quite eagerly. She might try to lick and even appear to want to eat any piece of loose clothing.

As the birth becomes closer, she may for a time stop pawing and stand, shifting her weight back and forth uncomfortably. She should still ruminate periodically during this time. When she is almost ready to push she is likely to lie down, get up, and lie down

again, changing sides, every few minutes. When a contraction (at this stage without pushing) occurs, if she has erect ears you may see her ears straighten (looking a bit like a rabbit's), her tail straighten and point upward, and her eyes grow large. During stage one, contractions are irregular and widely spaced. Unlike the case of human mothers about to deliver, though, counting their frequency isn't necessary or even helpful.

Observe the doe from the side: Her rump angle will look quite steep and her back will almost appear hunched. Observe from the rear: Her vulva will no longer look engorged and puffy but will instead appear stretched out and elongated. Both of these changes are due to the ligaments in her pelvis

Figure 8.12. This kid is being delivered in a normal presentation. The mucus that preceded it is tinged yellowish green (rather than being mostly clear) in what is called meconium staining. There are also small streaks of blood. Meconium staining is a sign that one or more of the babies has experienced distress.

Figure 8.13. Often the amnion of the first kid will bulge out of the vagina ahead of the kid. Sometimes called "the bubble," the amnion helps expand the pelvis, vaginal canal, and opening.

loosening and the distance between the spine and the pelvis increasing to open the birth canal for the babies' exit.

When delivery is very close, there will likely be a discharge of clear to amber-colored mucus. This may not start until stage two, when the doe starts pushing, but it may begin in stage one. Observe the mucus to make certain it is not any color other than amber. Greenish tinges (called meconium staining) might signal fetal stress; the color is caused by the baby releasing the first fecal matter, meconium, inside the uterus—something not normally done. Bloody tinges can be normal, but if there is dripping blood it might mean that an umbilical cord has torn or that there is bleeding in the vagina or uterus.

The Normal Delivery

When the doe is ready to push she may continue to lie down and get up, changing sides, but her back legs are likely to both be stretched straight out to the side when she lies down due to discomfort in her pelvis. She may lie down and then shift her body in a way that lifts it and moves it forward a bit. This is not pushing; she is simply trying to move away from the discomfort in her pelvis. Remember, the uterus is contracting and the babies are moving into position during the first stage of labor. If this continues for an extended period of time, and the doe does not ruminate or grinds her teeth, then there may be a problem. But if she occasionally chews her cud, say at least every 15 to 30 minutes, it is probably just taking a while for the cervix to dilate and prepare. If she seems tired, offer a pail of hot water and perhaps administer an oral dose of an energy supplement such as Nutri-Drench or Power Punch. If the doe is past her prime age, about five, an oral dose of calcium drench is a good idea because it can help with contractions. While the science isn't conclusive, it has seemed to me that giving the doe a handful or more of fresh or dried raspberry leaves to eat at this stage can be helpful in advancing the labor. Usually the goat will eat them eagerly. She can have as much as she likes.

A Normal Labor and Delivery

Early: The first stage may last many hours; the doe is not pushing.
Delivery of first kid: Once pushing begins, the first kid should be progressing through the canal or delivered within 45 minutes.
Additional kids: Additional kids might follow immediately or 10 to 20 minutes apart.
Delivery of placenta: The placenta might be delivered within minutes or hours.

Make a note of the time of the first push. A real push will involve the doe's entire body. Her head and neck will extend forward, and she may extend her top lip as well. The back legs, if she is lying down, will stiffen and extend. She may be very quiet or grunt during the first pushes. Some does stand while pushing, but most lie down. Some will even continue to eat and chew their cud at this stage. Interestingly, first-time does seem to be the most oblivious to what is happening.

If pushing begins and you do not see a bulging of the vulva as an amniotic sac, or bubble, is entering the birth canal, it is a good idea to check for the presentation of the kid. First, wash your hands and trim your nails. (You won't be probing deep for this simple check, so you do not need to sanitize your hands or put on gloves unless for your own protection.) Use your middle and index finger to gently explore the birth canal. This check does not cause any discomfort to the doe. If she is actively pushing, you should feel either the rounded bubble of an amnion (the sac around the kid) or, if it has broken, some part of the first kid's body. You may see, or may be doused with, a big gush of warm liquid if the amnion breaks. It takes some practice to understand exactly what you are feeling. The ideal presentation for birth is two front hooves on the bottom and, just behind that, the muzzle (see presentation 1, page 203). If you don't feel anything, or you feel something that is not the ideal presentation, prepare your supplies for intervention. You might not need them, but it is better

Warning Signs: Time to Help!

- The doe's "water" (amniotic fluid) has gushed or flowed out, but no pushing has begun after about 15 to 30 minutes.
- Pushing began, but then stopped for more than about 30 to 45 minutes.
- The doe is grinding her teeth.
- The doe is pushing hard, but there is no progress after 15 to 30 minutes.

- After one kid has been manually removed, and other kids are present but don't follow after 10 to 20 minutes, or the doe doesn't begin pushing again after 10 to 20 minutes.
- A lot of mucus is apparent, but the doe is making no attempt to push.
- Meconium staining of the mucus is visible (see figure 8.12).

Figure 8.14. Checking a doe for a malpresentation of kids. In this case the cervix did not dilate because two kids were trying to present at the same time. Manual dilation of the cervix for about 15 minutes then repositioning one kid to the side resulted in the successful delivery of quads.

Figure 8.15. Back-feet first is a normal presentation. Note that the hooves are upside down, which is normal for back feet, but rare for front feet. To verify whether protruding hooves belong to front feet or back, you must insert your fingers more deeply into the vagina and possibly through the cervix and feel farther up the leg to find either a hock or a knee.

to be ready. Within about 45 minutes or less of the first push, the first kid should have been delivered.

While most does don't need any help during delivery, it is perfectly fine to gently assist with a normal delivery. As the kid is delivered, you can clean off its nose with a clean terry-cloth towel. Then support the body of the kid as the doe pushes it out, and guide it around the side of the doe's body toward the doe's head so that she can reach back and inspect and start licking it. If the kid is back-legs first (also a normal presentation), support it and guide it out a bit more quickly than for a front feet and head first delivery. During a backward delivery, the cord often breaks before the head is out. If it does so while the kid's head is still inside the canal, the kid might try to take a breath, and it will receive no air. If this happens, the reflex that stimulated the attempt may not reoccur when the kid has emerged into the air. Support and guide the kid out in a downward arc that curves toward the doe's udder. Then dry its nose as quickly as possible so that a minimal amount of mucus is inhaled.

With a normal head-and-front-leg delivery the umbilical cord is usually still intact as the kid

emerges. With a back-leg or rear presentation, the umbilical cord may break before the kid is fully delivered. If it is intact, continue to move the kid forward toward the doe's head. This puts a bit of tension on the cord. It's this tension that signals the blood flow between mother and kid to stop. You will see the strands of the cord stretch and start to snap as you apply more tension. This is normal and the best way for the cord to separate. **Warning:** *Never* cut the cord with scissors while blood is still flowing through it! Once the cord breaks, you might see a bit of blood drip from it, especially if it breaks quickly as the doe pushes the kid out. This is usually perfectly fine, but you should still observe it for further bleeding. Usually a clot will form at the end that is still attached to the kid. You can use dental floss to tie a cord that continues to bleed. Umbilical cord clips are available from some supply stores, but I find them to be useless if the mom is allowed to clean and dry the kid—she simply licks and pulls off the clip. Although I have been present for hundreds of goat births, there has been only one occasion when I needed to use dental floss to tie off a cord.

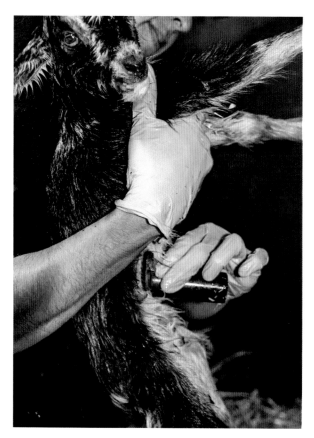

Figure 8.16. The small cup pressed against this newborn kid's belly contains strong iodine. Clipping and dipping the umbilical cord with strong iodine is always a good step for the future health of the kid.

Figure 8.17. This kid was born with a loose right rear hock—a condition that in my experience has never failed to correct itself within a day or two without any intervention.

After the cord breaks, you can leave it long or trim the portion hanging from the kid with clean, sharp scissors. Conventional advice is to leave a stub ½ to 1 inch (1.25–2.5 cm) long, but holistic vets recommend leaving it the length it breaks. Dip the entire stub, including the spot where it protrudes from the belly, in strong iodine (7 percent equivalent) or tea tree oil. To do this, pour the solution of choice into a small container with a small opening, such as a flip-top vial or pill container, filling the container about a quarter full. Hold the kid in one hand with its belly facing down, put the container over the umbilical cord, and press it firmly against the kid's belly. Then, holding the container in place, tip the kid backward so that the solution coats the entire cord and its base at the kid's abdomen. This can be repeated in a few hours or the next day.

If the doe is going to have more than one kid, the additional kids might emerge immediately after the first or a few minutes later. If the doe doesn't begin pushing after about 10 minutes, palpate her abdomen just in front of her udder. If another kid is present, you should be able to feel it. It will feel hard and knobby. Sometimes you will feel the uterus contract, and there will be a firmness that then goes away. It's easy to mistake that for a kid. (It still occasionally happens to me.) If there is only one small kid left, you might not feel it by palpating while the doe is lying down, because it could be up higher and partially engaged in the birth canal. If you are reasonably certain that there is at least one more kid, don't allow more than 30 to 45 minutes pass before intervening—the placenta might be starting to detach, and the remaining kid may lose its oxygen supply and die. Sometimes a doe will stop pushing because she is tired, she's out of shape, or there is a problem with the next kid. This is a time when intervention is imperative to save both the kid and the doe. Without help, it might take the doe hours to deliver the kid, and it will be born dead. Or she could experience complications and not survive.

The Assisted Delivery

Over the years I have become quite good at sorting out kids tangled up inside the womb and successfully delivering them. There's no way to develop

Presentation Possibilities

In this collection of illustrations you will see many of the possible presentations of kids—but not all of them! The possible combinations of presentation that can occur is huge. Because the scenarios can be quite complicated and because you will likely be anxious in the moment when a delivery is taking place, I recommend you take the time to study all the illustrations below well before kidding season arrives. Imagine each one, and mime the movements you would make with your hands to assist in the delivery. It can make a big difference should an emergency occur.

Other possible presentations not shown here include a normal front presentation, but with one front leg crossed over the neck; two twins presenting normally, but both trying to exit the uterus at the same time, and so blocking each other's way; a kid presenting stifle joint first (this is so hard to imagine, even though I've had it occur, I couldn't figure out how to draw it!); and back pressed against the cervix. And there are many more; too many for me to describe here.

Presentation 1: Normal presentation of twins, one in each horn of the uterus, before labor begins.

Presentation 2: Normal presentation of twins, delivery of twin in right horn, front feet and head first. No assistance should be needed.

Presentation 3: One front leg and head first, one front leg back. Usually no assistance is necessary. If help is needed, apply gentle pulling pressure to the presented front leg and around the kid's skull. If this presentation is determined *before* the kid's head clears the pelvis, the front leg that is back can be repositioned.

Presentation 4: Normal presentation, back legs first.

Presentation 5: When a small kid presents butt first and the back legs are forward, no assistance is usually needed. But in this instance, two more kids are present. The one in the left horn is a larger kid, with front legs and head first. In

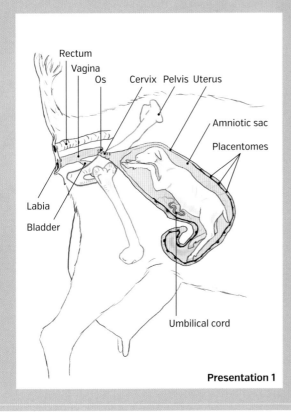

Rectum
Vagina
Os
Cervix Pelvis Uterus
Amniotic sac
Placentomes
Labia
Bladder
Umbilical cord

Presentation 1

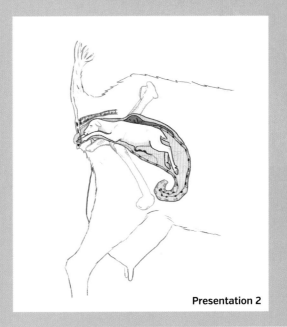

Presentation 2

this case, the larger kid may block the exit of the smaller, backward kid. You can gently push the smaller kid farther back into the uterus and deliver the larger, normal-presentation kid first if needed. The last kid is upside down in the right horn. As it is pushed by the contractions up toward the cervix, it may rotate on its own so that it isn't upside down. If not, you may need to gently rotate its head and front legs before the kid can clear the pelvis and be delivered.

Presentation 6: Butt-first presentation with hocks bent, causing the kid to lodge behind the pelvis. To assist, reach in and gently push the kid toward the mom's head. Then follow each back leg down to the hock, continuing to push the kid back in as each contraction pushes it toward you, and hook the lower legs up and toward the cervix. Once they are freed of being caught behind the pelvis, they will quickly extend backward into the vagina, and the kid can be easily delivered.

Presentation 7: This kid is almost in a normal presentation of head and front feet, but its elbows are back just enough to cause the shoulder blades to broaden and become wedged in the pelvis. This is sometimes called elbow lock. It can happen with the kid a bit farther out, too. When it does, push one leg backward a bit and pull gently on the other leg and the head.

Presentation 8: In this drawing twins are presenting with the left twin in a normal, back-legs-first presentation, but the larger twin in the right horn is head first with the front legs back and is blocking the cervix. In this case, the doe may not even dilate fully or perhaps not try to push. When you reach in, you will feel the larger kid's head and perhaps the back feet of the other kid. If you aren't sure if they are back feet and even which kid those feet belong to, follow the neck of the larger kid to its shoulder and then down to the front legs. Then you have two options: You can push the bigger kid back in a bit and hook up and bring forward its front legs, one at a time, and deliver it first; or you can push it down into the horn and deliver the other kid first.

Presentation 9: In this common presentation the front feet are first, but the kid's head is down. Help is usually required. Reach in and gently push the dome of the head back in toward the mom's head until you can reach under its chin and lift the muzzle up between its front legs. While you do this, you may "lose" a front leg and have to hook it back up and make sure it stays in place as the kid moves into the birth canal. Less common but with a similar presentation is front feet with the head turned to the side. In that case, when you reach in you will feel the neck at the throat. Push on the neck to move the kid back in a bit, then follow the neck to the head and swing it around. When it has been bent down or sideways, it will want to move back to that position, so keep your hand on the dome of the head until it moves into the vagina.

Presentation 10: The first kid presenting has its head down and its front feet back. A triplet in the left horn is presenting normally with front feet and head, and a triplet in the right horn, below the first kid, is presenting normally, but backward. Because the first kid is blocking the cervix with its forehead, the doe may not have fully dilated or may not be pushing. When you reach in, you will feel the forehead of the first kid, but you will also likely feel the front feet of the kid on the left and maybe even the back feet of the other kid on the right. You're likely to want to first raise the head of the kid, but be sure you determine which kid's feet belong to that head! Once the kid that has been blocking the exit has been delivered, the other two shouldn't need assistance if the doe is still strong and pushing.

Presentation 11: When the cervix hasn't fully dilated it might feel like a funnel when you palpate it. You will feel the other rings as your hand moves toward the uterus. The last ring will be the smallest. Because of the funnel-like form, you might feel the kid through the cervical tissue. Don't mistake that tissue for an amniotic sac and try to break through it to reach the kid!

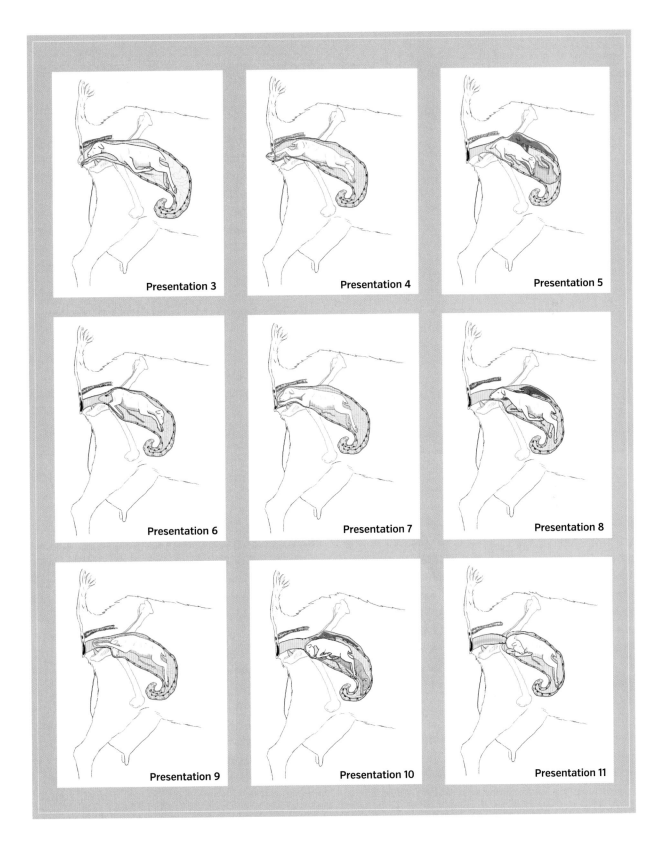

Presentation 3

Presentation 4

Presentation 5

Presentation 6

Presentation 7

Presentation 8

Presentation 9

Presentation 10

Presentation 11

competence at assisting during difficult deliveries, though, without having to go through some very stressful and challenging times. I learned early, when I still needed a vet's help during particularly difficult deliveries (*dystocia* is the proper term), that the essential attitude to cultivate for success is a combination of patience and persistence. If you are lucky enough to have a goat mentor who will let you be present during deliveries then you will gain a lot of useful knowledge. That same mentor might be willing to walk you through the steps over the phone. When I teach our "Goat Academy" here at Pholia Farm, I try to time it for a week when several does are due to kid just for that opportunity. And I hope this book can help you learn the skills needed to assist during deliveries.

I always err on the side of helping a doe earlier in the process, rather than regretting later. My perspective is that the animal is only in this predicament due to my choices, so she deserves my help. Another part of my viewpoint is the potential to decide not to breed a doe in the future if difficult deliveries are her norm.

When a doe is in labor and you've decided that further manual exploration is necessary, first take a few moments to gather your supplies—which should all be near at hand already. Fill a pail with a warm water and Betadine or other wash solution (follow the dilution instruction on the bottle). Be sure your fingernails are trimmed, remove any rings or bracelets, and put on gloves. You can use long OB gloves if you like. A trick I learned (from that patient vet I mentioned earlier) that leaves your fingers more agile than using an OB glove alone is to cut the fingers out of a long OB glove, put a nitrile glove on your hand first, and then put on the fingerless long glove over it. It's preferable to use your dominant hand for exploring the doe. This may mean you have to roll her over. If it's not possible to do this, you can use your non-dominant hand, but it will feel less efficient. If the doe is standing, ask an assistant to try to hold her head to keep her still.

Wash the doe's backside with a bit of the warm wash solution, then dip your gloved hand and forearm in the solution and shake it off. Dispense a generous amount of lubricant, either gel or powder,

Seven Things to Remember During a Tough Delivery

1. Keep calm!
2. Take your time. Move methodically and slowly inside the doe.
3. Give the doe and yourself a break every 5 to 10 minutes.
4. If the delivery requires intervention that increases the doe's pain to the point of causing her extreme distress, administer pain medication. (Banamine, which requires a vet's prescription, is a good choice.)
5. Medicate yourself! Have a helper bring you a quick shot of chamomile tea or even whiskey (if it's on your menu).
6. Give the doe a dose of high-potency oral nutritional supplement during the delivery, between kids or pushes, and when delivery is over.
7. Keep calm!

on your gloved fingers. You don't need to squeeze it all over your hand; it will spread as you enter the doe. Repeat this process whenever you reenter the doe.

With your other hand, lift her tail. Bunch your gloved fingers together, holding them straight, with your thumb tight to your palm, and then insert the fingers into the vagina. The doe is likely to begin pushing as the pressure of your hand in her vaginal canal stimulates that response. You will have to push against it. Slowly but steadily continue to move your fingers forward, and then your hand. As your hand moves forward, note the tightness of her pelvis and the dilation of the cervix. When the cervix is fully dilated you might find that as your hand goes into the vagina, the cervix will feel like a large funnel, with the small end toward the uterus. You will feel the rings of the cervix and maybe even feel parts of a kid through the cervical tissue. Do not mistake the tissue of the cervix for an amniotic sac and try to push through it! Instead, continue straight forward until you find the final cervical ring. Then your hand will enter the uterus. Sometimes the body of the uterus will feel large and spacious; other times it will be small and tight, with the kids obviously positioned in the horns.

UNTANGLING MALPRESENTATIONS

Now comes the tricky part, trying to distinguish—without being able to see inside the uterus and with very limited hand mobility—just what part of a kid's body you are feeling. Again, take your time. Close your eyes, and concentrate on trying to visualize the kid's position. You may be able to find benchmark parts that are easy to discern. See table 8.3 for some possible combinations of options.

Here are some tips and cautions to keep in mind as you assist with a delivery that involves malpresentations or other problems:

- Be calm!
- If you push a fetus in the direction of the doe's head, the uterus will expand. You may have to do this many times to create enough room to allow movement. Determine which horn the kid is in if possible, and push it back into that horn.

- As you push or move a kid, think about the natural curvature of its spine. Never try to bend the spine in a direction it wouldn't go naturally.
- Avoid bending legs in a direction that they wouldn't naturally bend or stretch—you can dislocate the kid's hip or injure a shoulder.
- If you aren't sure what body part you are touching or what to do next, explore beyond that part with your fingers until you find another part of the body that you can identify.
- If a newborn kid is available at the scene, you can use it as a "guide map." Ask a helper to hold that kid in a position similar to the position you envision the fetus is in. Close your eyes, and simultaneously explore how both kids' bodies feel. See if you can discern parts that match up.
- It is possible to completely reverse the direction of presentation of a kid if needed, but think carefully about all of the above considerations as you work. In addition, take care to avoid allowing the teeth of the kid to drag across the uterine wall, because this could rupture the wall.
- If you can find a head, always maneuver it to the birth canal. Also find at least one front leg—and be sure it's a leg attached to the same kid whose head you've moved into place. Getting one leg into place beside the head is important. Only if the kid

Table 8.3. Sorting Out the Unseen—Malpresentations

What You Feel Initially	What to Feel for Next	What Part of the Body It Likely Is
Pointed, soft	Is it one of a pair?	• If a pair, a hoof • If alone, a tail
Pointed, hard	Bones extending away from the point in a V-shape	Hocks
Pointed, hard, sharp	Teeth; you will also be able to feel the muzzle	Muzzle. Check to see if teeth are sharp side up or down. If down, the kid is upside down.
Rounded, hard	Eye sockets	• If eye sockets are present, a forehead • If no eye sockets, probably the stifle joint
Hard with ridges	Ribs or neck	Back or neck
Soft, rounded	Ribs	Belly

is very tiny can you easily deliver it head first with no leg in proper position alongside the head.

- The pelvic opening might feel awfully tight, but if your hand can fit through and the kid's head is not larger than your fist, assisting the delivery safely should be possible.

- If the presentation is rear legs first, the umbilical cord is likely to break before the head is out of the birth canal. You must not let the kid's head remain in for too long, or the kid will try to take a breath and fail to receive air, which could dampen its urge to initiate breathing.

- Avoid pulling both front feet out too far beyond the head. One foot is okay, but if you pull both forward, the shoulders may lock in the pelvic canal.

- If you are having trouble assisting a very large kid that is presenting normally with head and two front feet, try pushing one front leg all the way back into the uterus, and then pulling gently on the other leg while you help the head clear the outlet. You may have to use your fingers like a lever on the side of the head to help the doe deliver.

- Once the kid is in the canal, even if its head is out, don't let it remain there for too long. The tightness of the pelvic opening can stop the kid from breathing.

- On occasion, even after delivery and wiping the nose clear, a kid will not start breathing. If it has a heartbeat, try pinching very hard between the nostrils. This stimulates nerves that, like cool air hitting the nostrils, tell the kid's brain it's time to breathe. If you are certain that the kid is not carrying any zoonotic disease organisms, you can also cover its entire mouth with yours and puff gently to provide air. You can repeat this a few times.

MANUALLY DILATING THE CERVIX

A fully dilated cervix has an opening about the size of a fist. But in some cases, if the cervix is not fully dilated, and you are sure the doe needs assistance, then you will probably need to help manually dilate the cervix before you can help with delivering the kids. Sometimes this situation is referred to as ring womb. It seems to most often occur when there is a malpresentation or two babies trying to get out at the same time. There isn't enough internal stimulation of the cervix to promote dilation. Before you attempt manual dilation, be sure that you have gone over all of the other signs, such as seeing the amniotic fluid from the rupture of one of the amniotic sacs (seeing the water break), to make sure the doe is actually ready to deliver. The cervix can usually be manually dilated over a period of time; in my experience it takes 15 to 30 minutes of intermittent sessions. Use your gloved fingers to gently massage and open the cervix by running them in a circle around the inside of the ring. Massage for a minute or two and then give the doe a break. The mere presence of your hand will stimulate her to push, and over time this will tire her out. Be sure to resanitize your gloved hand before each session, and use plenty of lubricant.

Smith and Sherman suggest that in this type of situation, a C-section is required. However, I have dealt with an undilated cervix many times, and in all but one case (described in the Ohmnom sidebar on page 209), I have been able to manually dilate the cervix in time to deliver healthy kids without too much stress to the mother.

CESAREAN SECTIONS

Cesarean sections are sometimes unavoidable, especially when the time for manual intervention has passed or hasn't been effective.

For there to be any possibility of the doe's surviving and recovering from the surgery, a veterinarian must perform it. The farmer will not have the necessary skills, tools, and medication to do it humanely. Most does will enjoy a full recovery from a cesarean, as long as the procedure wasn't performed as a last-ditch attempt to save either the doe or the kids. Whether to breed the doe again is a choice that must be assessed based on not only the C-section, but also her age and health.

If a doe in labor is dying, then the producer can perform a cesarean in order to save the life of the unborn kid or kids. It's assumed that the doe cannot recover. If the doe is unconscious, no pain medication or sedation is necessary, but if she is

Ohmnom

Up until the spring of 2016, I had never lost a doe due to complications from kidding. I always say that every year something new and challenging will probably happen to keep me humble, and this time it was a doozy. Ohmnom, usually called Nommie, was a five-year-old fourth freshener with no history of kidding troubles. (As goats age, the odds go up that something will go wrong during delivery.) When I was called down to the barn by my intrepid intern, Amanda, at 1 AM, Ohmnom seemed ready to deliver; she was pushing but making no progress. Upon checking her cervix, I found it barely dilated. She had been leaking amniotic fluid for a few days, so there was concern over a dry delivery, where the fluid around the kid has been lost.

Ohmnom's pelvis was incredibly tight and narrow. I was unable to make much progress dilating her after an hour of trying, but she was progressing a bit and was handling my manipulation well. However, if it had been daytime, when my vet was available, I would have taken her in for a C-section. As it was, the only emergency vet care available was a small-animal hospital, which would not have been helpful. Right or wrong I made the choice to keep working on her.

We gave her some Banamine for pain, calcium drench, Nutri-Drench, and hot water to drink. During breaks, we offered her feed. After three hours, her cervix had dilated enough that I could pass my fist through. I could feel a very large kid, with its head down, presenting. After another hour, I had been unable to manipulate the kid into the canal. Its head was so large, and she was so small. I decided that the kid, who had stopped moving, could be lost, but not Ohmnom. I was able to get a loop around the kid's head and with one front leg out (two would have been even more crowded) I worked it out of her, dead. After a break, I felt her abdomen and palpated another kid. I went in and found, far down in the right horn, a small kid that was delivered alive. We broke down in tears, in gratitude for the live kid and the fact that it was over.

Ohmnom was in shock but got up and nursed the kid, a doe. She seemed as good as you would expect given all she had been through. I put her on antibiotics because of the invasive work I had done. Twelve hours later, she hadn't delivered all of the placenta, so I administered oxytocin. I think that was too much for her. I believe that the contractions that it caused (I gave another dose eight hours later as recommended) were too hard on her. Two days later she perished. In hindsight, I should have ignored the retained placenta and simply continued the antibiotics to cover any retained material as it broke down. I also could have waited until daytime and taken her in for a C-section. She may have lost the kids, but we wouldn't have lost her. I also wish I had known about the condition called uterine torsion. I could have ruled that out, as it might have been the problem. We named her kid Lil Nommie.

alert, then the humane choice is to administer pain medication. Lidocaine is effective for local numbing of pain at the incision site and xylazine as sedation. (Xylazine is not a recommended sedative for a cesarean if there's a chance the doe will recover, unless the reversal drugs are also used. Without them, there can be serious complications for the pregnant goat.) If pain control isn't available, the doe should be euthanized using a bullet or captive bolt (see the Euthanasia section on page 174 for instructions), after which the C-section must be quickly performed. Any kids alive and healthy at the moment of their mom's death can be removed safely if the procedure is done immediately.

Before you begin, gather materials: You will need a sharp scalpel, skinning knife, box cutter, or X-Acto knife. You will also need disposable gloves, wipes, and clamps. To perform the cesarean, lay the doe on her right side, with her left side facing up. The cut through the animal's flank should be made at a right angle to the spine just in front of the hind leg and pelvis. Make the incision in several shallow cuts so as not to accidentally cut too deep and hurt a kid. The first layer will be skin, the second the abdominal wall, and the third the lining of the abdominal cavity (called the peritoneum). You can then reach in and find the uterus. Make a careful cut through the uterine wall and locate and remove a kid. Quickly clean the kid's nose. Then clamp or tie the cord in two places and cut (using the knife, scalpel, or scissors) the cord between the two clamps. Immediately check for other kids. There will be a lot of amniotic fluid and blood. Once all the kids are delivered, the kids should be dried and fed colostrum. You may be able to harvest colostrum from the doe, although she may not have produced much or any.

It's also important to quickly attend to the sad task of disposing of the doe's body. Options for this were described in the Field Necropsy and Liver Sampling section on page 175.

Dead Kids and Deformities

If you are present for enough births, sooner or later you'll encounter some sad and odd results, even when you have done everything right. You may never know what happened to cause the premature death or the deformity, which can make it feel even more frustrating.

Typically the body of a living fetus in the uterus will feel resilient when you touch it, although it may or may not move on its own while still inside. Kids that have recently died in the uterus will be in varying states of breakdown. If they die during delivery, they will look normal but be so limp that repositioning them in the uterus can be surprisingly difficult. A kid dead for more than about 12 hours will have sunken eyes, and the jaw will look pointed and small. Even longer, and hair will start to come off.

When a fetus dies earlier in the pregnancy, it will often mummify. Delivering a mummified kid or finding one in the afterbirth (they are often very tiny and can easily be missed if the afterbirth isn't carefully inspected) can be a bit disturbing. In retrospect, you might realize that the doe had a difficult point in her pregnancy, or even seemed to struggle for the entire time. When a fetus dies, it puts stress on the doe, and her body struggles to retain the other kid or kids. In most cases, even if a doe delivers a dead kid, you don't need to treat her any differently during her postpartum time.

Many birth defects are possible in goats, but creating a list of all of them would be pointless and discouraging. The only deformities we have experienced here at Pholia Farm, among hundreds of newborns, are one living anencephalic (missing a part of the brain and skull) kid and one kid whose pelvis was fused (or at least the stifle joints were unable to bend). A neighbor who keeps goats assisted with the delivery of a hair-covered ball with a gel-like texture. Another friend had a goat that bore a kid whose neck was fused turned backward. A veterinarian friend of mine once delivered a lamb that had developed "inside out" with its organs and musculature outside and the hide inside. Whatever the defect, you will have to decide based on your own philosophy whether or not the kid will have a good quality of life. At the beginning of their careers, many goat producers allowed kids to live that later had to be euthanized as their life became too difficult. Over time, you learn to look into the future in an informed way, based on your experience, and make a different decision earlier.

Postpartum Baby Care

Immediate care for all newborn kids includes checking for deformities, clipping and dipping umbilical cords, drying the kid, and ensuring that it receives adequate colostrum in the right amount of time.

Non-life-threatening deformities include the presence of more than two teats and undescended testicles (for details about this condition, cryptorchidism, see chapter 16). You will be able to tell the gender of a kid immediately after birth by feeling

What to Do About Extra Teats

In most breeds teat deformities are relatively common. Extra teats are those that are separately delineated from the clearly normal two teats. Forked or double teats are those that are joined at the udder and then split into two of roughly equal size. Teat spurs are those that exist as a pronounced bump or protrusion on an obvious normal teat.

How you choose to manage these issues is, for the most part, your own decision. The first consideration is how the extra growth might interfere with milking or suckling. A double or forked teat isn't easy to hand-milk and is impossible to machine-milk. A teat spur can also interfere with either. A small extra, or supernumerary, teat will usually not cause problems with either. Most breed associations and registries rightly consider teat deformities a flaw and undesirable. But since the genetics that create them aren't easily weeded out, their importance as criteria for culling animals must be considered on a case-by-case basis. For example, if the animal in question is a doe kid from superior parents born with a small spur or extra teat, the best course of action might be to snip off the offending tissue, and then keep her and breed her in her first year. You will not want to keep the kids as breeding stock, but you can assess her milking quality and mothering abilities and decide whether to keep or cull her based on her performance. Buck kids with teat deformities, on the other hand, should likely not be used as breeding stock.

One of our best milk goats had a teat spur that was unnoticed by the breeder or me when I first picked her up as a newborn kid. Once I discovered the spur, I had our vet remove and suture it while she was very young. She has kidded many times over the years and has never produced any offspring with any teat deformities. However, I also purchased her full brother. He did not have any teat deformities, but he sired many offspring with extra teats. He ended up in a friend's freezer.

If you decide to remove a teat spur or extra teat from a doe kid yourself, do it when the kid is very young. Sterilize a pair of super-sharp scissors or a scalpel, cleanse the area with Betadine, and then cut off the spur or teat. You can numb the area a bit with ice just before making the cut, but if the teat is very small, the cut is unlikely to cause much discomfort. If you sell the kid, be sure to tell the prospective owner about the removed deformity.

carefully for testicles and also by checking the area under the tail for the presence of an anus and vulva (a doe), or an anus only (a buck). Intersex kids, also known as hermaphrodites, are difficult to distinguish at birth. It's not unusual for kids, especially from large litters, to show some characteristic that appears to be a major deformity but is actually a temporary condition that is simply related to development in an overcrowded uterus. They may be born with "loose" hocks in which the joint can move both forward and backward, making the leg appear broken. This usually corrects itself within a few days. Such a kid may have a bit of trouble getting good traction for nursing its mom, and it may need a bit of help with that task.

Figure 8.18. A successful delivery with healthy kids and mom.

Smaller kids often "knuckle over," meaning their front and/or rear fetlock joints remain bent instead of straightening, so that the kid walks on its fetlocks This can last a few days but should also remedy itself with movement.

THE IMPORTANCE OF COLOSTRUM

Kids must receive the right amount of colostrum within a short time after being born. Their systems are able to absorb the invaluable antibodies in the colostrum only for a few hours. Also, the dam's colostrum will quickly become diluted as other kids suckle and milk starts replacing it. Ideally kids will receive their first feeding of colostrum within an hour of being born and no later than six hours after birth. Without the antibodies from colostrum, the kids are extremely susceptible to illness due to infection by bacteria that are naturally present in the farm environment. In addition, colostrum delivers a powerful dose of nutrients to the kid's system.

Kids being raised on dams should be observed to be sure they are nursing. I suggest feeling their bellies at birth, then comparing it with a later time—if they have nursed, they will feel fuller—as a way to determine whether they've eaten. They can also be weighed at birth and again a few hours later and the two weights compared. The rule of thumb when bottle feeding is 1 ounce of colostrum for every pound of

Steps for Tube Feeding

I've tube-fed many a tiny kid, including Niblet, the little doe shown in these photos. Fortunately, even though she had a tough start, she did well and grew into a normal-sized kid. Here's a step-by-step explanation of how to tube-feed a kid. Note: For an adult goat, a similar procedure is followed, but using an adult-sized tube and a speculum or piece of PVC pipe placed in the goat's mouth to prevent it from biting down on the tube.

Step 1. Hold the tubing so that the tip of the catheter is at the same level as the area between the kid's front legs. Then position the tubing to follow the length of the neck up to the mouth. Mark this distance with a piece of cloth or masking tape.

Step 2. Place the catheter in hot water to warm and soften it.

Step 3. Warm an appropriate-sized dose of colostrum. The recommended dose is 1 ounce for every pound of body weight (or 50 ml/kg)

Step 4. Remove the plunger from the syringe, and set it aside.

Step 5. Hold the kid in your lap or lay it on the floor—depending on how active it is. (If the kid is really active, you probably don't need to tube-feed it!)

Step 6. Insert the tip of the warmed catheter into its mouth, and guide it gently to the back.

Step 7. Lift the chin a bit, and continue to feed the tube down the kid's throat. With your non-dominant hand, palpate the skin of the neck. You should be able to feel the tube as it moves down the esophagus, a bit to the left of the center of the neck.

Step 8. Once the tape mark on the tube is positioned at the kid's lips, the catheter tip is in the right spot. To verify, place the open end of the catheter near your ear; you may hear some gurgling, but no breath sounds. Additional reassurance comes from the fact that if the tube had entered the lungs, it wouldn't have been able to penetrate the full distance—you would have felt resistance.

Step 9. Hold the tube in place at the kid's mouth, and insert the big syringe into the open end of the tube. You may need a helper to pour the colostrum into the syringe. Add all of the colostrum at once, keeping the syringe and tube at a low level, not raised above the kid's body.

body weight (or 50 ml/kg). Another guideline is to feed 20 percent of their body weight over a 24-hour period. If you are present at the birth, it is helpful to clean the udder and remove a squirt of colostrum to clear the wax plug from the teat end. Eager, strong kids will certainly have no problem clearing the plug themselves, however.

It's always a good idea to keep some backup colostrum on hand. If you are new to goats, you won't be able to put any away until your first doe kids. But when she does, try to steal a bit; even a few ounces can make the difference between life and death for a future kid. Colostrum from the first day, less than 12 hours from the time of delivery, is ideal, but even

if collected 24 hours later it will still provide some benefit. If the kids have nursed only from one side of the udder, take the colostrum from the other side. I suggest heat-treating it (see Pasteurizing Milk for Kids on page 236), cooling it, and then labeling it in a ziplock bag and storing in the freezer. If you have a deep freeze, it should last a full year.

Sometimes, usually a day or so after birth, some kids will produce feces that are pasty and sticky enough to form a plug that adheres to the tail and anus. The outside surface of the material will dry and harden. Don't try to pull off the stuck clump, because skin might be torn off or irritated in the process. Instead, take the kid to a warm-water faucet

Step 10. Carefully raise and then lower the syringe to allow the colostrum to flow through the tube and enter the kid's stomach a bit at a time, slowly.

Step 11. Once all of the colostrum disappears from the syringe, continue to keep the syringe elevated for about 10 seconds so that the fluid in the

catheter can drain completely before you pull it out. This helps prevent any colostrum remaining in the tube from accidentally being inhaled by the kid as you pull the tube out past its airway.

Step 12. Remove the catheter in a smooth, continuous motion.

Step 13. Place the kid in a warm spot.

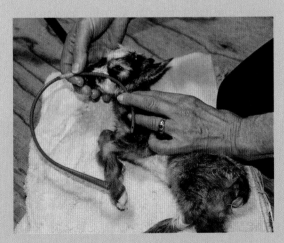

Figure 8.19. Measuring the length of the tube needed to reach the kid's abomasum. (Note: The abomasum is not located in the chest cavity, but the distance shown is equivalent.)

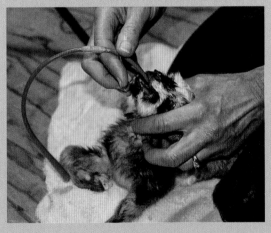

Figure 8.20. Inserting the tube to the marked length.

or bucket of warm water and soak and rinse the plug until it loosens and the skin is exposed. These plugs can occur when babies drink a bit more rich colostrum than needed.

Sometimes a mom will reject a kid. This can happen if the kid curls up with other kids and picks up their scent. It can also happen after disbudding, when kids smell like burnt hair. Try to anticipate when a new mom might have trouble accepting a kid that doesn't smell like her. There are some simple tricks you can try to encourage the mother to accept the kid. If the pen where she delivered has not been cleaned out yet, you can rub some of that bedding on the kid. Try rubbing the kid's bottom on the doe's backside, and squirt a bit of the doe's milk or colostrum on the kid's muzzle and face. A stronger scent disguise can be created by using mint or eucalyptus oil and dabbing some on the mom's nose and the kid's backside. These techniques can also be helpful when you are trying to "graft" a kid that has lost its mother onto another mother. (Grafting is usually done only in meat and fiber herds; bottle-raising kids is the norm in dairy herds.)

WARMING AND FEEDING A WEAK KID

Occasionally a kid will be born too weak to suckle, or the mother will neglect one as she focuses on others. A kid that was under stress during the labor, that isn't dried properly, or that is smaller and less competitive for the teat than siblings might become chilled and weak. If these kids aren't given extra care, they may die within hours, be accidentally crushed by the mother, or perish within a few days from a lack of colostrum.

If the kid is dry but chilled it may simply need to be warmed up in order to be strong enough to nurse or be bottle-fed. You can check the kid's body temperature to confirm that it needs your intervention, or you can feel its nose or inside it's mouth, which should be warm. The fastest way to raise a kid's body temperature is to submerge all but its head in a bucket or sink of 100°F (38°C) water. Within minutes the kid will go from limp and barely responsive to bright-eyed and kicking. When the change happens, you can towel and then blow-dry it to keep it from

cooling off while it dries. In some cases, the water will wash the mother's scent from the kid, causing rejection. I haven't ever had this happen, but as a precaution if the kid is from a first-time mom, you can put the kid's body, with its head sticking out, in a plastic garbage bag and then submerge it in the warm water that way. Be sure to let the air escape from the bag so that the heat can quickly reach the kid's skin.

If the kid is dry and warm but too weak to take the teat or bottle, tube feeding can quickly provide nutrients and a valuable dose of antibodies. The first time you insert a stomach tube on a weak kid, it is a bit scary, but it's surprisingly simple and easy. A tiny tube, or catheter, purchased from a vet or goat supply company must be used. Basically these are urinary catheters packaged for this new use. You will also need a 60 to 120 ml syringe with a tapered tip to insert into the fat end of the catheter. Colostrum will flow into the kid's stomach through the force of gravity.

Postpartum Doe Care

Delivering kids is a major physical stress on the doe. Even when delivery is normal and uneventful, she will be tired and somewhat depleted afterward. If it was a complicated delivery, you will have offered nutritional and hydration support during the experience, and the does will need additional support after it as well.

A doe that has experienced a normal or mildly complicated delivery should be up and about within minutes after the last kid is born, assuming she hasn't given birth standing. If delivery has been traumatic, such as requiring intervention to rearrange the kids in the uterus, her pelvis is likely to be quite uncomfortable, and she may not want to stand. A dose of pain management medication can be a kind thing to offer. Then assist the doe to stand so that her kids can nurse or you can milk out colostrum.

After delivery, most does will drink warm water with relish. Provide a pail full of water that is hot to the touch—about 110°F (43°C). Some breeders add a bit of molasses (a tablespoon or less) to the hot water. Be careful of overdoing it with molasses or diarrhea

Figure 8.21. Before the placenta is delivered, you will often see one or more amnions hanging out.

might result. Alternatively, give the doe a dose of a liquid nutritional supplement, such as Nutri-Drench or Power Punch, separate from the water (to ensure she ingests it all). Make sure her hay feeder is stocked. If the doe was receiving grain before delivery, offer her usual ration. If she wasn't, you can offer a small amount, a couple of handfuls or so.

If the kids are remaining with the doe, they will provide the stimulation to her udder that will cause the release of the hormone oxytocin. This hormone plays several roles, but at this point in time it is important in order to stimulate uterine contractions that will help expel the placenta. If the kids have been removed, massaging the udder, milking out the colostrum, and stimulating the teats will help release the hormone. If you milk out the doe immediately after delivery, you should still return to her a few times to massage her udder or take other action that

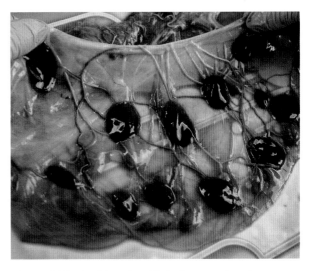

Figure 8.22. Healthy afterbirth tissues, showing the fetus-side portion of the placentomes, the cotyledons, and the blood vessels that lead to the umbilical cord.

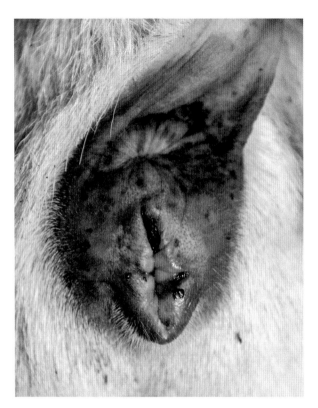

Figure 8.23. It's not unusual for a small tear in the vulva to occur during delivery. Typically it won't need any extra care, but ointment can be applied.

will stimulate the release of the hormone, until the placenta is expelled. The doe may eat the placenta, but if you can retrieve it before she does, you can dispose of it. We bury them in the compost pile.

The goat's placenta, which attaches the fetus to the uterus, is made up of multiple attachment points called placentomes, instead of a single placenta as is the case for humans. (Even so, it is usual to call a goat's afterbirth membranes, which include the amniotic sacs and placentomes, simply "the placenta.") A placentome is a two-part structure that functions and looks much like a snap on a piece of clothing. The cup-shaped caruncle portion of the placentome develops on the wall of the uterus, and the button-shaped cotyledon forms on the fetus side of the membrane. Blood flows through the placentome to multiple blood vessels, through which it then travels to the umbilical cord or cords.

Ideally, you will observe that the placenta has been passed from the uterus and inspect it, but in the real world this isn't always possible. The time it takes for a doe to pass her placenta can range from a few minutes to a few hours. It shouldn't take longer than 12 hours after delivery of the final kid. If you are not present when the doe passes her placenta and she consumes it right away, then the evidence is gone. Don't automatically assume that a placenta has been retained just because you don't see it anywhere in the pen.

Before and after the placenta is delivered, it is normal for reddish, viscous liquid to drip from the vulva. Bright-red blood is seen only for short moments. The source is an umbilical cord that has separated from the kid before the placental blood supply has stopped. This type of bleeding should be brief. If the bleeding is heavy, a vaginal or uterine tear might have occurred. A tear in the uterus is a dire emergency situation. The doe may survive if a veterinarian can perform an emergency removal of the uterus (hysterectomy). A minor vaginal tear should clot and heal, providing it is not major. Either way, encourage the doe to consume as much liquid as possible. Some will even be eager to drink a dose of their own colostrum—which provides quick nutrition, antibodies, and liquid. If you have the equipment and supplies, IV fluids can be given if the situation is severe.

Many does will experience a small to medium tear in the skin of the vulva as a result of delivering a kid with a large head. Such a tear might look quite tender for a few days. The vulva may also be bruised. You can apply some aloe gel or other ointment to help with healing, but the tear will likely heal fine on its own. Brownish discharge from the vulva, called lochia, is normal for several weeks after kidding.

Training a Doe to Be Milked

The time to begin training a doe for milking is the day she kids—or ideally, a week or more before kidding. Does that have been handled a lot and are tame will be much easier to train for milking, but even those

The Proactive Producer

To set the stage for training goats to be milked, the proactive producer can take one or more key steps, such as introducing does to being on the milk stand early in life, even before the first pregnancy, and getting them accustomed to having their udders touched. Another tip is to be present when the kids are born, because then the mothers often bond with the producer as well as with the kids.

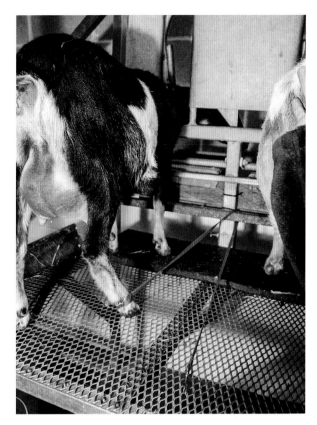

Figure 8.24. A webbed dog leash works well to make a simple hobble for use when training a reluctant goat to be milked. A second leash can be used on the other leg if needed.

that are gentle will likely protest unless you make them understand from the beginning that they have to share their milk with you.

Even if you are planning on dam-raising the kids, and milking after they are weaned, you should handle the doe's udder daily. You can massage the udder and milk her a bit, or pretend to milk her by squeezing the teats without forcing out any milk. This helps her learn what to expect when milking begins in earnest. If you leave a doe on her own with her kids nursing for the first few days, she will usually put up quite a protest if you then try to begin milking her. The milk will be ready for human consumption three to four days after kidding, providing both sides of her udder are being emptied of colostrum by her kids or by you.

If you have been slowly introducing grain to the doe before kidding, that's a great time to start touching her udder. Some does have very sensitive teats, so don't interpret a bit of stomping at this time as bad behavior. Others seem to enjoy the experience. Picture the young, first-time doe coming up on the milk stand for the first time. She's experiencing anxiety at several possible levels—leaving her kid that she must protect, having her head restrained, having a "predator" that she can't see well approach

her hindquarters, and then having milk removed by a being that is definitely not her own baby. She may kick. She may do a great impression of an Irish dance number. Or she may simply just sit down. It's pretty hard to avoid becoming exasperated! The best approach is extreme patience, calm demeanor, and persistence. This is just one example of many times that the prey instinct will dominate the experience of working with goats. To be helpful for their stress level, as well as your own, you must keep this instinctive behavior in mind.

If possible, begin bringing the doe to the milk stand within a few days after delivering. There she can eat her grain, if you are feeding it, and be handled. If you weren't able to start conditioning the doe in advance or if you've purchased a goat that's in milk but has never been milked, then prepare for a battle, but one you can win. Here are some tips for success:

- If the doe has a kid that's still nursing, have someone hold it in front of her while you milk. Usually after doing this for a few days, she will adjust to sharing.
- If the doe doesn't have kids, try to make the time on the stand as pleasant as possible—don't feed extra grain, though! That simply becomes another bad habit.
- Hold a small bowl or pail in one hand and place your other hand on the back or side (depending on which direction you're milking from) of her udder. Try to keep it there while she stomps and protests. Once she stops, glide your hand—still touching the udder—down to a teat and hold it there for a moment. If she doesn't kick, put the bowl under her, still holding it, and try to milk her.
- If necessary, you can use a hobble to help a goat learn to not kick. A hobble isn't painful or distressful, except in regard to its preventing her from doing what she wants to do! I don't like any of the hobbles that I've seen available commercially. Instead, I use a nylon web dog leash that has narrow webbing, making a slip loop from the handle end, slipping the loop around her fetlock (as shown in figure 8.24), and then clipping or tying the free end of the leash forward (if you are milking from the rear) or back and down on one side. After you have applied the hobble, retry the technique described above for making contact with the udder and starting to milk. I've occasionally had to apply two such hobbles, one on each rear leg, but rarely for more than a few milkings. Goats are smart, and they learn fast.

CHAPTER NINE

Offspring Management

Although goat farmers may not agree on the perfect approach to raising goat kids, we all agree on one thing: Kidding season is the most chaotic, exhausting, and challenging time of the year. Of course, it's one of the most fun and invigorating times as well. Other than puppies, goat kids are arguably the most entertaining and adorable animals we humans interact with. This chapter covers the many options

Figure 9.1. No matter what their ultimate destination, be it meat or breeding stock, Boer goat kids are adorable. At Sospiro Goat Ranch in North Carolina, these kids are wearing goat coats to compensate for cold weather and minimal housing facilities.

for raising goat kids, including options for feeding for optimal health and management strategies to make these goals sustainable for the farmer. I also describe the intermittent and important management tasks that you will be faced with, such as castration, identification, and disbudding. Again, remember that there's no one-size-fits-all scheme for rearing kids. Over time, and with some experimentation, you will develop a holistic approach that best complements your farm and goals.

Kid-Rearing Realities and Philosophies

Unless you have a very small farm and your goats are more pets than working livestock, you'll have to sometimes make difficult management decisions in order for the kid crop to become an asset or, at the least, less of a liability. Goats are prolific breeders and are more likely to have twins and triplets rather than a single baby. Some breeds handily have litters of four, five, and sometimes more. Breeds with ancestors that evolved closer to the equator, such as Boers, Nigerian Dwarfs, and Nubians, are the most likely to produce large litters. The more kids are born, the more natural mortalities will occur. When survival rates are high, though, there is greater stress on the entire kid population. Although it is a tough task to undertake, creating a plan for how to deal with the inevitable deaths is something that a proactive goat farmer accepts as important and necessary. In addition, the health of the kid crop depends on managing the population in a way that minimizes stress on the kids and makes management more efficient for the farmer—in other words, having a plan for how you will disperse extra kids that aren't going to be an asset for the farm.

Mortality

Even the best of farms will lose a percentage of the kid crop to difficult deliveries, illness, disease, or injury. Also, a few kids will be born dead. Most goat farmers feel that an acceptable rate is 5 to 10 percent, depending on whether the mothers are managed more closely or unobserved out on range during pregnancy and delivery.

For example, if 100 kids are born on your farm in a year, the loss of 5 to 10 from birth to after weaning might be considered acceptable. One year we lost 20 percent of our kids, but we usually average a loss of 3 to 4 percent. The reality on some farms, however, might be grimmer. If the management approach is extensive, with does delivering unobserved, the rate of loss is likely to go up, as least in the short term

The Unsavory Reality of Unwanted Bucklings

It wasn't all that long ago that unwanted newborn buck kids on dairy farms were regularly dispatched—that is to say, killed at birth. This is shocking to many who are new to the world of dairy goats. The harsh reality of a dairy farm, however, is one of thin margins and the need to allocate often slim resources to the animals that have the potential of helping the farm succeed. Historically, some buck kids were destroyed by drowning. This practice should never be considered an acceptable option. There are only three humane means of euthanasia: use of a captive bolt, use of a gun, or use of barbiturate overdose. Even the newborn goat with no viable or valuable future deserves one of these methods. As an alternative, I encourage the development of goat meat options by both goat farmers and entrepreneurs (see the Hill Farm Dairy profile on page 54) for these kids. Death is still the result, but the kid's life will have had an outcome that makes the resources—namely, the food and energy required to support a pregnancy—purposeful.

until the genetics of the herd are strong for mothering skills. Extensively managed farms may also have greater predation losses. They may, however, have fewer losses due to illness from diseases such as coccidiosis that are common in intensive management conditions. However mortalities occur on your farm, have a plan for how the deaths will be managed. As I mentioned in chapter 3, at Pholia Farm we deal with these losses through composting.

Kid Population Numbers

How you view your kid population will depend entirely on what kind of goat farm you are managing. A goat dairy will see most buck kids as a drain on resources (see The Unsavory Reality of Unwanted Bucklings sidebar on page 220), while a meat goat operation will see them as assets. Goat mothers that have large litters may not be able to adequately care for all of the kids. On an extensively managed farm, this problem is likely to resolve itself, because the less viable kids will more likely perish (and indeed, this might be considered a good management practice—the farmer likely only wants to allow the strongest genetics to survive). Farms with more hands-on management might intervene and hand-raise the surplus kids. When not raised extensively, the kid crop must be managed in a way that prevents overcrowding and the conditions of stress that would otherwise occur—in other words, providing ample pen space, a rigorous pen cleaning plan, and a feeding schedule that limits gastric stress and gorging (discussed later in this chapter).

Deciding how many kids can healthily occupy a pen involves balancing the animals' needs with the resources you have available—including space, time,

Figure 9.2. Angora kids are about as irresistible as it gets, making it fairly easy to find new homes for those not retained in the herd, a management advantage in most cases. *Photo by Ignite Lab/Bigstock.com*

and budget. Ideally kids will be raised in peer groups (also referred to as contingents) in pens that allow them adequate space to behave like normal goat kids—meaning they have room to run and cavort, can be kept clean, have exposure to sunlight and protection from the elements, and have good ventilation. If space is limited, then the kidding season should be structured so that kids are born in waves that can move through the available facilities without stressing the resources. You will have to think backward when considering this. For example, if you know that only five to seven kids can comfortably occupy the newborn pen where the kids spend their first week, then you should aim to have only three or four does kidding per week. As an example of pen size, a 12 × 12 feet (3.7 × 3.7 m) pen with no outside yard might only accommodate 10 standard-sized kids age one or two weeks, and 3 or 4 up to weaning age. With an outside run available (and remember that runs should ideally be longer than they are wide so that the kids can work up a good head of steam) it can probably accommodate 10 until weaning. At Pholia Farm we are lucky enough to have a person who takes, for a very small purchase price, all of the pet-quality doe and buck kids when they are a few days old. This immediately opens up more space, and more of my time, for the dairy-quality kids that add value to our farm and program.

Kid ID and Tracking

Many large goat farms don't keep track of the lineage of their goats; they find it too daunting a task, and it would require a big change in how the farm is managed. I believe that the results of tracking parentage—the ability to breed for improvements—pay for themselves, but I understand the difficulty in such a paradigm shift. For most small to midsized farms, however, keeping track of the kids is paramount. You must have a system of identifying them before they are separated from their dams, which usually means before they are permanently identified with a tattoo, microchip, or ear tag. Depending upon the size of your herd, there are many options for accomplishing this. One is some sort of neck band, such as a colored

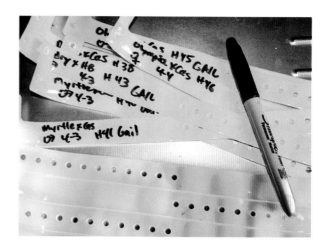

Figure 9.3. Vinyl party wristbands make great temporary IDs for kids. The bands fit well around the kids' necks. They last for several weeks and then tend to break off. (These kids weren't all named "Gail" but were being picked up by a woman with that name!)

string that is coordinated with the same color tied to the mother's collar. Some companies manufacture kid collars made of adhesive paper or Velcro with a section that can be written on. I purchased a box of party wristbands that are sturdy enough to last until kids are at least a month old, by which time they will have been tattooed.

As soon as kids are born it is helpful to jot down the statistics of their birth. If you don't have very many kids during a season, a calendar might be all you need to keep track of the important info. I have devised a chart with spaces for all the information I find vital, including vaccinations, registration number, and the name of the person who purchased the kid. I print out the chart (there's a copy in appendix C) on sturdy cardstock paper. I keep the previous year's charts in a file, and they have helped many times over the years when someone calls saying they have a goat from our farm, which by then I likely have no memory of! If a kid loses a collar, the color description can be extremely helpful in identifying it, along with its tattoo.

Permanent ID

Goats can be permanently identified by ear tag or tattoo. Microchipping is also an option but usually must be done in conjunction with one of the more easily visible forms of ID to comply with federal and registry requirements. Federal requirements in the United States are becoming more stringent regarding goats being moved across state lines, and export regulations usually require extensive documentation, including registration paper and identification. Most countries, even Canada, allow goats to enter only if the goats originate from a farm that is part of the US scrapie program. (See the Scrapie entry on page 301 for more information.)

Ear-tagging works well for goats on range and in less intensively managed situations. Ear-tag numbers that comply with the rules for transporting animals will include your facility number and an individual animal ID number. They can be supplied by the Animal and Plant Health Inspection Service (APHIS)—a part of the USDA—or your state veterinarian office as a part of the scrapie control program. The USDA tags have a unique herd number (by state), a unique animal ID number that the producer requests, and the USDA logo.

You can also purchase tags, specifying custom features, including color. (See appendix D for suppliers of tags.)

Richard Johnson of Lookout Point Ranch (see the profile on page 26) shared the farm's preferred ear-tag method: "We double-tag, with a USDA tag in one ear and a plain non-USDA tag in the other ear. Both have matching numbers. The non-USDA tag is easier for us to read, as the numbers are larger." Richard notes that if an animal loses its USDA tag, which tends to happen every few years, the tag must be replaced with one that has a new unique number, because the USDA doesn't issue replacements. "We add an 'A' at the end of the former number on the replacement, making it unique," Richard explained. "We now clip a radio chip tag [USDA FDX EID tags are available online] on the bottom side of the plain tag for use in the corral with a wand reader." He added that each year, they change the color of the tags they use for doelings. That way they can easily tell the age of does even from a distance. They use black tags on all males.

Ear tags are usually attached just after birth utilizing a special tool. Many ear tags have a distinct front and back face, containing different information. The two faces snap permanently together when applied. Before applying tags, clean the kid's ear with rubbing alcohol and a cloth or cotton ball. Affix tags well into the ear and between the ribs of ear cartilage.

Most dairy goat owners do not like using ear tags, and goats in the show ring never sport ear tags. Dairy producers, whose goats are usually more intensively managed, find that goats chew on the tags and will even pull the tags out of other goats' ears; even in extensively managed situations tags can break or be torn from the ear. For kids that will be registered with the ADGA, a series of digits that identify the farm is inked into one ear (or side of the tail webbing for LaMancha-type goats), and a series of digits unique to that kid is inked in the other ear (or other side of the tail webbing). The series for the farm must be accepted by and recorded with the ADGA. It is typically inked on the goat's right side. The animal's ID is on the left. Usually the animal ID will begin with a letter suggested by the registry to be used for any animals born that year, followed by numbers indicating the order of birth. These guidelines are optional, though. You can use any combination of letters and numbers, as long as each animal's ID is unique.

Goat kids can be tattooed as soon as their ears are big enough. I prefer doing it at the same time as disbudding (see Disbudding and Dehorning on page 225), because I sedate and provide pain medication—getting the entire uncomfortable session over at one time is quite appealing to me, and I would like to think it's better for the goat. It's best to tattoo a goat before you register it. If you register an animal and simply assign the tattoo that you intend to use and then make any error in how the tattoo is completed—even reversing ears—you will have to resubmit the registration papers. Trust me, it will happen!

Tattoo equipment can be purchased through most goat supply companies. Even if you have very small goats, I don't recommend the smallest tattoo size available, because it's difficult to read in the adult goat. If possible, buy two sets of pliers and keep your herd sequence set up permanently in one. That way you don't have to change it out every time you apply an individual goat ID tattoo.

STEPS FOR A SUCCESSFUL TATTOO

If you are new to tattooing, practice first using the pliers on paperboard, such as a cereal box. It feels about the same as tattooing an ear. When you feel ready, follow these steps to tattoo an ear.

Step 1. Administer tetanus antitoxin unless the kid is already covered by a previous tetanus toxoid vaccine.

Step 2. Clean the ear with a cotton ball or piece of paper towel lightly saturated with rubbing alcohol to remove any dirt before tattooing.

Step 3. Double-check the accuracy of the tattoo numbers by first clamping the pliers on a piece of paper—some people clamp it on the kid information card as a way of checking the number and documenting that the kid has been tattooed.

Step 4. When the rubbing alcohol on the ear has dried, line up the tattoo sequence with the long edge of the ear, making sure to avoid the ear's ridges. See figure 9.4.

Step 5. Clamp down hard and fast, then release.

Step 6. Examine the holes. They should be clearly visible and slightly red; there might even be a bit of blood. Occasionally a hole will have opened through to the back of the ear and there will be blood there. That is okay! Better too deep than too shallow.

Step 7. Place a bit of tattoo ink on the handle of an old toothbrush, your finger, or the back of a spoon.

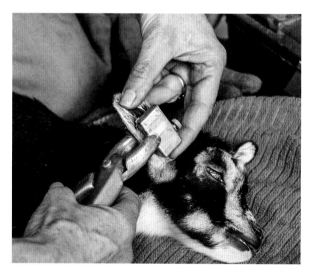

Figure 9.4. Tattooing, step 4.

Figure 9.5. Tattooing, step 7.

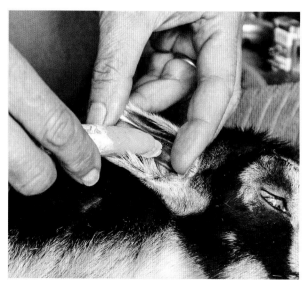

Figure 9.6. Tattooing, step 8.

Green ink shows up better than black or white ink (trust me, I've tried them all). See figure 9.5.

Step 8. Rub the ink thoroughly into the holes. If blood is present, wait about five seconds, and then rub it again to make sure the blood doesn't wash out the ink.

Step 9. Sprinkle the inked area with a bit of talcum powder or cornstarch. This absorbs the extra ink so that it doesn't rub off in unwanted areas—such as on your clothing or on another goat's nose. See figure 9.7.

Disbudding and Dehorning

Horns on goats are beautiful and give the goat a noble, classic look. They also serve a purpose—they help goats radiate excess heat and can be useful when defending themselves against predators and scratching their backs. Goats that are managed mostly on open range such as many herds of meat and fiber goats can benefit from having horns for those reasons. Also, horns can serve as handy management accessories for the farmer—you can grab them to handle the goat. Pack goats are often left horned so that they will be more able to keep themselves cool when working. But unless every member of the herd has horns, they can be a hazard and a liability. (The benefits of horn and scur control on adult goats was discussed in chapter 7.)

Disbudding is a procedure in which the early horn growth, called the bud, and the cells surrounding it are destroyed in order to prevent the growth of a horn. Disbudding is usually done using a heated tool called a disbudding iron. Chemical pastes, called caustic pastes, should not be used, because they are likely to end up smeared into the eyes or mouths of other goats as the kids interact.

Dehorning refers to the removal of horn growth that is too advanced for disbudding treatment to work. Dehorning can be done on an adult goat, but it brings the risk of infection and shock from the surgery. Banding of the horn—placing an elastrator band designed for castration on the base of the horn—isn't recommended. The band easily slips up the horn, rendering it useless. When it does stay put and successfully cuts off the circulation, there is likely to be pain over the long period of time that elapses before the horn comes off.

Typically, registered dairy goats that have horns cannot be shown at registry-sanctioned events in the United States. While you may not plan on showing your goats, should you ever wish to sell a horned goat, a prospective buyer may consider the horns a flaw because options for showing the animal will be limited. Although adult goats can be dehorned, it is a high-risk procedure for the goat. For most dairy goat breeders, disbudding—the destruction of growing horn tissue at a very young age—is the best choice.

Disbudding must be done with a properly heated iron. The greatest risk of damage comes from an iron that isn't hot enough and must therefore be applied for too long, which can cause brain damage—this is called heat meningitis. I have seen this happen only one time, when someone who was just learning over-applied the iron. When it does occur, the kid seems fine until a couple of days after the procedure, when it may have trouble swallowing and become lethargic and disoriented. Afterward, the animal may recover and become normal again over time (the one I observed took about 10 days to return to normal with

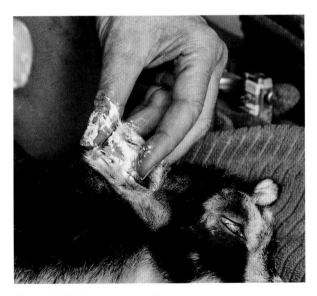

Figure 9.7. Tattooing, step 9.

Avoid Rejection After Disbudding

When a kid is returned to its mother just after disbudding, some does find the scorched odor confusing and disorienting, sometimes to the point that they will reject the kid. To avoid that, when you return the kid to the mom, distract her attention—a handful of grain can help—and then place the kid so that it's under the udder, facing away from the doe's head. When she reaches back to smell the kid, she'll be able to identify it from its hind end odor. By the time it's done nursing, it will again have the doe's scent on its head.

treatment). Antibiotics and IV fluids, or tube feeding, are likely to be needed.

To gauge when the iron is hot enough, you can carefully hold it about 6 inches (15 cm) from your face; at this distance, it should feel almost hot enough to burn your skin. The technique I use is to apply the iron in a continuous application of 5 to 10 seconds and then check for the telltale copper ring, which indicates that the burn has been of sufficient time and heat. Another technique is to apply and lift, apply and lift, checking as you go. Either way is effective.

To Drug or Not to Drug

I have been using sedation and pain control when disbudding and tattooing goats for over a decade. During that time I have administered a drug called xylazine to at least six or seven hundred kids. This drug, which is legal for a vet to prescribe, is quite dangerous if not dosed properly, and in rare instances humans will have a severe reaction to the drug if they accidentally inject themselves. For these reasons, most veterinarians are reluctant to prescribe it to producers. In addition, many people feel that the risk is too high and it shouldn't be encouraged. I am sure that my background as a nurse, which requires a course in how to calculate and properly dose medications, was of great help in my confidence, and probably in reassuring the vets I have worked with as well!

Why do I like using this method? Yes, it takes longer and has more risks, but I feel much more at ease because the animals are not expressing pain. I can absolutely do a better job tattooing and disbudding on my own. Many producers believe that the stress and pain experienced by a kid during these procedures are limited to a brief time and as much due to being restrained as having intense heat applied to their heads and sharp needles piercing their ears. However, all research (on calves) indicates that stress hormones are the most elevated and heart rates stay elevated longer when no pain management is used.

In chapter 7 I included a quick course in pharmacological math and how to calculate dosages. If you are interested in convincing your vet to let you try sedation, please do study up on how to give tiny dosages very accurately. You should also know how to administer the reversal medication (given IV) for xylazine. I have never had to use it, but it's important to keep it on hand. It is routinely given to older, ruminating animals, because staying sedated for too long brings about its own risks. For young kids who aren't yet ruminating, the risks are different and not related to bloating of the rumen.

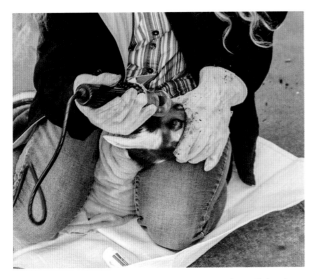

Figure 9.8. When sedation isn't used during disbudding, kids need to be restrained. This kid has been wrapped in a blanket and then straddled.

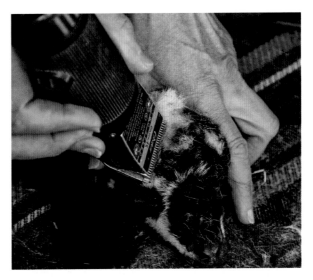

Figure 9.9. This sedated kid is getting a buzz cut around the horn buds to make them more visible. Shaving the area also reduces heat loss from the disbudding iron, and reduces the smoke produced during the procedure.

Figure 9.10. The hot iron is applied with firm pressure, and the pressure (not the iron itself) is rotated in a circular pattern, without lifting the iron, for seven seconds. Then the iron is lifted and its edges are used to burn a cross onto the top of the bud. Afterward, an ice pack is applied to the bud. Another approach, also correct, is to apply the iron firmly and without rotation for a few seconds, lift and inspect for the copper ring, then apply again, repeating until you observe the copper ring.

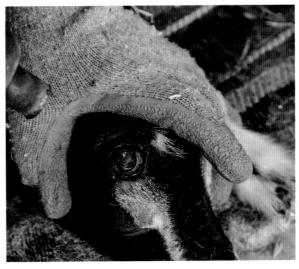

Figure 9.11. This is a proper disbudding burn pattern. Notice the desired copper-colored ring.

Kids must be restrained when disbudding to prevent the risk of burning and severely injuring them or yourself. There are several ways to do this. For those working alone with an unsedated animal, the use of a *kid box*, into which the kid is placed with its head protruding, is common. Kid boxes can be purchased or made from easily available instructions online. Fully awake kids can also be wrapped in a blanket (as in figure 9.8) and then straddled. If a helper is present, the kid can be held in the helper's lap with the person doing the disbudding kneeling in front. Be sure the helper's lap and hands are protected from an accidental burn by the extremely hot iron.

Unless a goat kid is polled (genetically lacks the ability to grow horns), it should be disbudded as soon as the buds on the forehead where the future horns will grow can be detected. For buck kids this is almost always within a few days of birth and for doe kids just under a week, although some will grow more slowly. When done properly, disbudding will prevent the growth of any horn tissue in does, but bucks that remain intact (capable of breeding) may develop some growth of horn, called scurs, even after disbudding. Such scurs are generally not difficult to manage. (Instructions for cutting off a scur can be found in the Scur or Horn Control section on page 172.)

In some countries, laws require that disbudding be done with the use of pain management drugs, and the procedure is often done with the animal entirely or partly sedated. In the United States producers of all horned livestock often perform the procedure (along with castration) without any pain control or sedation. This approach is falling under more scrutiny, and I am a proponent of utilizing medications to reduce pain and the stress of the procedure (see the To Drug or Not to Drug sidebar on page 226). If you have only a few goats you might be able to budget for a vet to do the entire procedure, but because the timing of disbudding is very important to its success, this may not be realistic—if your vet isn't available and your goat is at the proper stage of development, then the task might fall to you.

Castration

Unless a buck kid is going to be kept for breeding or butchered at a young age, castration is an important, if unpleasant, procedure that must be performed. The timing of the castration will depend on the destination of the animal—pet, plate, or pack. Pet wethers and pack wethers can be castrated later, between six and eight weeks of age, in the hope of allowing for additional growth of their urinary system so that urinary stones, calculi, will more likely be able to pass out of the animal without difficulty. When a kid is destined for the plate it's important to consider the market in order to make the decision. If the kid is to be butchered before eight weeks of age, there is no reason to castrate. Goats destined for ethnic markets where intact males are preferred don't need to undergo the procedure. If kids will be grown for the meat market as wethers, then castration can take place between four and fourteen days to minimize stress on the animal and maximize growth.

There are three main ways to change a buck into a wether: surgery or "cutting," banding, and emasculation. I have taken part in all three of the methods and feel that each has equal points for and against it. The age of the kid, pain control, and protection against tetanus should all be considered. Before your first batch of kids has been born, you should decide how you'll approach this and be ready with the right tools—and resolve.

Surgery

Sometimes called simply cutting, this method is usually done at a very young age to minimize the risk of bleeding. If performed on an older animal, the likelihood of unstoppable bleeding increases; such a procedure should be done only by a veterinarian, who can also give the goat proper pain control.

The scrotum is cleaned with a Betadine surgical scrub; the bottom third of the scrotum, with the testicles pushed up above it, is cut off with a sterilized scalpel (there will be a little blood); and the testicles are pulled down and exposed. The spermatic cords are then either scraped with a sterilized knife until they separate or crushed with an emasculator (see the section below). A third option: Grasp the testicles, and then quickly pull them while you hold a finger at the inguinal canal—the small opening in the abdominal wall where the spermatic cords emerge (to keep from putting pressure on the opening that might make the opening bigger)—to break the spermatic cords. This last method of breaking the cords with a quick pull is only done on very young kids.

If the kid's mother was immunized for tetanus before the kid's birth, then the kid will have immunity. If not, administer tetanus antitoxin. If the weather is warm and flies are present, apply fly repellent around the wound.

Banding

In the United Kingdom, banding is not allowed because of concerns for animal welfare. In my opinion, though, when banding is done at the right time—namely, when the scrotal size is neither too large nor too small—there is minimal pain associated, especially when pain medication is given.

Standard-sized goats are usually ready to band by about three weeks of age and Nigerian Dwarfs, five to six weeks.

There are two types of banding tools. One applies a one-size-fits-all heavy-duty elastic ring around the scrotum between the testicles and the abdomen. Because it is one size for all, the age at which it will properly fit the animal depends entirely on the size of the scrotum, making the timing of the procedure a bit less flexible if you're trying to delay the castration for maximum urinary system growth. The ring stops the blood flow to the scrotum, causing the testicles to die and, after a period of weeks, fall off. The second type of tool uses an adjustable band that you can custom-fit to the animal; Callicrate is one brand. This allows for delaying castration until the last possible moment—before the animal is capable of breeding but when the urethra has a larger diameter. It also works for castrating adult bucks.

Before banding a kid, be sure he is up to date with tetanus vaccination. Giving antitoxin isn't useful because it's effective only briefly and the wound from

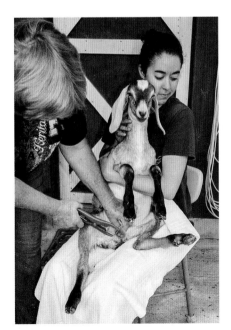
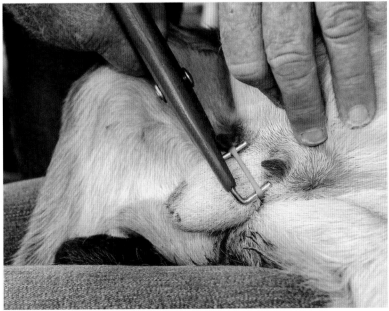

Figure 9.12. Daniella Rojas holds a buck kid about to be banded by Gail Swanson (*left*). The band is positioned so that the testicles are well below it and the teats well above before the clamp is allowed to close and the band slid off the prongs (*right*).

castration is present over a longer period of time. Aspirin can be given half an hour ahead of time to help with pain, and the area can be numbed with ice. A veterinarian will use lidocaine to numb the testicles. Clean the scrotum just above the testicles with a Betadine solution and soak the band in the iodine or alcohol. Have a helper hold the kid on their lap with the kid's back to their chest, grasping a front leg and back leg in each hand. The kid will struggle a bit and fuss but will quiet down fairly quickly. The band is then slipped over the prongs of the applicator and stretched wide so that it can pass over the testicles. Once over them, manipulate the testicles to verify that both are in the scrotum, and then release the band by allowing the tool to close. Now you can slide the band off the prongs. Be sure you are just above the testicles and not too close to the abdominal wall. If the band is too close to the kid's belly, a teat might get caught in it, or the size of the wound might be increased by pulling more skin from the belly into the band—rather than just the neck of the scrotum. Once the band is off the tool, open the tool again and remove it from around the scrotum.

Even when the band is applied properly and at the right stage of development, the kid typically shows signs of discomfort until the loss of blood supply numbs the area. When the band is tight enough, numbing occurs rather quickly, often within minutes. There is still discomfort due to the pressure of the band, though, so banded kids will usually walk "funny" for a day or so.

Emasculator

The word *emasculator* brings thoughts of torture chambers or derogatory, man-bashing remarks. And indeed, it is quite dramatic in appearance and effective in its use. Burdizzo is a brand made specifically for goats. The tool is used to crush the spermatic cords, along with their blood supply, at the point where they exit the abdomen and enter the scrotum. The fact that the procedure doesn't create an open wound assures a much lower risk of tetanus.

This procedure can be used on bucks of any age. You can convert a buck formerly used for breeding to

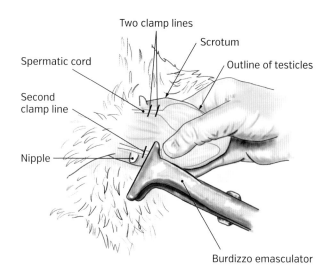

Figure 9.13. Proper position of the emasculator so that it will crush the spermatic cords. Two clamps are made on each side.

a wether if you like. The disadvantage of this method is that the scrotum does not disappear, but instead shrinks slowly if the procedure has been done properly, meaning that for some time the animal might be mistaken for an intact male, at least from a rear view. The older the buck is, the more tissue is left from the larger scrotum. The mature buck's secondary sex characteristics, his extensive beard growth, smell, and behavior will gradually cease.

Anesthesia is not usually used when this procedure is performed. A local lidocaine injection, usually administered by a vet, can provide pain control. A producer can give the buck aspirin if desired. Fortunately the procedure is fast, and goats appear to recover quickly. If the goat is young, position it as for banding (see figure 9.13). If it's older, restrain the animal on its side. You may need to use one or more tethers to hold the front and back legs on each side together.

The emasculator is applied to one side of the scrotum at a time. Manipulate the testicles so that they are well down in the scrotum and then locate the spermatic cord for one testicle and push it out to the side of the scrotum. Continue holding it to the side, and line up the tool so that the cord is located in its jaws; clamp the tool shut. Recommendations vary about how long to leave it in place, but a few seconds should suffice. Make a second clamp ⅜ to ¾ inch (1–2 cm) above the first. Repeat the process, clamping twice, on the other side. When you're done, you should be able to feel the indentation and separation of the cords. Clean the site with an iodine solution.

Vaccination

Vaccinations are given to adult goats and kids to prevent the occurrence of deadly and costly diseases. There aren't many vaccines specifically designed and approved for goats. In the United States the only routinely given and approved vaccine for goats is for *Clostridium perfringens* types C and D (the cause of enterotoxemia, also known as overeating disease) with tetanus toxoid, often referred to as CD&T. Vaccines labeled for sheep and dairy cattle can be given to goats when labeled and approved by a veterinarian. The most common of these are for contagious ecthyma (soremouth or orf), caseous lymphadenitis (CL), rabies, and chlamydia. In areas where rabies or other diseases are a concern, those vaccines can be given. A veterinarian should be able to help you make the right choice.

On most goat farms, vaccination is a regular part of management. There are many diseases for which I would vaccinate if a vaccine were available, such as Q fever and paratuberculosis (Johne's disease). We have chosen to not give our goats CD&T, but we do use tetanus antitoxin (not a vaccine but a treatment) and tetanus toxoid (a vaccine) when appropriate. As with other management decisions, please make your vaccine choices based on your vet's input and your own comfort level. (See the Enterotoxemia entry on page 258 for more on that disease.)

If the vaccine for *Clostridium perfringens* and tetanus is to be a part of your program, ideally does should receive their booster shot four weeks before giving birth to provide immunity for the newborn kid. Kids are then given a booster at one and two months of age, then annually. For other vaccines, consult with your vet to decide whether your herd is at high risk if not vaccinated and to learn about the pros and cons of vaccination—including the possibility that vaccinated animals may test positive for some diseases because of the presence of antibodies. This may be an issue if someone else (such as a potential buyer) tests for that disease and isn't aware or convinced that the animal has been vaccinated.

Feeding Kids— Newborn to Weaning

Decisions regarding feeding kids often hinge on multiple factors, some within and others outside of your control. For example, organic and holistic ideals may indicate that dams raising their own kids is the most natural approach, but presence or danger of a disease such as CAE would mean that dam-raising is, in fact, an inhumane choice because it might well result in the kids contracting that serious disease. You need to find a method that works well for you—in all aspects. At Pholia Farm, we experimented over the years with many methods of feeding kids, and I still haven't settled on the perfect way! (Visit the Pholia Farm website to read more about the many combinations of feeding methods I have tried.)

The main choices to consider include dam-raising, bottle-raising, free-choice feeding (a form of bottle-raising), and combination feeding. As you consider what type of feeding system to try and when and how to wean kids, it's important to first understand how the digestive system of a kid works, because it is not the same as that of an adult goat. It's a system in transition and needs to be supported as such to create the best possible healthy working goat.

No matter what feeding approach you take in regard to milk, young kids should have access to water, minerals, free-choice grass or legume hay, and ideally fresh forage. In addition, they should have clean bedding material—which they will spend a good deal of time nibbling on. They rarely will start drinking water until their milk source is limited, but it's a good idea to provide it anyway. I don't usually offer baking soda

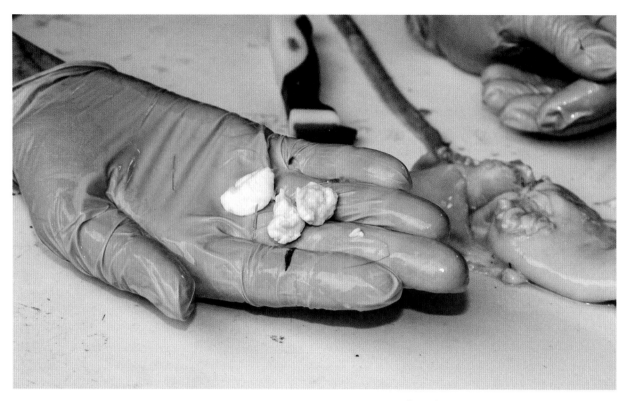

Figure 9.14. The baby goat is truly the first cheesemaker. This abomasum [*right*] from a two-week-old kid contained these amazing cheese curds.

until they are eating a lot of solids, usually by age four to six weeks. To encourage consumption, hay should be put out in tiny amounts and replaced daily.

The Kid's Unique Digestive System

When a goat is born, its rumen is tiny and non-functioning. In the adult goat, the rumen is the largest stomach chamber, but in the newborn, the abomasum or true stomach is the largest. When the kid nurses its mom or a bottle, or even drinks milk from a pail, a reflex occurs that diverts the milk from the rumen and reticulum and sends it directly into the abomasum. The tube, also called the esophageal groove, forms from tissue in the reticulum and basically extends the esophagus so that it bypasses the first two chambers (see figure 4.9 on page 74). This is important, because the non-functioning rumen would not be able to process the milk. In fact, milk cannot be processed well even in a mature,

Why Do Kids Shiver After Feeding?

Kids that are being bottle-fed or intermittently fed on their moms will sometimes shiver after filling their tummies. If you're bottle-feeding, you might think that perhaps you didn't warm up the milk enough. But more likely, it is the simple response of the goat's physiology to suddenly having a full stomach. This demands a shift in the blood supply so that the stomach can go to work processing the meal. The result can be the kid feeling a bit cold and shivering. More frequent feeding of smaller meals can prevent the shivering response.

functioning rumen. If a kid overeats and fills its abomasum to the point that milk backs up into the rumen and reticulum, that kid is risking death—the milk will basically rot in the rumen.

The milk flows quickly through the omasum into the abomasum, where the baby goat's stomach quickly turns it into a coagulated gel. You might even say that the baby ruminant is the first cheesemaker. Enzymes in the kid's stomach, chymosin and pepsin, are responsible for this gelling. (These same enzymes, especially chymosin, are harvested and used in traditional cheesemaking.) Coagulating the milk in this way allows more time for the kid's stomach to slowly absorb the nutrients. A premature or runt kid lacking the proper enzymes, or enough of them, may suffer with white, thin diarrhea. If the situation isn't corrected, either by adding enzymes to the milk or feeding yogurt (which has already been partly broken down by bacteria and is therefore easier to digest), the kid will starve.

In a natural setting, goat kids nurse in frequent small feedings. However, most farmers who raise kids apart from the dams feed the kids only two to three times in a 24-hour period, simply because of logistics. This can lead to gorging, overfilling of the abomasum, and stress.

Within the first week of life, a kid will begin to nibble and chew on bits of feed, bedding, and even dirt. Bacteria from these sources as well as from the mom's teat, udder, and saliva all begin to contribute to populating the upper digestive system with the bacteria needed for the rumen to function. Rumination usually begins around the end of the second week. As the rumen grows, it pushes the abomasum from the center left of the abdomen over to the right, where it is positioned in an adult goat. The enzymes also begin changing in the stomach; pepsin becomes dominant.

Until its rumen begins to work, a kid depends on vitamin B from its mother's milk, which is present in the milk in a decreasing supply from the time she kids. It disappears from the milk at roughly the time that a kid is ruminating well.

Vitamin C is also present in raw milk and is not made by a kid's body until it is several weeks old. These things are very important to consider if you aren't feeding a kid raw milk from its own dam, whose milk supply will be perfectly balanced for its needs.

Dam-Raising

When herds are free from diseases that can be passed through the milk or by contact with the feces of adults, it's safe to allow kids to be fed by their mothers, or dam-raised, for the most holistic approach to nutrition and rearing. Kids raised with the herd often learn how to browse and forage better as they watch and copy the adult goats, especially their moms. They might have an increased activity level and exposure to sunlight (vitamin D). They are also less stressed, because frequent feeding keeps the abomasum and intestinal tract working and prevents kids from getting so hungry that they gorge at the next feeding (risking overflow of milk into the rumen). It's also quite beautiful to see the relationship that develops between most mothers and their kids.

Dam-raised kids are much more likely to be a bit wild than hand-raised kids. If you have only a few goats and some time to spare, you can usually socialize dam-raised kids to accept humans. Depending on the kids' future careers, skittishness might be a liability. For example, dairy goat farmers need to work calmly with their goats on a daily basis. And wild goats are hard to accommodate if a producer counts on being able to catch animals easily and lead them by the collar. On some meat and fiber goat farms, less-than-tame goats can be managed through the use of chutes and pens that allow skittish animals to be moved about and allow farmers to perform necessary maintenance tasks, such as hoof trimming, without too much stress to either the farmers or the animals.

Pens that are shared by adults and kids must be "child-proofed" to prevent kids from escaping or becoming injured or trapped. Kids are likely to treat feeders as sleeping boxes, so dealing with

> Kids raised with the herd often learn how to browse and forage better as they watch and copy the adult goats, especially their moms.

Homeopath and Veterinarian Jovita Jokubauskienė
LITHUANIA

I met veterinarian Jovita Jokubauskienė in 2014 when I had the good fortune to visit Lithuania and tour the country, interviewing cheesemakers and visiting small dairy farmers. Jovita and her husband, Audrius Jokubauskas, raise dairy sheep and cows in a national park in southern Lithuania. Jovita has been a veterinarian for five years; she began practicing homeopathy shortly after becoming a vet. She works at a small-animal clinic and also regularly treats the sheep and cows at their farm and other farms. Goats aren't common in this small nation located in Central Europe along the Baltic Sea, but one of the wonderful things about homeopathy is that its principles apply to all species.

I asked Jovita what convinced her that homeopathy is an effective form of treatment. "I was looking for an alternative treatment for myself when I discovered homeopathy," Jovita said. "The results beat all skepticism about homeopathy that had been created for me during my veterinary studies. In addition, at the beginning of my career I had a case of a sheep mismothering after birth. She would push her lamb away and wouldn't let him suck. That meant I had to wake every two or three hours, walk to the barn during winter at night, and force the sheep to feed the lamb. But the homeopathic remedy Chocolate [*Theobroma cacao*, a remedy that is very hard to find here in the United States] solved the problem. Literally within seconds—just after giving one dose—the mother started calling and licking the lamb and let him suck. Wish I had a camera then. That was a miracle."

Jovita received her homeopathic accreditation through the British Faculty of Homeopathy and passed the exam (LFHom Vet). In Lithuania, as in many countries, there's no official recognition of homeopathy; nor are there official requirements to practice it. Also similar to other countries, it is small-animal vets and their owners who are much more likely to turn to homeopathic remedies. Jovita

Figure 9.15. Dr. Jovita Jokubauskienė with one of the dairy sheep at her farm in southern Lithuania.

is convinced of its usefulness on all livestock, though. She told me about how it had helped alleviate mastitis on a cow she was treating.

I asked her if a goat farmer can use homeopathy without much training and, if so, what some good remedies to start with might be. She said all goat producers can absolutely get started with just some simple recommendations. She suggested five emergency remedies to be used during kidding season or times of trauma for the animal:

- *Bellis perennis* is ideal for pelvic bruising, and pain, especially after delivery. (This remedy is not usually included in basic kits.)
- Carbo Veg can be used for collapsed, weakened animals and for cold animals. It's also good for newborns with breathing problems after birth.

- China (*Cinchona*) helps restore body fluids after loss due to diarrhea, giving birth, giving lots of milk, or dehydration.
- Arnica is for inflammation and pain of all kinds.
- Apis helps after spider and tick bites. It reduces edema, swelling, and pain. Use for mastitis when swelling is present.

You can order pet-focused remedy kits directly (online, by phone, or by mail) that are much more reasonably priced than buying single remedies. Some practitioners believe remedies packaged in glass vials are better than plastic, but Jovita says it's far more important to simply protect the remedies from light and heat. That's good news, because not only are plastic-packaged remedies more commonly available, but they are much more reasonably priced as well. Her last piece of advice is the best: "Find a homeopathic vet that you trust as well as like!"

feces in the feeders is a reality for the multigenerational goat pen. Predator protection is also a must, because kids are at much more risk from predation.

Although dam-raising avoids the emotional (and often auditory trauma to the farmer) of separating kids from their moms after birth, it only delays it until weaning time, when the kid's lungs are stronger and the bond between mother and offspring is tighter. Kids raised with their dams will remember them for many months after weaning. Although some moms won't allow the kids to nurse after a few weeks or more of being apart, others will continue to be available for nursing almost indefinitely. It's not so endearing to see a mom nursing her kid when you can't tell from the size which one is the kid! Meat goat operations typically send kids off to butcher concurrent with weaning, meaning the kids don't go through weaning stress. The mothers will experience some stress, but since they won't hear their kids calling, it will fade more quickly than if the kids were separated and within hearing distance. I discuss weaning options in more detail later in the chapter.

Hand- or Bottle-Raising

Even when dam-raising is the preferred choice, bottle-raising kids might be an intermittent or required option. Dairy farmers who for financial reasons might require the milk for themselves as soon as possible, those raising pet goats (including those who have purchased a kid to raise), and farms where milk-borne or direct-contact illnesses are a possibility are more likely to bottle-raise (also called hand-raise) their kids. Kids are most commonly raised in part or whole being fed from a bottle, bucket, or group feeding system. When kids are fed from a group feeding system, it's sometimes referred to as artificial rearing. As noted above, as long as the herd is free of disease the kids can nurse colostrum directly from their mothers and then be switched to a bottle. The length of time the kid might be left with the mom varies, from a day or so to more than that, depending on the needs and approach of the producer and what the milk will be used for. Producers selling milk for drinking won't want to use the milk, which may have some colostrum in it, for at least three days after kidding, while those using or selling it for cheesemaking are more likely to wait four to six days. (Colostrum isn't dangerous, but the natural antibodies it contains interfere with cheesemaking.)

WHAT TO FEED

If your herd is not on a milk-borne illness prevention program (previously known as a CAE prevention plan), you can feed a kid its own dam's raw milk. Simply save and refrigerate the milk, then feed it back to the kid. I advise never feeding raw milk from another dam, or mixture of other does, to a kid. Not only will it not have the right vitamin profile, but it could contain undetected microbes that will cause illness

Pasteurizing Milk for Kids

If you plan to feed pasteurized colostrum or milk to kids, follow these guidelines.

Heat colostrum to 138°F (59°C) for one hour. One way to do this (for small amounts of colostrum) is to preheat a thermos with hot water, then dump out the water and pour in the colostrum.

Heat milk to 145°F (63°C) for 30 minutes or 161°F (72°C) for 15 seconds.

later in the kid. The microbe of biggest concern in this regard is mycoplasma. Although you can't reliably test an asymptomatic goat for mycoplasmas (see the Mycoplasma entry on page 290 for more about this disease), limiting feeding of raw milk will at least limit the spread of mycoplasmas to only the goat's own offspring.

Bulk or comingled goat milk from an entire herd can be heat-treated using pasteurization times and temperatures and fed to kids. Heat treatment can be accomplished on a small scale on the stovetop or by using a home pasteurizing unit. I use an electric turkey deep fryer. I set a 5-gallon (19 L) pot filled with milk in the fryer, add water to the unit, and turn it on. You can vary the heat setting depending on how often you are able to stir the milk and verify that it isn't becoming too hot or heating too unevenly. The fryer heats the water, which in turn heats the milk in the pot. At a larger scale, a calf milk pasteurizer (or any pasteurizer) can be used. Calf milk pasteurizers are the least expensive type because they don't have to meet regulatory requirements for legal pasteurization.

As I mentioned earlier, keep in mind that pasteurization destroys some vitamins, so you should add supplements to pasteurized milk. Vitamin B and C

are both water-soluble vitamins that are present in raw milk and important until kids start producing their own. Because both of these are water-soluble vitamins, there's no risk for overdosing. I'm not very scientific when it comes to administering them. I typically add ground-up vitamin C tablets (equivalent to 500 to 1,000 mg of the vitamin) and 1 to 2 ml of vitamin B complex liquid injectable to a quantity of milk (or in our case yogurt) sufficient to feed 10 kids over the course of a day. I only add the vitamin B complex until the kids start ruminating at about two weeks of age.

Powdered goat's milk replacer is widely available and in use. Milk replacers aren't allowed in organic milk production in most countries. There are supposedly "organic" milk replacers on the market, but if you read the fine print on the label, you'll find statements such as "there may be an ingredient in this product that does not meet the National Organic Program standards established by the NOP, and should not be used on organic farms unless specifically authorized by your certifying agency."

I have yet to talk to any producer who has been totally happy with milk replacer. The most common complaint is that kids suffer from bloat (of their abomasum) as a consequence of consuming it. Some producers mix at least half pasteurized milk and half replacer with better success. The likely cause of bloating from drinking milk replacer is the presence of concentrated lactose from dried whey, usually the first ingredient listed on the label. Dried whey product, on the other hand, which is often the second ingredient, has had the lactose removed.

Nutritionally, milk replacers are formulated to meet the basic requirements of kids and supply the vitamins that are missing in pasteurized milk. Replacers are also rather expensive, however, so don't choose them simply because you think it will save you money.

I feed yogurt to our kids. I learned this from Jennifer Bice, the inspirational founder of Redwood Hill Farm and longtime goat breeder. Her company is famous for its goat's-milk yogurt. In their yogurt-making process at Redwood Hill, inevitably they end up with some quantity of yogurt that isn't

Getting Kids to Accept a Bottle

Once a kid has nursed its mom, it will be quite reluctant to take an artificial nipple. The same thing can happen in reverse—kids that are given bottle on the first day of life might not take as readily to their moms afterward. This happens with human infants as well and is known by the rather amusing term *nipple confusion*. It can be a very frustrating experience.

It takes some time and experience to become good at converting a kid to the bottle. Kids who have been handled a lot by humans are always easier to get on the bottle, simply because they aren't afraid of you, but you'll still need some strategies. Even if you don't plan on bottle-feeding kids, sooner or later one or a few will need it, because mother goats sometimes reject kids or don't produce enough milk to feed all of them. Here's the bottle-feeding process we use at Pholia Farm:

1. Give the kid a "diet day" during which you handle it frequently, but don't offer a bottle. If it's very hungry, it will be less picky about the nipple.
2. During the diet day, frequently check the kid for the "rooting reflex," which is when it wants to butt your hand when you rub or cup that hand over the top of its muzzle. This is a good indicator that it's ready to feed. This may take 12 to 24 hours.
3. Make sure the milk is very warm, about 104°F (40°F).
4. If possible, offer the kid a bottle containing milk from its own mother, to avoid flavor differences that might confuse or upset it.
5. Try kneeling on the ground with the kid positioned between your thighs and its head facing away from you.
6. With your non-dominant hand, hold the kid's head so that your palm covers the eyes and your fingers are on one side of the mouth, your thumb on the other.
7. If the kid starts rooting and lipping, work the nipple into its mouth using an in-and-out motion—don't just thrust it in as far as it will go. Believe it or not, kids are more eager to latch onto something that is moving away from them!
8. If the kid starts sucking right away, great. If not, try steps 6 and 7 over again. Sometimes a little taste of milk will prime them, sometimes it takes more waiting.
9. Above all, if a kid is upset, avoid forcing open the mouth. This only seems to make kids more stubborn.
10. If you get desperate, syringe-feed the kid enough milk (the same way you would give an oral medication) to satisfy it for the night and then try again the next day. It won't let itself starve to death! You can also offer a bowl of milk to see whether the kid will drink on its own.

---

salable, although it is still perfectly good for consumption. They make good use of this yogurt by feeding it to goat kids at their farm.

I like feeding straight yogurt to kids because it contains probiotics, which pasteurized milk lacks. Also, yogurt is acidified, so it can be fed in a free-choice bucket system without much concern for contamination by spoilage bacteria. And its thick consistency slows down consumption by kids on the free-choice method, which helps prevent digestive problems. During the winter, I keep enough goats in milk to make enough yogurt to stockpile in several deep-freeze chest freezers for use in the spring. When needed, I thaw or partially thaw the yogurt and feed it to kids without diluting it. Remember, yogurt might be thicker than plain milk, but not because any moisture has been removed. Diluting it with water would not be a good idea because the kid would become full before it could take in all of the nutrients needed.

BOTTLE AND BUCKET OPTIONS

Bottles of many kinds work well for feeding kids. Which type to choose depends on the type of nipple you're using, which in turn depends primarily on the size of the kids. Nigerian Dwarf goats, some small miniatures, and runts are likely to do best on a Pritchard or other flutter-valve-type nipple, which is designed to screw onto a plastic water or soda bottle. Some brands seem to fit better than others. I find that bottles from Coca-Cola products, including Dasani water, fit well. (To test the fit, attach the nipple and upend the bottle. If it leaks, try tightening the nipple more. If it still leaks, choose a bottle from another brand of water or soda.) These nipples must have the end snipped off before use. Be careful not to cut too

Figure 9.16. Kids on free-choice bucket feeders don't experience the stress that comes from getting hungry between feedings. Free-choice feeding has allowed us to do away with the use of coccidiosis preventive drugs, thanks to the reduced stress and the constant supply of food in their digestive systems.

far: 1/16 inch (1.5 mm) is usually plenty. Another commonly used type, usually labeled a caprine nipple, fits over the outside lip of many glass bottles that have unthreaded necks, such as soda, beer, and even wine bottles. This size works great for bigger kids.

Using a bucket feeder (see figure 9.16) is convenient for feeding larger numbers of kids. Sometimes called a lamb bar, a bucket feeder is easy to make, or you can purchase different configurations, sizes, and systems. Open-top bucket feeders are meant to be used for intermittent feeding and washed between uses. Closed-top buckets and automatic feeders can also be used for free-choice feeding, also called ad libitum (ad lib) feeding. If you plan to use a bucket feeder for scheduled intermittent feedings, be sure the bucket has more than one nipple slot per kid. Otherwise, the kids will crowd and bully, often butting nipples and the bucket and making a mess. For free-choice feeding, usually one nipple for every three to five kids is sufficient.

FREE-CHOICE BUCKET FEEDING

The free-choice, cold-milk method not only saves a great deal of labor but also prevents kids from overeating, avoids encouraging them to be noisy and pushy to humans, and keeps milk in the kids' stomachs, which prevents mental and physical stress and is more natural for their digestive systems. Cold milk (or yogurt, milk replacer, etc.) is used in these systems, for two reasons: First, it prevents the kids from continuing to suckle just because the warm nipple and milk are comforting. Second, it helps prevent the milk from spoiling or experiencing the growth of contaminating bacteria.

With this method, the buckets must never run dry. If the milk runs out, the kids may get quite hungry. Then, when you refill and rehang the bucket, the kids will gorge on the cold milk. This can cause a kid to go into shock and die (this happened once on my farm). When the bucket runs dry, or if a nipple gets plugged up, kids will chew the nipples out of frustration. Plan on checking the level of milk in the bucket multiple times during the day, until you get an idea of how quickly the kids are emptying it. It's also a good idea to check all nipples every time you refill or wash the bucket to make sure they are in good working order. Disassemble the bucket or feeder at least once a day for cleaning and sanitizing. In cold weather—below 40°F (4.4°C)—the milk will stay cold enough, but if the weather is warmer, add capped bottles of frozen water to the bucket to keep the milk chilled.

Bucket feeders with valve-controlled nipples installed at the bottom are a good choice, but be sure the nipple style (often designed for lambs) will work for your goats. Despite the claims of some companies about being able to modify a Pritchard-style teat to work with this system, I have not found it to be successful. I did, however, create a hybrid in which I use the ball valves installed above the level of the milk, extra gasket rings, a Pritchard nipple with the flutter valve sealed with ABS cement (used for gluing plastics), and a latex tube over the valve stem and extending down into the milk. This improvised system works very well.

Getting kids started on free-choice bucket feeding requires some extra thought and work. If they have not been nursing their dams, it is relatively easy to start them on warm colostrum in the bottle first, then gradually switch to room-temperature bottles while they are in a pen with kids just a bit older that are already feeding from the bucket feeder. The new kids are quite likely to respond to the sight of other kids nursing the bucket feeder by trying it themselves. You can help a bit by holding a kid (by the body, not the head) so that the nipple touches the sides of the kid's face.

If there are no kids already trained to the system to serve as role models, I do as follows: When kids eagerly seek out an offered bottle, I switch to cooler and finally room-temperature milk at each feeding over a day or two. Then, at a morning feeding, I give them only half their usual amount in the bottle. Every hour or so after that, I take them to the bucket feeder and encourage them to eat a bit. They usually figure it out after just a few times. The secret is to have them hungry enough to be motivated to try it but not so hungry that they overeat.

Kids on the free-choice system grow well and don't usually become overweight. And because they are not overfull, they are willing to start eating forage and a bit of grain at an early age, which is a must for rumen development and preparing for weaning.

Combination Raising

Combination rearing can offer the best of both worlds for some farmers. Such approaches offer the kids the wholesomeness of the mother's milk while giving the farmer reduced labor and a greater likelihood that the kids will be easy to handle. There are two main ways to share "custody" of goat kids with their mothers, but you can hybridize your own method as well.

One approach is to separate kids from their mothers for a portion of the day so that the dams can be milked, and then the kids nurse the rest of the day. This works as long as there is enough milk for everyone, which depends on how many kids a doe has and how many of them remain dependent on her until they are weaned. For example, you may sell some of the kids early in the season, in which case there is likely to be plenty of milk to share. And of course the dam's productivity is also a factor to consider.

Another approach is to separate the kids from their mothers and supplement by bottle-feeding. This reduces demand on the doe and encourages the kids to bond with humans. This is my preferred approach. It does require more labor, but for a small to midsized herd, it offers the best of both worlds.

Weaning

Weaning is the time when kids are transitioned to a diet that no longer includes milk. Kids are usually given milk until they are 8 to 12 weeks old, although some dairies wean earlier, and some extensively managed operations leave kids with a dam until she is dried off for her next kidding. Weaning is a stressful time for kids, both physically and psychologically. The psychological stress results from separation from their mothers or

> The stresses of weaning combine to increase the likelihood of medical issues such as coccidiosis, gastrointestinal parasites, and pneumonia.

from the act of suckling. Physical stress is due to the switch from a milk-based diet to one of solids only. Both of these stresses combine and increase the likelihood of medical issues such as coccidiosis, gastrointestinal parasites, and pneumonia. To help you reduce the odds of things going poorly, it's important to consider the age of weaning, how the kid's rumen adjusts to the change in diet at weaning, and the health risks associated with this time of the goat's life.

On farms where milk is the main income source, farmers tend to wean the kids at a much younger age than meat goat operations do. Some dairies wean kids as young as four to six weeks of age, but most shoot for six to eight weeks. At Pholia we begin weaning at eight weeks, but allow the transition to occur over the course of two to four weeks, during which time the kids receive less and less milk. If you recall, kids are born with a non-functioning rumen that is gradually "jump-started" by ingesting bits of dry matter and bacteria that begin the process of fermentation. They begin chewing their cud at about two weeks of age, but the rumen is still quite small. Over time it expands due to the amount of feed taken in as well as the gases produced by the bacteria. A diet high in roughage is believed to be the best for developing rumen function, but a diet that includes concentrates

Reducing Separation Anxiety

No matter what approach you take to weaning, unless you let the goats decide when it happens, there will be separation anxiety on the part of the mother and the kids. There are several things you can do to reduce separation anxiety, but be aware that with the stress comes an increased risk for health problems. Kids being weaned should be carefully watched for several weeks for signs of illness, especially coccidiosis and pneumonia.

- If kids are separated from their mothers at birth, there is a brief period of obvious stress on the mother's part, especially if she can hear her kids' voices after separation. The kids don't experience anxiety with this approach. If separation at birth is important (usually for the prevention of communicable disease), then ideally all kids should be moved to a separate barn, because anxious mothers will respond to the sounds of other moms' kids as well.
- If you plan to separate kids after a few days of nursing (so they get colostrum and initial care by the mom), limit the one-on-one time to 24 to 48 hours. Then return the mom to the main pen; this length of separation is brief enough that she shouldn't be picked on by the herd after returning. If you want to continue having the kids nurse, take the kids to the mom or vice versa several times a day for feeding. Or create a setup where the mothers can check the kids as they wish. We put the kids in large galvanized tubs placed at openings in the shared fence line inside the barn. The moms can visit and inspect their kids as they choose and then go back to the herd activities. After a few days of this, most moms seem content when nursing stops and you switch the kids to a bottle.
- If you plan to allow kids to nurse for several weeks before weaning, it's less stressful if you wean an entire peer group of kids at the same time and then move them to a barn or facility out of earshot of their mothers.

However you approach weaning, the biggest mistake is to give in and reunite a mom and her kids. If you allow the sound of the mom and kids calling to pull hard at your emotions and wear down your resolve, the end result is that you are teaching the animals to keep calling, because it gets them the results they want.

(grain) will provide more energy. This is sometimes the preference when early weaning is the goal so that the kid will be getting enough high-energy calories from grain and forage combined to make up for the loss of milk. Often a pelletized feed that contains both forage and grain along with minerals and vitamins is fed. I favor feeding a very small amount of grain after rumen function is already well established and just before and during weaning.

If possible, it's best to reduce milk consumption gradually. This allows the kids to adjust their intake of feed to fill the gap in nutrients left by the withdrawal of milk. You will likely notice an increase in the amount of alfalfa the kids decide to eat, and they will begin consuming a mineral mix (which should be available free-choice when they are a week or two old). They will also begin to drink water, if they haven't already. Making sure the water is clean and palatable and even warm can help with the transition.

The risk of coccidiosis increases during weaning, due to the stress and the fact that as kids increase their intake of solid feed, they are also likely to increase their intake of coccidia oocysts. Other diarrheal upsets might also occur. Be watchful and ready to treat if needed. I like to give probiotics during this time. Access to plenty of clean space and sunshine is helpful to the process. When kids are transitioned to pasture, they will also be exposed to other internal parasites that may be present on pasture grasses. Fecal testing and watching for signs of anemia through checking inner eyelid tissue color (see the Inner Eyelid Color and FAMACHA section on page 126) at this time is a good idea.

Pneumonia, a respiratory disease caused by either viruses or bacteria, is a risk for any animal in a weakened and stressed condition. The risk of contracting it can be reduced by making sure that kid pens are not overcrowded and that no kids are being bullied, as well as by keeping the pen clean and free of harsh

Figure 9.17. The design of keyhole feeders does a good job of keeping the hay clean and preventing goats from wasting it as they eat. It's important to provide enough keyholes to prevent too much competition for access.

ammonia odors from urine. Good air ventilation and sunshine are essential.

Feeding After Weaning

No matter what the future career of the goat, whether for meat, milk, fiber, or other, the quality of the diet from weaning to age one is critical to success. You can track the young goat's progress in several ways: by tracking weight gain (see table 8.1, page 182, which lists weight goals for breeding stock), monitoring body condition, and comparing the kid with its peer group. It's possible to calculate rations for young goats, especially if you have growth-rate ideals, but this method of determining what to feed is more likely to be a part of intensive management than the holistic approach. As I mentioned in chapter 5, you can access a helpful online ration calculation tool from Langston University (see appendix D for details).

Figure 9.18. One way to ship kids to a new home is by commercial airline, in an approved kennel. I like to zip-tie plastic containers, such as for yogurt, to the door of the kennel to hold water and feed pellets. Set the containers high enough that they won't likely get pooped in.

From weaning to maturity, young does still have a great need for minerals to sustain bone growth. As they near age one, their height growth will slow and continue a bit until about age two. If they are pregnant during this time, they will also be supporting the growth of offspring. High-quality pasture and browse will fill many of these needs, along with dry forage, including alfalfa for calcium as needed, and of course access to a properly balanced mineral and salt source.

Although excess phosphorus and calcium are a concern (risk of urinary calculi) in the diet of adult bucks and wethers, the young buck is still growing and in need of these minerals. Until about 18 months, he should be supplied with them through high-quality forage that might include alfalfa (which is high in calcium). Phosphorus needs are usually met without the need for concentrates.

Young castrated male goats and unbred female goats, whether they are working or pets, will be prone to obesity as they head toward maturity. This is from both the low demands of their lifestyle, unless they are hardworking pack or cart goats, as well as their owners' tendency to pamper them with treats. Once maturity is neared, their diet should be restricted to graze and browse and low-energy forages such as plain grass hay. Of course, high-quality mineral and salt sources must be present as well. But they will rarely if ever need alfalfa or legume hay, grain hays, or concentrates. The risk for urinary stones is quite high with wethers and should be anticipated with a strict diet.

PART V

Solving Goat Health Problems

A-to-Z List of Disorders

The following is an A-to-Z listing of all the disorders covered in this part of the book. If you want to find information on a particular disorder, you can check this list to quickly find the page where the disorder is discussed.

The health of your goats will be almost completely sustained by your day-to-day management and holistic approach—from feeding and managing genetics to understanding and providing for the animals' psychological needs. However, homeostasis is occasionally disturbed in even the most well-managed herd, and some animals will dip below the ideal and experience a health crisis that requires intervention.

In chapter 6 you learned how to troubleshoot some common health problems by assessing goats' demeanor; posture; eyes, coat, and skin; body condition; feces; and vital signs. This part of the book digs deeper into diagnosing and treating goat diseases and metabolic and nutritional disorders. Each of the eight chapters focuses on a particular body system, or pair of related systems, and the health issues that can affect those systems. Within each chapter you'll find individual entries on particular goat health problems, such as bloat and mastitis. Each entry includes a description of signs and symptoms, possible treatments, and prevention measures. In some cases you will be able to adequately treat your goat based solely on the information this book provides, but in many cases you will need the assistance of a trained practitioner. You can greatly assist your practitioner by clearly describing to them the signs you have observed and the symptoms your goat seems to be experiencing.

Medical professionals always troubleshoot a patient's illness by assessing each body system and its particular symptoms, knowing of course that the systems work together. By starting with the symptoms and considering how they relate to the system, however, you can more quickly troubleshoot the problem, pinpoint the possible cause or causes, and provide treatment yourself or with the help of a veterinarian. (See the A-to-Z listing of disorders on page 244.)

Each chapter in part 5 begins with a table of signs and symptoms related to that particular system. For example, the digestive system table in chapter 10 includes signs and symptoms such as loss of appetite, slobbering, and abdominal distension, with an alphabetized list of possible causes for each. A good way to use part 5 is to begin with these tables. Figure out which body system the signs or symptoms of concern relate to, and turn to the related chapter. Review the table and the lists of possible causes.

Keep in mind that a goat may experience a symptom that is not actually related to the underlying cause. For example, if a goat is breathing rapidly (which is a respiratory system symptom), the initial assumption could be that the goat has a respiratory problem. But in fact, the goat's respiratory system may be perfectly fine, and she's breathing heavily due to the fact that she is in labor.

By process of elimination, you may be able to zero in on the cause of the problem (if you can't, your next step may be to call your vet). Keep in mind that the tables do not cover every possible condition; nor are the lists of signs and symptoms exhaustive. I have focused on the most common diseases and disorders likely to be seen in North America or throughout the world.

That said, if you are reasonably confident you have settled on the cause of a problem, you can read the treatment information in the relevant entry and decide whether to take action on your own or, again, to first consult with your vet or practitioner.

CHAPTER TEN

The Digestive and Metabolic Systems

Goats have a paradoxical digestive system and metabolism. On the one hand they are hardy, efficient consumers of forages that many other livestock find unpalatable and non-nourishing. But on the other side of the equation, goats can easily undergo unhealthy changes in the digestive system. In fact it's not uncommon for seemingly straightforward problems, such as a brief loss of appetite, to lead to a change in goat rumen function that cascades into a crisis situation and may even lead to death.

For this reason there are a huge number of disorders that are associated with this body system, which includes the goat's upper and lower digestive tracts and the liver, gallbladder, and pancreas. (See chapter 4 for an overview of how the upper digestive system works.) In addition, digestive disorders are some of the most common ailments faced by the goat producer. For these reasons we're starting this section of the book with the digestive and metabolic system.

ACIDOSIS

OCCURS MOSTLY IN ADULT GOATS

Acidosis is a condition in which a goat's rumen acidity has increased beyond a healthy level (the lower the pH below 7.0, which is neutral, the more acidic the condition). If the rumen remains too acidic, the negative effects spread throughout the

goat's other systems. Acidosis can be subclinical and chronic—barely noticed by the producer—or it can be acute and cause death within a day of onset. Floppy kid syndrome (see the entry on page 260) is an example of metabolic acidosis, a state of acidity throughout the body (blood is normally slightly alkaline, not acidic at all). The cause of floppy kid syndrome is unknown.

The goat's rumen is designed to break down fibrous plant matter, such as woody browse and dry forage. The normal pH (acid level) of the rumen allows bacteria that digest fiber and cellulose to thrive. The desirable level is 5.5 to 7.0; the closer to 6.5, the better. If you recall from chapter 4, a goat's saliva (which contains acid buffers) is the key ingredient that keeps the rumen pH at its ideal. When goats eat grain or concentrates, which don't require as much processing by rumination (chewing the cud), less saliva is produced, and that can lead to acidosis. In addition, the simple carbohydrates in grain favor rumen bacteria that make acid, which also raises the pH level. Remember it this way: The more grain is eaten, the less saliva is produced, and the farther the acid level rises.

Acidosis can also result from anesthesia, because when a goat is sedated it may not swallow; instead, saliva will drool out of the mouth.

As the acid level in the rumen increases, the desirable rumen bacteria start to die and acid-producing

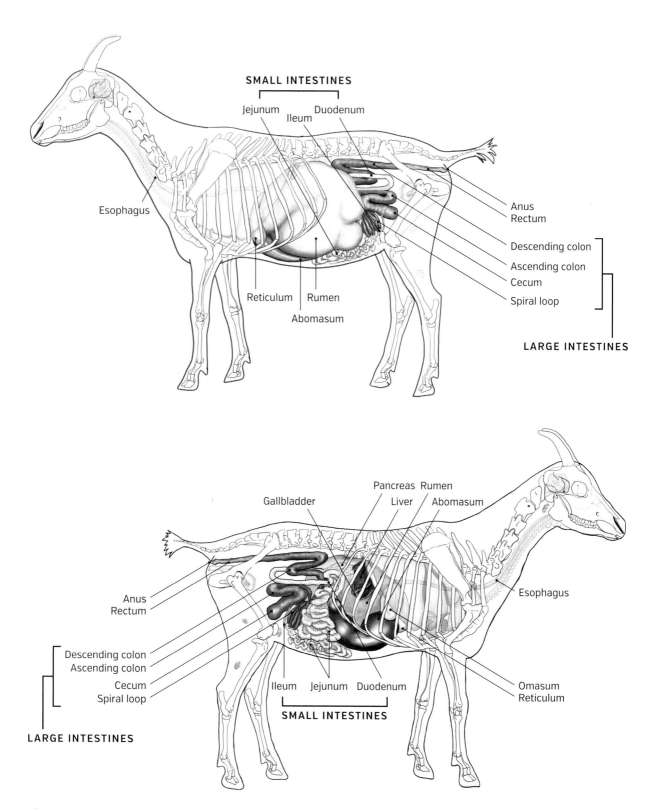

Figure 10.1. The right and left side of the goat's digestive system, including pancreas, liver, and gallbladder.

Table 10.1. Common Digestive System Signs and Symptoms

Sign or Symptom	Possible Causes*	Sign or Symptom	Possible Causes*
Abdominal distension	Abomasum bloat (more common in kids on milk replacer) Bloat (rumen) False pregnancy (hydrometra) Hernia (rare in goats) Inflamed gut/intestines (enteritis) Inflamed lining of abdominal cavity (peritonitis) Parasites Rumen or abomasum blockage (impaction)	Lack of rumen movement (atony)	Anesthesia or sedation (side effect) Dehydration Fear or excitement Fever Indigestion Ketosis (toxemia) Low-fiber diet Milk fever (hypocalcemia) Pain Rumen acid imbalance Urea poisoning
Abdominal pain or discomfort (see chapter 6 for additional information)	Coccidiosis, acute Inflamed gut/intestines (enteritis) Inflamed lining of abdominal cavity (peritonitis) Ingestion of cold milk or water Overeating disease (enterotoxemia) Poisoning Twisted gut (torsion) Urinary stone (urolithiasis)	Loss of appetite	Acidosis Poisoning
		Slobbering	Blue tongue Foot and mouth (not found in North America) Soremouth (contagious ecthyma) Vesticular stomatitis (rare)
Diarrhea	Copper deficiency Indigestion Intestinal infection, bacterial or viral Overeating Parasites Poisoning Selenium toxicity	Vomiting	Blocked trachea or esophagus Copper toxicity Poisoning
Difficulty swallowing (dysphagia)	Inflammation of gums and inside of the mouth (stomatitis) Neurological problem Teeth problems	Weight loss	Behavior problems CAE (caprine arthritic encephalitis) Chronic bacterial infection Chronic wasting disease (Johne's disease; paratuberculosis) CL (caseous lymphadenitis) Enterotoxemia MVV (maedi-visna virus) Nutritional problems Obstruction in digestive tract Plant toxicities Salmonellosis TB (tuberculosis) Teeth or oral problems
Frothing or foaming at the mouth	Poisoning		
Lack of or reduced feces	Blockage Coccidiosis, subclinical Constipation Dehydration		

* Causes listed include the more common possibilities, but the lists are not exhaustive or intended to replace the diagnostic services of a licensed veterinarian.

bacteria become the dominant microbes, leading to even more acid production. Once the microbial population undergoes this shift, a goat may struggle with a chronic state of acidosis even after its diet is changed back to one that supplies adequate fiber. Thiamine deficiency, polioencephalomalacia (goat polio), chronic or acute laminitis (founder), and pregnancy toxemia (ketosis) can result from rumen acidosis.

SIGNS AND SYMPTOMS

Mild: Loss of appetite, decreased rumen movement (motility) and rumination, decreased milk production. Diarrhea and grinding of teeth also possible. Standing with back hunched, discomfort when walking.
Severe: Loss of appetite, no rumen motility, mild bloat, muscle tremors, teeth grinding, groaning or crying out, mild fever, diarrhea, increased heart rate and breathing rate. Lameness and laminitis.

TESTING

Test results can indicate the severity of acidosis. A fecal pH test result may be below 5.5 (normal range is 5.5 to 7.0, with the ideal closer to 6.5). Urine pH will fall below 7.0. A sample of rumen fluid may be whitish gray and sour smelling, with a pH below 5.0. A sample of rumen contents may contain a high level of gram-positive microbes. Blood calcium levels may be low. For advanced tests such as these, consult your veterinarian, who will help you determine the right course of treatment.

TREATMENT

Mild: Offer high-fiber feeds such as good grass hay or fresh-picked tree greens. Give a B complex injection (to prevent polioencephalomalacia). Give antacids such as an oral dose of 1½ tablespoons (20 g) baking soda dissolved in warm water or 15 ml milk of magnesia (this is for an adult goat). Administer the rumen contents from healthy goats via rumen transfaunation (see chapter 7).
Severe: When rumen pH is low but within normal range, administer antacids (as above) and perform rumen transfaunation. When rumen pH is below 5.0, surgery to remove the rumen contents and clean the rumen lining may be necessary for the goat to have a chance of survival.

PREVENTION

Remove grain rations as long as the goat is clinical. Provide plenty of forage, ideally browse. When reintroducing grain, do it slowly, increasing the daily amount over a period of two to four weeks. Feed forage before feeding grain, and divide the grain ration among several feedings per day. Make sure clean, fresh baking soda (or another buffer) is available free-choice.

BLOAT

Bloat is the extreme distension of the rumen with gas. There are two types: The more common is frothy bloat; the less common, free gas bloat. (Yes, the names are similar, and it's hard to keep them straight!)

Frothy Bloat

The name *frothy bloat* describes the appearance of the rumen contents (see figure 10.3) of goats suffering from this condition. This type of bloat is also known as primary bloat, because it is the most common form. It's also called nutritional bloat due to the fact that the condition is caused by something a goat has eaten. Another name is ruminal tympany, which basically means the goat's rumen becomes so taut it makes sounds like a drum when tapped.

Frothy bloat is caused by the overconsumption of fresh, leafy greens (especially legumes) or by the consumption of grain, especially when finely ground. Both types of feed have the potential to form compounds that trap gas bubbles in the rumen contents. When this occurs, gas doesn't rise to the top of the rumen, where it can be belched out by the goat; instead, gas accumulates in the fluid. This causes the fluid level in the rumen to rise, which ultimately leaves no room for gases to escape. This puts intense pressure on the heart and lungs, enough to kill the goat. Goats that have experienced a mild case of grass tetany or milk fever are also more at risk for this type of bloat.

SIGNS AND SYMPTOMS

Distension of left side of abdomen out- and upward. Drum-like sound when distended area is tapped. Skin will feel tight when pressed, with no give. Signs of abdominal pain, including grinding of teeth, kicking at belly, shifting weight, crying out or moaning, difficulty breathing. Sudden death. **Note:** When any goat is necropsied more than a few hours after death, frothy bloat is likely to be present from the natural activity of the rumen microbes, which continues even after a goat has died. Frothy bloat may or may not have been the cause of death.

TREATMENT

Administer 100 to 200 cc vegetable or mineral oil (an ounce—about 30 ml—of dish soap can be added) or a purchased anti-frothing bloat treatment. (Therabloat contains poloxalene, which is approved in organic production for emergency bloat treatment.) Both the dish soap and the poloxalene combine with the small bubbles to form larger bubbles, which the goat is able to belch out. You can add a few drops of peppermint or spearmint oil to improve the flavor for the goat. After giving, encourage the goat to walk; if a gentle slope is available, walk up that. If the goat is lying down and can't rise, roll her over a couple of times and massage the rumen to help mix the oil into the foam and break down the bubbles. Be sure the goat is belching gas and that the rumen distension is diminishing. If not, a stomach tube should be passed to release the gas. (Follow the process described in the Steps for Tube Feeding sidebar on page 212, but using an adult-sized tube and a speculum or piece of PVC pipe to keep the goat from biting the tube.) In an extreme emergency situation, a large-gauge (16 or 14) needle, a knife, or a specially designed tool called a trocar can be be used to puncture the rumen through the abdominal wall (as shown in figure 10.4) and release the gas. This procedure presents other risks to the goat's health; it should be performed only as a last-ditch attempt and followed with proper veterinary care. Note that this treatment is more successful in frothy bloat when the goat has also been given a detergent to combine with the tiny bubbles.

Figure 10.2. Angelica, a 10-year-old Nigerian Dwarf doe, is not suffering from bloat. She has a naturally distended abdomen, or dropped stomach, a condition that is thought to be genetic and is seen in all breeds.

Figure 10.3. These rumen contents, photographed during a necropsy, exhibit the multitude of tiny foam-like bubbles typical of frothy bloat but that also form naturally several hours after the animal's death.

PREVENTION

Avoid sudden changes in the diet. Provide goats with access to dry forages, such as hay, when fresh greens are also available, and make sure they have eaten some before you give them access to the green pasture. Limit grazing time on legume-rich pastures. If legume-heavy pastures are unavoidable, non-certified organic farms can provide an anti-frothing agent in a mineral mix or block. (The active ingredient, poloxalene, can be used in organic production only for emergencies.)

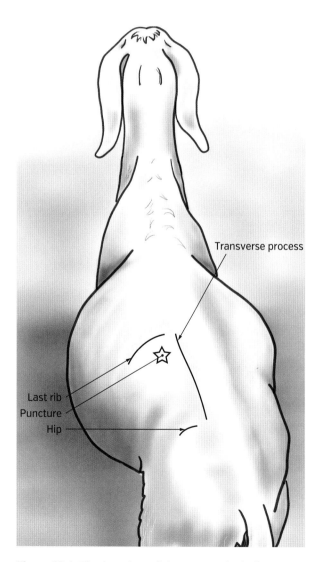

Figure 10.4. The location of the rumen, including where to puncture for emergency bloat relief.

Free Gas Bloat

This type of bloat is also known as secondary bloat, secondary ruminal tympany, and sometimes as choke (see the Choke entry on page 253). It is a result of a blockage in the esophagus that prevents rumen gases from escaping. The blockage might be cud or food chunks, or an abscess or a tumor that puts pressure on the esophagus. If gas production is greater than belching, free gas bloat can occur. This type of bloat is rare in goats.

SIGNS AND SYMPTOMS

The same as for frothy bloat. In addition, possible visible swelling of the neck if a tumor, feed wad (as in choke), or abscess is present; signs that the goat is having difficulty swallowing—including not wanting to eat and inability to ruminate.

TREATMENT

Administer 25 to 50 cc vegetable or mineral oil or a purchased anti-frothing bloat treatment. The oil may lubricate the esophagus to help the goat clear the obstruction. Administer it slowly; if the goat cannot swallow well she may aspirate the oil into her lungs. In one case of esophageal blockage without bloat, I was able to massage a lump in a doe's neck to help break it up and allow her to swallow. Inserting a stomach tube might help dislodge the blockage as well but carries a risk of puncturing the delicate esophageal wall. As for frothy bloat, releasing gas from the rumen by inserting a needle or trocar can be done in a life-threatening emergency.

BLUE TONGUE

Blue tongue, a reportable disease, is caused by a virus that's spread by carrier insects; namely, little flies called midges. The virus can't be spread from goat to goat. There are different species of blue tongue viruses (BTVs) throughout the world. Blue tongue is more likely to be a problem in warm, moist areas—the same type of climate favored by biting midges. It was previously thought that BTVs could be introduced to regions free of them, and thus restrictions for goat, semen, and embryo transport (all of which can harbor the virus) were put in place. Many of these restrictions have now been relaxed as more has been learned about how the virus is spread.

Goats are almost always symptomless, even when carrying the virus. When symptoms are present, there is no specific treatment, and the disease is self-limiting, usually resolving after less than two weeks.

SIGNS AND SYMPTOMS

Usually asymptomatic. May include loss of appetite, fever, mild depression lasting a few days. Subsequently, sores on the gums, lips, and tongue, swelling of the face, slobbering.

TREATMENT

No medical treatment is needed. Isolate the animal to allow it time to eat without competition. Sunlight can aggravate skin lesions, so shade is helpful.

PREVENTION

In the United States one vaccine is available, but it protects against only one type of BTV. It must be administered off-label by a vet. In the UK, an eight-virus vaccine for BTV is available. Using insecticides may limit the presence of midges, but it takes only one bite to infect a goat, so such controls are not effective. Because this disease rarely makes goats sick, focusing on midge control as a preventive measure is not a high priority for most producers.

CHOKE

There is a big difference between choking and choke. Choking refers to an obstruction in the trachea or bronchial tubes that leaves the subject unable to breathe. Choke refers to an obstruction in the esophagus, which leaves the subject unable to swallow. Choking is life threatening, but choke is not immediately so. If the obstruction remains in the esophagus for long, though, then free gas bloat (see Free Gas Bloat on page 252) may develop and can cause death.

SIGNS AND SYMPTOMS

Shaking the head, slobbering, coughing (not with every case), possible visible or palpable lump on left side of neck.

TREATMENT

If the goat is not in too much distress, see if you can feel for the obstruction along the left side of the neck. If you can, gently massage the blockage upward, not down. If you can't locate the obstruction, try giving 20 to 50 ml of vegetable or mineral oil if the goat will swallow. If the goat is able to cough and shake her head and is not in any other distress, such as bloat, you can simply wait to see whether the obstruction clears on its own.

PREVENTION

Goats are more likely to choke when fed dry pelletized food and when they're in a rush to eat. Pre-wetting pellets or ensuring that no goat feels pressured to eat quickly can help prevent choke. You can slow consumption down by placing medium-sized rocks in a feeder so the goat has to move them about to expose more pellets.

COCCIDIOSIS
OCCURS MOSTLY IN YOUNG GOATS

Coccidiosis, the infection of the intestinal lining by microbes in the genus *Eimeria* (a protozoan), is the bane of all livestock farmers. There are several species that can infect every species of mammal. The greatest losses occur in the young stock, whether that is baby chicks or baby goats. It is the leading cause of diarrhea in young goats between the ages of three weeks and five months. *Eimeria* species (usually simply called coccidia) exist normally in adult goats without causing disease, but certain conditions cause the coccidia to reproduce rapidly, which leads to a flare-up of disease.

Coccidia don't lay eggs but instead form spores called oocysts that are shed in goat feces. Oocysts are extremely tough and can reside in the environment for a long time. When shed, they change from oocysts to sporocysts, which are viable and able to infect the goat. The change to sporocysts occurs when moisture and ideal temperatures are present, from about 54°F (12°C) to 90°F (32°C). During these temperature ranges and in humid climates, infective coccidia will be present in greater numbers. This is why coccidiosis is more of a problem in moist, warm climates than in cold, dry ones. Oocysts shed during less-than-desirable conditions will remain dormant and change when the conditions are right. Because of this, there can be a sudden change from no cases of coccidiosis to a serious outbreak.

Goats ingest oocysts present on feed and soiled bedding and around water troughs. In the intestine, and sometimes in the abomasum, the sporocysts proceed through several other phases of reproduction and invade the tissue of the intestinal lining, where they produce more oocysts. While the coccidia reside in the mucosa of the intestines, they cause extensive damage and blood loss. Anemia results. The goat's intestines respond by flooding the intestine with liquid. Diarrhea results, along with dehydration, electrolyte imbalance, and metabolic acidosis. Goats that recover from a severe case can have lifelong scarring in the intestines that prevents the ideal absorption of nutrients.

Figure 10.5. On necropsy this young kid, which I suspected had perished from coccidiosis, showed the intestinal damage that is typical of this disease. Small, whitish nodules are present, and the lining is red from blood loss. Parts of the intestines had clotted blood where the damage had caused the kid to bleed.

In order to become and remain disease-free, goats must be exposed to low levels of coccidia in their gut from the time they are young and continuing throughout their lives, which allows them to develop and maintain resistance. Immunity is usually developed by five months of age and continues until the goat is older. In goats over age seven, immunity starts to decline even with regular exposure.

Several things affect the likelihood of a goat succumbing to coccidiosis. They are:

- The stress the animal is experiencing, which is often related to specific life situations—weaning, other diet changes, pen changes, herdmate changes, unfavorable weather events.
- The number of eggs in the environment and therefore consumed by the goat.
- Previous exposure to that particular species of *Eimeria*—moving a goat to a new herd, or bringing a new goat in, can expose animals to a new strain to which they have no immunity.

Fecal flotation may or may not be useful in determining whether the number of oocysts present indicate a potential coccidiosis case. Healthy kids may have egg numbers from 1,000 to 1 million per gram of feces; clinically ill kids may have 100 to 10 million.[1] This may not seem logical, but the interpretation of the count's importance depends a lot on the texture of the feces. When a goat has diarrhea, eggs may move through the digestive system rapidly. However, the feces are also more diluted, so the gram-per-gram comparison to normal feces is not the same. Still, if a symptomatic goat has a count of over a million, you can be fairly certain that coccidiosis is the culprit. *Goat Medicine* authors Smith and Sherman suggest that the modified McMaster technique for fecal floats (which was described in chapter 7) is the best choice when testing for coccidiosis. It's my preference as well.

Two types of conventional mediation are available for the control of coccidia: coccidiostats and coccidiocides. If you look at the last part of each of these words, you will see that one ends with *stat* and the other *cide*.

You will recognize the root word *cide* from the words *homicide* and *suicide*. It basically means "death." *Stat,* on the other hand, is used as a suffix meaning "to prevent growth." I remember it by thinking of the word *statue*—something that doesn't change. Coccidiostats keep the numbers of coccidia under control without killing them all, and coccidiocides are meant to wipe out the entire population. Depending on the dosage, some medications can serve as either. In most cases, a coccidiostat is the preferred drug, because an animal needs exposure over time to the parasite in order to develop immunity to it.

In the goal of limiting coccidiosis and animal suffering while at the same time practicing or transitioning to a more holistic approach, coccidiostats might still be needed, either routinely or periodically. For this reason I've included table 10.2 (see page 257), which summarizes those currently used for treatment of the disease in goats. I believe their use can be eliminated in most cases, but it isn't realistic to expect a producer to simply stop their use before implementing other controls.

SIGNS AND SYMPTOMS

Peracute: Sudden loss of appetite. Behavior that indicates abdominal discomfort, such as standing apart from other animals, hunched back, moaning. Anemia. Death within a day or less.

Acute: Loss of appetite. Behavior that indicates abdominal discomfort, such as standing apart, repeatedly getting up and lying down, moaning. Feces changing from normal to soft, then brown diarrhea with a foul odor (due to blood in the feces); sometimes frothy looking with a golden color.

Post: Animals that recover from coccidiosis may suffer reduced intestinal efficiency, resulting in less-than-ideal nutritional status for their entire life. Such animals may be stunted, rough-coated, and pot-bellied. I've had a few animals that recovered from a severe case, however, with no noticeable difference in their health over their lifetime.

TREATMENT

Peracute: Try treatments recommended for acute, but prepare for death. Consider euthanasia, because goats with peracute coccidiosis seem to suffer a great deal of pain as they approach death.

Acute: Separate the goat, and hydrate it in any means possible. Force-feed an oral electrolyte mixture or administer IV fluids. Do not give milk for a day or two. Give probiotics. If agreed to by your vet, administer a non-steroidal anti-inflammatory (such as flunixin,

Diarrhea

A multitude of microbes, bacteria, protozoans, and viruses can cause diarrhea in the goat—most often in the young kid. Many cases of diarrhea that occur in a herd will never be attributed to a specific cause. Fortunately, the treatment for most cases is similar, whatever organism is to blame.

The two most common causes of diarrhea are coccidiosis and cryptosporidiosis—both protozoans. In young calves, *E. coli* is also a common cause of diarrhea. Research seems to indicate that *E. coli* is uncommon in goats—yet it is still often blamed when

diarrhea crops up. Salmonellosis is not too common in goats. Cryptosporidiosis and salmonella are also zoonotic diseases—they can be transmitted to humans.

No matter the cause of severe diarrhea, it's important to isolate affected goats. Wear gloves when handling the animals, and wait to handle them until after necessary handling of healthy goats. Thoroughly clean out the pens where the infected animals were kept. Note that many of these microbes are resistant to most sanitizers that might be used to clean pen surfaces, such as chlorine, iodine, and chlorhexidine.

The Year I Felt Like Quitting—Coccidiosis and Kid Loss

In my original, first-edition copy of *Goat Medicine*, the section that covers coccidiosis is the most worn and highlighted part of the book. About three years after founding our commercial dairy, we had a horrific year during which we lost over 20 percent of our kids to the disease. At that time we split the kidding season into spring and fall groups, so that we could milk and make cheese year-round. The mortalities started in the spring. I thought I had gotten a handle on the problem and went into the fall season feeling confident. Even so, more deaths occurred.

In all but one case, the kids died of the peracute form—without diarrhea and perishing within 12 to 16 hours of first showing symptoms. I sent off the bodies of the first kids that died for necropsy to rule out other issues, such as enterotoxemia. At the time of these losses, I had yet to try a more holistic approach to this disease, and we were using a popular coccidiostat, sulfadimethoxine, commonly sold as Albon. Our experience confirmed the loss of effectiveness of sulfa drugs to treat coccidiosis worldwide. I had also been using an herbal dewormer at the dose recommended by the maker to prevent coccidiosis. We raised kids separately from adults in pens with fresh bedding that were cleaned frequently.

After the first cases I changed to a coccidiostat from the ionophore class called lasalocid. That seemed to clear up the problem for the rest of the spring season. When the fall kidding began I started the kids on lasalocid right away. Then when they were between four and six weeks old we lost almost half of them.(The fall crop of kids was smaller in number than the spring group.) Again, no diarrhea to speak of. I treated the entire crop with the highest dose of sulfadimethoxine recommended and started doing fecal floats. While float numbers for coccidia can be misleading, ours did seem to correlate with the signs and symptoms we were observing.

The next spring I switched to a quinolone drug called decoquinate and started doing fecal floats weekly. And had a good year.

Fast-forward a few more years. I was determined to find an organic approach that worked. I had switched to raising the kids on free-choice cold yogurt. With the reduced stress, constant supply of feed to help keep the intestinal contents moving, and possibly some protection from the probiotics in the yogurt, I felt comfortable not giving the kids any preventive coccidia treatments. It has now been three years since I stopped using preventive treatments. I still have work to do, as I typically lose one or two kids each year after weaning and before immunity is established. I feel that most of these cases could have been caught sooner or stress reduced during weaning even more. It's a great improvement from our bad year but, like I said, still more work to do.

aka Banamine) for pain and its anti-inflammatory, gut-calming effect. If your operation is not certified organic, consider treating all goats of the same age with a coccidiostat (see the Prevention section following). Continue to hydrate as needed. Offer high-fiber feed—grass hay, fresh straw, tree greens. Administer B vitamin injections daily until the animal is ruminating again. If symptoms persist more than a few days, reintroduce milk slowly in small amounts, because lactose intolerance seems to occur during the disease,

increasing gut stress. Another option is to feed lactose-free milk or yogurt (in which the lactose will be naturally broken down by the bacteria in the yogurt).

PREVENTION

Detect problems early: Perform routine fecal floats (modified McMaster method) on combined samples from each pen or age group, especially high-risk groups—age three weeks to five months. If oocyst count are high, no matter what the symptoms, take

Table 10.2. Coccidiostat Usage Recommendations

Drug	Brand Name	Status for Use in Goats	Notes	Treatment Dosage*	Preventive Dosage*
Sulfadimethoxine	Albon, Di-Methox	Extra-label	Widespread resistance	75 mg/kg body weight for 4–5 days	Not recommended by the author
Amprolium	Corid	Extra-label	Because of risk of polioencephalomalacia, only use with thiamine supplementation; not controlled by VFD[†]	50 mg/kg for 5 days or 10–20 mg/kg for 3–5 days	25–50 mg/kg body weight daily from age 2 weeks to several months
Lasalocid	Bovatec	Extra-label	An ionophore, not controlled by VFD[†]	—	1.1–2.2 mg/kg daily
Monensin	Rumensin	Approved	An ionophore, not controlled by VFD[†]	—	15–20 g/ton feed
Decoquinate	Deccox	Approved	Not controlled by VFD[†]	—	0.5–1 mg/kg daily
Toltrazuril	Baycox	Not in US for goats	Highly effective treatment being used outside the US on goats and approved in the US for use in horses and poultry	20 mg/kg single dose	20 mg/kg once every 3–4 weeks

* Dosages as recommended in Smith and Sherman, *Goat Medicine*.

† VFD (Veterinary Feed Directive): FDA regulations effective January 1, 2017, changed the availability of antibiotics deemed important for human health from available over the counter to available only under the direction (prescription) of a veterinarian. The guidelines also control their addition to feedstuffs.

a second look at the next steps and make sure you are addressing all these needs. Then consider treating with a botanical such as oregano oil (which was described in the Botanicals section on page 153) and making high-tannin feeds available.

Optimize environment: Increase pen cleanliness. If the floor of a kid pen is dirt, at the start of each kidding season remove several inches and replace it with fresh soil or other material that has never been exposed to goats. Consider converting goat pens to slatted flooring or mats that can be scraped clean. Increase exposure of pens to sunlight by adding windows or other openings. Keep floors dry (don't rinse mats with water or sanitizer). Keep feed off the ground, and be sure areas around bucket feeders or waterers stay very clean.

Avoid stress: Avoid sudden feed changes, sudden weaning, overcrowding, and bullying. Group kids by age if possible. Increase bottle-feeding frequency (decrease the interval between feedings) or switch to a free-choice feeding system. Frequent feeding can help keep the gut healthy as well as limit stress. Give Rescue Remedy and frequent probiotics whenever stress occurs.

Drugs: Coccidiostats have long been used to limit *Eimeria* numbers in goat kids, but in many cases coccidia are growing resistant to the more commonly used drugs, particularly the sulfa-based coccidiostats. If you feel their use is warranted, it's a good idea to consult your veterinarian in order to choose an effective treatment and also address the other issues that can help improve immunity and prevention.

CRYPTOSPORIDIOSIS

CONTAGIOUS FROM GOATS TO HUMANS

Cryptosporidiosis is caused by a protozoan. It is relatively new, with the first cases reported and identified in 1981. This disease is zoonotic, and both animals and humans with decreased immune function are at particular risk for contracting the disease. Kids are most likely affected and are at greatest risk between the ages of 5 and 21 days, but usually less than 14 days old. The protozoans have a life cycle similar to coccidia, and the risks of spreading and concentration of oocysts is also similar (see the Coccidiosis entry on page 253). Oocysts can be seen on flotation, but they are easy to confuse with yeasts, so I recommend having your vet run the fecal test if symptoms seem to fit this disease, rather than coccidiosis. Cryptosporidiosis has a death rate of up to 40 percent.[2]

SIGNS AND SYMPTOMS

Severe, watery white to yellow diarrhea lasting or recurring for several days to two weeks. Depression, loss of appetite, rough coat.

TREATMENT

Isolate affected animals from the rest of the herd. Rehydrate with electrolytes given IV or forced orally. Reduce the volume of milk fed, because lactose intolerance seems to occur during the disease. Or feed lactose-free or -reduced milk or yogurt (see Make Your Own Lactose-Free Milk at left).

PREVENTION

Prevention measures are similar to those for coccidiosis: Increase herd hygiene and manage for cleanliness, reduced stress, and nutrition. Make certain kids receive ample colostrum, keep them grouped by age, and reduce stress.

Make Your Own Lactose-Free Milk

Goats can develop several diseases that result in the bacterial population in the gut being disrupted and becoming sensitive to lactose, thereby contributing to further diarrhea.[3] If you want to feed a kid suffering from one of these conditions milk from its mother or from other animals in your herd, it's important to make sure the lactose content is reduced. There are several options for accomplishing this: First, you can ferment the milk into yogurt, which won't be lactose-free but is lactose-reduced as well as easy to digest. Second, you can buy lactase-producing bacteria from a cheese supply company. Third, you can pulverize tablets or drops that contain the enzyme lactase, such as the brand Lactaid, and mix the powder into the milk. I've had good results with this last option, as well as yogurt. If you're using the enzyme, simply follow the directions for dosage on the package. You can find these tablets at grocery stores and pharmacies.

ENTEROTOXEMIA
(Overeating Disease, Pulpy Kidney)

Enterotoxemia is caused by the bacterium *Clostridium perfringens*. The disease is sometimes called pulpy kidney, which describes the texture of the interior of the kidney in some cases (observable if necropsy is performed immediately after death). It is also called overeating disease, because a sudden change in feed, especially gorging on grain, often precedes illness. Enterotoxemia is thought to be very common, but because it is difficult to definitively diagnose, even with necropsy and lab work, its real prevalence is unknown. The bacteria that cause the disease are common in the soil and are found in the gut of healthy goats. The disease is more likely to occur in intensively managed goat farms.

Clostridium perfringens can release several toxins in the gut of the goat that are activated by an enzyme naturally present there. The toxins are then absorbed into the bloodstream and damage the circulatory system and vital organs, including the brain and kidneys. When feed intake is normal, it's believed that the goat moves the bacterial population through its system, along with the digesting feed, at a rate sufficient to prevent the bacteria from overpopulating the gut. The theory is that when the goat eats too much grain or other lush feed, the meal may move too quickly through the upper digestive system, leaving ample nutrients to feed the *Clostridium*. Excess grain intake is also thought to slow peristalsis (normal intestinal movement), and the slowing allows even more time for the bacteria to multiply to levels that will release enough toxin to be harmful.

The peracute form often leads to sudden death. It is more common in kids, particularly fast-growing, healthy, aggressive eaters. It is almost 100 percent fatal. The acute form usually affects adults. It occurs over a few days, and the goat may recover with treatment but usually not without. The chronic form comes and goes and is most common in adult goats. It is the most difficult to recognize because the symptoms overlap so many other disorders.

The vaccine for this disease isn't effective for long in goats, and even when recently vaccinated a goat can still develop enterotoxemia. Antitoxin is available, but it's not usually effective in the peracute version. It can be in the acute, but large, expensive doses are required. In the chronic form antitoxin is at its most effective.

SIGNS AND SYMPTOMS

Peracute: Sudden loss of appetite, depression, abdominal pain as evidenced by standing with back arched, kicking at belly, crying out. Copious diarrhea—at first yellow-green and then turning watery with blood and mucus. Death—when a goat is found dead, the head is usually in thrown-back position.

Acute: Same as peracute but less severe and occurring over three to four days. Dehydration; acidosis is likely.

Chronic: Intermittent episodes of illness over several weeks. Listlessness, loss of appetite, decreased milk production, progressive weight loss, and intermittent soft stools.

Seen on necropsy: Intestinal lesions, pulmonary edema, hemorrhages on the heart muscle, diaphragm, and abdominal muscles. If necropsy is performed immediately after death, the kidney centers might be pulpy—described as looking a bit like cotton wool—although this might be rare in goats. Presence of toxin in intestinal contents may or may not indicate enterotoxin.

Why I Don't Vaccinate Against *Clostridium perfringens*

Although vaccination is common, studies show that goats retain active antibodies for only about 28 days after the injection, and some recommendations call for revaccination every three to four months.[4] In the past 10 years, I have never given the vaccine to my goats. I have never had a known case of enterotoxemia, either. Given the intense vaccination protocol, I have preferred to treat this as a herd management issue. That doesn't mean I won't lose an animal, but at this time I feel my time and dollars are better spent on other health issues that I believe are of greater importance.

If you decide to not vaccinate for this disease, you may receive some skepticism and even criticism from some veterinarians. If they are more familiar with sheep and cow vaccine effectiveness, this will be understandable. I photocopied the page about the limitations of *Clostridium* vaccination of goats from *Goat Medicine* and gave it to one vet that scolded me.

Please don't copy what I do without consulting your vet and analyzing the health of your own herd, as well as your own comfort level with risk! Remember, all vaccinations should be addressed based on risk.

TREATMENT

Antitoxin can be given, but in a peracute or acute case it may not be effective. Treatment focuses on hydration and dealing with shock—by keeping the animal warm and dry. Give pain medication. Give probiotics, yogurt, and charcoal. The chronic case can be treated with antitoxin.

PREVENTION

Avoid sudden feed changes. "Goat-proof" all grain bins and make storage areas inaccessible for escaped goats in order to prevent accidental gorging. A vaccine is available, but effectiveness is questionable (see Why I Don't Vaccinate Against *Clostridium perfringens* on page 259).

FLOPPY KID SYNDROME

OCCURS MOSTLY IN YOUNG GOATS

This rather mysterious problem, which causes metabolic acidosis, is restricted to kids at the age of 3 to 14 days. The kid's blood pH falls from a normal of about 7.4 to a low of 7 to 6.8. The major sign is total limpness of all limbs—hence the name. Floppy kid syndrome (FKS) was first noted in the late 1980s. Although the cause is still unknown, the cure is fortunately quite reliable and easy. If not treated, kids are almost certainly doomed to death. Floppy kid syndrome can occur in kids regardless of whether they are fed on dams or bottles, milk or milk replacer. According to Smith and Sherman, it does not occur in kids on pasture, but only confinement.[5] (Keep in mind that during kidding season, even herds that are on pasture might be self-confined due to the weather.) Overfeeding of milk, whether by bottle or by a high-producing doe, seems to be implicated in most cases. In addition, there are more FKS cases later in the kidding season than early, leading some to speculate that it might be related to increased numbers of pathogens in the kids' environment.

The only case I have ever had to deal with occurred minutes after dosing a kid with an oral herbal dewormer; it is impossible to conclude if there was a correlation or simply coincidence. I do suspect,

though, that somehow the herbs quickly put the kid in a state of metabolic acidosis. The treatment listed below completely revived the kid within an hour.

SIGNS AND SYMPTOMS

At age 3 to 14 days, profound muscle weakness or paralysis, unable to suck but able to swallow. Shivering or shaking possible.

TREATMENT

As soon as symptoms are noticed, immediately administer a lightly rounded ½ teaspoon (1.5–3 g) baking soda dissolved in as much cool water as it takes to dissolve the baking soda. Use a drenching syringe, and give slowly while you hold the kid's chin and head tipped up slightly to help it swallow. If one dose does not completely restore the kid, consider whether other problems may be causing the symptoms. If after 12 hours the symptoms haven't completely disappeared, give a second treatment. Occasionally a kid might need repeated treatments every 12 hours for up to four days. Following a case of FKS, restrict milk intake for 24 hours. Instead give the kid a bottle of warm electrolyte solution (see chapter 6 for more on electrolytes) and also give a dose of nutritional supplement.

PREVENTION

Since the precise cause of floppy kid syndrome is not known, it is impossible to know how to prevent it. Preventing overeating should always be the priority but is not always possible. Vigilance to symptoms and quick treatment are musts.

GASTROINTESTINAL PARASITES

Intestinal parasites, often collectively called "worms," are common throughout the world. (Not all goat parasites are classified scientifically as worms, however. For example, the parasite that causes coccidiosis is a protozoan.) While I cannot provide specific information about each type of parasite that can infect goats, I can emphasize some key points that will apply to any situation where parasites of the digestive system are involved. Preventing parasites becoming a problem in your herd requires a proactive approach to the following: the importance of nutrition and the goat's immune function; climate; grazing patterns and pasture management; the utilization of anemia assessment (for certain parasites only); an estimation, through fecal counts, of the severity of parasites already present in the goats; and finally a plan for culling animals that have chronic parasite problems even when these other management factors are being addressed.

The goat's immune system is its best defense against infestation. A well-nourished goat with access to good feed and all of the macro and micro minerals has a better chance of remaining trouble-free when it comes to internal parasites. However, a goat that is carrying a large parasite load will not be able to absorb the proper nutrition from even the best of feeds, so an improved diet may not help. Young animals do not have a fully developed immune system, so they're at greater risk of infestation and will subsequently fail to grow well—continuing the cycle. Genetics also play a role, and resistance to parasites should be an important consideration when choosing breeding stock.

In general, the colder and drier the climate, the fewer gastrointestinal parasites are present in the environment. Pastures that experience a deep winter freeze will have egg loads decreased but not totally eradicated. Goats in warm, wet tropical regions will suffer the greatest parasite infestations.

Goats that live in dry-lot pens that are well managed will have fewer issues, as will goats that spend more time browsing upright forage rather than grazing in open pasture. When pasture is utilized, it must be managed in a way to limit exposure to eggs and larvae. Maintaining sufficient pasture height along with other pasture management techniques such as rotational and interspecies grazing can all help reduce parasites that are a threat to goats. Diets high in tannins, such as from oak and many other plants, act as natural anthelmintics (dewormers). Therefore, there is a twofold benefit of browsing.

Although not all parasites feed on the blood of their host, the most problematic worm does. The barber pole worm, *Haemonchus contortus*, is a nasty, aggressive, and prolific parasite that bites into the wall of the abomasum and draws blood from the host. It takes very few *Haemonchus* to kill a goat. Fortunately the anemia they cause can be detected by observing the goat's mucous membranes. The easiest place to check for anemia is by looking at the inner eyelid of the goat. The procedure is known by the acronym FAMACHA and comes with a copyrighted score card; it was developed in South Africa for the evaluation of parasitic anemia in sheep and goats. FAMACHA was described in more detail in chapter 6 (see the Inner Eyelid Color and FAMACHA section on page 126). The best advice is to routinely check all goats for these signs of anemia when your herd is at risk for *Haemonchus*.

Gastrointestinal parasites should be evaluated by routine fecal floatations, especially when accompanying signs of parasite load are present. You can send samples out for testing or do this procedure yourself, as was explained in chapter 7. As I mentioned in chapter 7, however, fecal egg count numbers shouldn't be your only criterion for managing internal parasites.

Figure 10.6. Occasionally some worms can actually be seen leaving a goat's digestive system through the anus. These tiny whip- or thread-worms were caught on camera at Oats and Ivy Farm, California. *Photo courtesy of Amanda Nuñez*

The most common gastrointestinal worms affecting goats in North America are the barber pole worm, *Haemonchus contortus*, and the brown stomach worm, *Teladorsagia circumcincta*, both found in the abomasum; the bankrupt worm, *Trichostrongylus colubriformis*; the long-necked bankrupt worm, *Nematodirus* spp.; the nodular worm, *Oesophagostomum* spp.; the whipworm, *Trichuris* spp.; and tapeworms, *Moniezia* spp., all found in the intestines. Of these, the barber pole is of the most concern.[6] The most promising research regarding the control of internal parasites with natural remedies focuses on the use of plants or feeds naturally high in tannins and the use of copper oxide wire particles (aka COWP or copper boluses). (Chapter 6 reviews each of these in detail.) Many wild plants, such as oak and grapevine, as well as their domesticated versions, are high in tannic acid. Studies into the use of concentrated tannins (CT) have focused on plants and materials that are more readily available for widespread commercial use, including the legumes *Sericea lespedeza* and bird's-foot trefoil; grape pomace (a residue of the wine and grape juice industry); and pomegranate hulls. All have been shown to reduce parasite egg counts and improve FAMACHA scores in small ruminants.

SIGNS AND SYMPTOMS

Anemia, but only for *Haemonchus, Oesophagostomum, Trichuris*, and the protozoan coccidia (see the Coccidiosis entry on page 253). For tapeworms, visible segments that look like rice grains in feces or on rectum. For all types of worms: failure to thrive, depressed attitude, loss of appetite, rough dry coat, potbelly.

TREATMENT

Refer to the Internal Parasite Control section on page 150 for detailed information on treatment. Provide supportive care, including easy access to feed, water, and a low-stress environment. Treatment with anthelmintic appropriate for the situation—check with your vet for known resistance and dosage for treatment rather than simply giving the recommended dosage stipulated on the product package. That dosage is intended for prevention, not treatment. Tapeworms

are fairly common in goats but have few negative effects, and the general approach now is to not treat them. Resistance to tapeworms develops with age.

PREVENTION

To prevent parasite problems, provide nutrition and environment that promote immune function, follow feeding and housing practices that limit exposure to eggs and larvae, and provide natural browse or access to plants high in tannins. You may decide to use copper oxide wire particles (COWP). Conduct routine fecal egg counts and eyelid anemia (FAMACHA-type) assessment. Possibly most important, create a plan for culling animals who despite your efforts at prevention still suffer from parasite problems.

HYPOCALCEMIA (Milk Fever)
OCCURS MOSTLY IN ADULT GOATS

There are two variations on the condition of hypocalcemia (low levels of calcium in the blood), also known as milk fever. One occurs after a doe gives birth; the other, later in lactation. Hypocalcemia is common in dairy cows, but rare in goats as a standalone condition. In fact, it is so uncommon that it's poorly documented in the scientific literature, and most of what is written is simply the rewriting of what is known about dairy cow milk fever.

Cows come into heavy milk more quickly than do most dairy goats, and it's thought that this may put them at higher risk for milk fever. Having lost one cow to this problem myself and having had another who experienced it every freshening, I am very glad to not have to worry about it much with goats! However, hypocalcemia can often develop if a goat is also experiencing ketosis (see the Ketosis entry on page 263), and it's important to remember that hypocalcemia is common as a secondary condition.

After a normal birth, a doe's blood calcium level will naturally drop, but she will not show symptoms of milk fever. Heavy-milking dairy goats occasionally experience hypocalcemia with the same symptoms as milk fever type, but it occurs one to three or more weeks after kidding, instead of immediately after.

Traditional tactics for reducing the risks of milk fever in cows included removing alfalfa (high in calcium) in late pregnancy. While this might seem like the wrong way to prevent low blood calcium, the tactic was thought to encourage the cow's system to switch to the mobilization of bone calcium, making it more available for the sudden drop in blood calcium at calving—thanks to the cow's drop in dietary intake during calving and the sudden demand of milk production. Smith and Sherman state that a low calcium diet in goats is *not* needed and may in fact lead to hypocalcemia (as it evidently does in sheep).[7]

SIGNS AND SYMPTOMS

Early: Loss of appetite, slight tremors and twitching, weakened uterine contractions (if hypocalcemic during labor), poor milk production (later in lactation). **Advanced:** Lying down, first on sternum, then on side, with muscle spasms and yelling.

TREATMENT

For both types of hypocalcemia, subcutaneous 23 percent calcium gluconate (borogluconate) can be given if symptoms are mild. Oral calcium supplements can also be given in addition to the subcutaneous dose. If the condition is advanced, IV calcium is given, but it should be administered by a vet, because the goat may abruptly die of cardiac distress if blood calcium levels rise too quickly. In an advanced case it won't hurt to administer a sub-Q or oral paste dose while you wait for a vet to arrive; it can be effective and helpful.

PREVENTION

Based on the rarity of milk fever in goats and the uncertainty about how to treat it, I recommend focusing on providing a balanced diet as a preventive rather than removing alfalfa hay toward the end of gestation.

INTESTINAL TORSION (Colic)

Colic is a term often used to describe a condition in horses in which there is a twisting or kink in the intestines, as well as a general term for unspecified abdominal pain. In the case of the goat, torsion of the intestines is usually related to other causes of pain and is more a result than the primary problem. If you do a field necropsy of a goat, you might observe this twisting or a section of the intestines that is deep red because of the blockage of blood flow due to the twist.

KETOSIS (Toxemia)
OCCURS MOSTLY IN ADULT GOATS

Ketosis is a term that describes a metabolic crisis. One type of ketosis occurs just before kidding and is often called pregnancy toxemia. The other occurs after kidding and is known as postparturient toxemia or, more commonly, lactational ketosis. The signs and symptoms of both are similar. However, because different stresses—pregnancy versus lactation—lead to the condition, the treatment and odds of recovery are quite different. Thus, it's helpful to think of them as separate disorders. The pregnancy version is much more common and also more deadly than the lactational version. Lactational ketosis is much more likely to respond to treatment than pregnancy toxemia. Both variations are sometimes called fatty liver disease or syndrome, and both versions often are accompanied by severe hypocalcemia, which will need to be addressed during treatment.

The word *ketosis* refers to the fact that animals in this state have switched from using glucose (sugar) as their primary energy source and are instead metabolizing fat through ketones. All forms of ketosis occur when the goat's daily needs for energy are not being met. In the case of the pregnant doe somewhat paradoxical situations lead to this: A very thin or poorly fed doe will not be able to keep up with the energy needs of her body and the fetuses; an overweight doe will not have sufficient room in her abdomen for both her internal fat deposits and a litter of kids. In both cases the intake of energy isn't enough, and the body switches to metabolizing fat.

Pregnancy Toxemia

Pregnancy toxemia can be fatal for the doe and the unborn kids. Ketones are not able to supply energy to the fetus, so once the doe switches to metabolizing mostly or all fat for her own needs, the fetuses begin to basically starve. If the kids die, they release toxins that poison the mother, causing her to decline even more rapidly. As ketones increase in the blood, the pH changes and acidosis will result, eventually causing coma. A mild state of ketosis can exist toward the end of pregnancy without causing any symptoms. If acute pregnancy toxemia is not treated early, it can easily be fatal. It typically develops slowly over 3 to 10 days and in the last six weeks of pregnancy.

SIGNS AND SYMPTOMS

Early: Vague signs of dullness of behavior. Isolation, reduced appetite, slowness to rise.

Midstage: Loss of appetite, grinding the teeth, staring upward (stargazing), muscle tremors and weakness, stumbling, mild diarrhea, drooling. An odor of ketones (a sweet acetone, or nail-polish-remover-like, odor) on the goat's breath is possible.

Tricks for Getting a Goat to Pee

Goats often urinate just after standing up. You can take advantage of this opportunity to collect urine by putting the goat of interest in a dog crate that is a little too short. The goat will usually lie down. Try to ensure that her head is turned toward the crate door when she does lie down—you can encourage that by placing some feed at the door (if the doe has an appetite). Don't place bedding in the crate—goats are also drawn to urinating where the ground is soft. Place a pile of shavings or fresh straw outside of the crate. When you are ready to collect the sample, encourage her to get up and exit the crate. Stand her in the deep pile of bedding and be ready with a specimen cup.

Advanced: Rapid breathing; feces small, hard, and mucous-covered; inability to rise; finally coma and death.

Seen on necropsy: Possibly multiple fetuses, often dead for longer than the doe has been; enlarged liver with yellow fat deposits (from body fat that has been mobilized in an attempt to meet nutritional needs but has instead become lodged in the liver) and crumbly texture; enlarged adrenal glands.

TESTING

Ketones in urine and blood can be detected using diagnostic strips (Keto-Test strips is one brand). You may smell ketones on the breath. Tests may reveal low blood sugar and low blood calcium level.

TREATMENT

Early: Administer an oral dose of propylene glycol, because it will go directly from the digestive system to the liver to be transformed into glucose (this is not the case with molasses or other sugars). Dosage is 60 ml two to three times daily. Propylene glycol is the main ingredient of both Nutri-Drench and Power Punch remedies, which can also be given for early or mild cases (see the label for dosing instructions).

Administer a B vitamin complex injection containing the equivalent of 1 g of niacin per day. Because hypocalcemia is a risk associated with ketosis, a sub-Q injection of 60 ml of a 23 to 25 percent calcium gluconate solution or an oral calcium supplement is also advisable. Offer a bucket of hot water, 105 to 110°F (41–43°C)—or, even better, raspberry or nettle tea—to help prevent dehydration. Give sub-Q fluids if needed. Offer high-quality forage and increase the grain ration. Separate the doe so she can eat without competition or stress.

Mid- to late stages: Try the measures described above plus forcing feed of the highest quality (Smith and Sherman recommend making a mash of a complete, pelletized, horse feed). Seek veterinary advice, which is likely to include IV glucose. If recovery isn't rapid and the doe is within a week of delivering, consider inducing labor to immediately remove the physical demand of the pregnancy. If she is too advanced to withstand the rigors of labor, cesarean section may be needed to save her life or that of the kids. If the fetuses are living, corticosteroids can be given to induce labor, which will occur two days after administering. After delivery, stimulate the doe's appetite by offering small amounts of high-energy feed to help the doe switch back to metabolizing glucose.

PREVENTION

Prevention is far more likely to be successful than treatment, so assessing the feeding needs of the pregnant herd and individuals is imperative. Does might begin pregnancy overweight or increase in weight as the pregnancy progresses, putting them at high risk. Grain should be slowly added during the last few weeks before delivery, or earlier for does that are carrying larger-than-normal litters and are either underweight, low in the herd order, or obese. Make sure that high-quality hay and forage are always available. Moderate exercise and low stress conditions are also important.

Lactational Ketosis

In the case of lactational ketosis, the energy demands of producing milk are simply too high, driving the doe's body to begin breaking down fat. The released fat infiltrates the animal's liver and accumulates there. If the animal is necropsied, the liver will be found to be enlarged with fat deposits, a yellowish color, and a crumbly texture—hence the fatty liver disease moniker.

SIGNS AND SYMPTOMS

Decreased milk production, refusal to eat grain, loss of body condition, reduced rumen activity. Ketones in urine and odor on breath. If hypocalcemia accompanies ketosis, and it often does, symptoms will progress as for advanced pregnancy toxemia above.

TREATMENT

Give propylene glycol as recommended for pregnancy toxemia. Also, offer high-quality forage, grain, or soaked, pelletized, complete high-energy horse feed or other high-energy pelletized livestock feed. Your vet may prescribe corticosteroids to help the doe's body to switch back to metabolizing glucose and as an appetite stimulant. Treat for hypocalcemia.

PREVENTION

High-producing does during their peak lactation need special attention to prevent lactational ketosis. Provide high-quality forage and include grain in early lactation, and increase it every few days to maintain body condition until peak milk production is reached, usually two months into lactation.

LIVER FLUKES

Flukes are a type of parasite. They are flatworms that usually require two or more hosts to complete their life cycle. In North America and Europe there are three categories of liver flukes, each with some differences in how they interact with and harm the goat. Goats of all ages and type are equally at risk; those in damper regions are more likely to encounter liver flukes. In all cases upon necropsy, damage to the liver will be visible and flukes should be found in the bile duct and liver. In some parts of the world there are also flukes that invade the pancreas of the goat.

Fasciola hepatica and its larger relative *F. gigantica* (limited to Asia, Africa, and the Middle East) are the main cause of liver fluke disease in goats. In some respects they are also the easier fluke to control because they require moisture and a particular species of freshwater snail to complete their life cycle. Limiting access to areas that are wet or have recently been wet greatly reduces the likelihood of infection. If your land is perpetually marshy, you may have to keep the goats off pasture and manage them in a more confined dry-lot situation.

Adult *F. hepatica* reside in the bile ducts, where they lay their eggs. The eggs pass out of the goat through the feces, hatch, and then infest their secondary host, the snail. Inside the snail, they grow to the larval stage and then leave the snail in moist to wet areas. The larvae swim to plants and attach themselves, ready to be consumed by an unsuspecting herbivore. Once inside the digestive system of the goat, they burrow through the intestinal wall and move through the abdominal cavity into the liver. Occasionally they miss their target and end up in the lungs or kidneys instead. Inside the liver they cause the damage that leads to the goat's demise. In acute cases, which are rare in goats, the liver ruptures, leading to internal bleeding and likely death. In chronic cases, the flukes still cause damage and bleeding, but in a slower, long-term fashion.

Fascioloides magna, also known as the large American liver fluke, shouldn't be confused with *Fasciola gigantica*. *Fascioloides magna* is the largest of the liver flukes and causes more disease in goats in North American than does *Fasciola hepatica*.

There are several reasons that it's such a problem. First, it has a wider range of habitat in which it will thrive; second, goats are not the intended host. *Fasciola magna*'s preferred hosts are deer, elk, and moose. When goats ingest these larvae, the flukes still migrate to the liver, but instead of finishing their life cycle, they continue to probe and burrow, seemingly aware that they are not in the ideal quarters. This wandering causes extensive damage to the goat's liver, and infestation by even a single fluke can kill a goat.[8] The fact that they rarely lay their eggs inside goats means that no eggs will be found in fecal samples. Because it is the immature fluke that causes the damage and even a few flukes can cause serious issues, drugs targeted to kill them must be almost perfectly effective. Short of fencing out all deer and elk (which is virtually impossible), the only way to

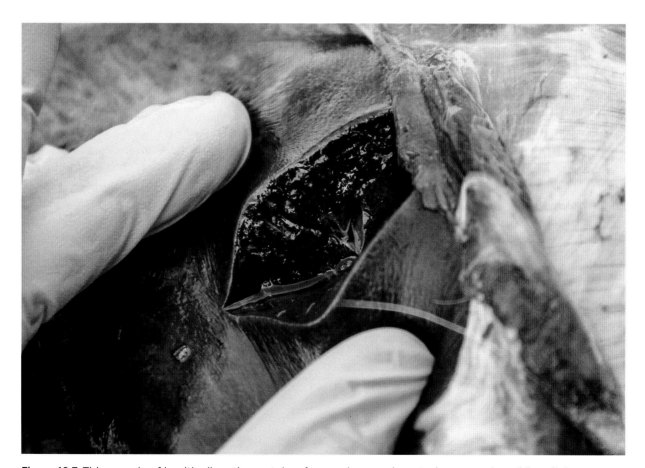

Figure 10.7. This sample of healthy liver tissue, taken from a deceased goat, shows no sign of liver flukes.

limit the likelihood of infection by flukes is to keep goats confined or to allow them browse and graze only on tall, dry grass.

Dicrocoelium dendriticum (lancet liver fluke) is quite small by comparison with the others and typically causes less damage and fewer deaths. *D. dendriticum* is the only liver fluke to use two intermediate hosts. Land snails consume fluke eggs and release the larvae in their slime. Ants ingest larvae in the slime. When the larvae develop inside the ant, they damage the nervous system in a way that paralyzes the ant. Paralyzed ants on plants are then eaten by the goat (or cow or sheep). Because the intermediate hosts don't require a wetland, this liver fluke is very common and more difficult to prevent. See table 10.3 for fluke treatment and prevention information.

PARATUBERCULOSIS
(Johne's Disease)
CONTAGIOUS FROM GOAT TO GOAT

Paratuberculosis or Johne's (*YO-knees*) disease, named after the scientist who first identified the bacteria that cause it, is a very discouraging and confounding problem. It is difficult to detect in the lab; transmittable through milk, through feces, and in utero (in the uterus); able to survive in soil for years; untreatable; and possibly contagious to humans.

This reportable disease is caused by the bacterium now called (it has had other names in the past) *Mycobacterium avium* subsp. *paratuberculosis*, or *Map* or MAP for short. An infected animal may remain symptomless for several years, but once

Table 10.3. Common Flukes in North America and Europe

Type	Eggs Detectable in Feces	Environment	Intermediate Host	Symptoms	Size of Flukes	Treatment
F. hepatica	Yes; sedimentation method better than flotation	Wet or frequently wet areas, temperature 79°F (26°C)–50°F (10°C)	Water snail	Acute cases rare in goats Chronic: weight loss, poor appetite, depressed, poor body condition, anemia, bottle jaw	18–32 mm long × 7–14 mm wide	Different drugs are effective at different stages of the fluke's life cycle; consult a vet
F. magna	Rarely	Wide range of habitat tolerated	Various snails	Anemia, peritonitis, death	23–100 mm long × 11–26 mm wide	Albendazole (Valbazen) extra-label
D. dendriticum	Yes	Dry forest grazing	Snail, then ant	Least severe of all liver flukes Weight loss, anemia, bottle jaw	6–10mm long × 1.5–2.5 mm wide	Several drugs are available; consult a vet for proper dosage for *D. dendriticum* as it may be higher than recommended on the label

symptoms begin, *Map* causes a debilitating, fatal wasting disease.

Symptoms usually don't manifest, no matter when the goat was infected, until age two or three. Symptoms often begin after a stressful event, such as kidding or being moved to a new herd. Once symptoms begin, animals progressively lose weight over weeks or months. Their coat may become rough, and they become increasingly inactive and depressed. Unlike the same disease in cattle, goats don't suffer diarrhea until perhaps the very end. Infected animals, whether symptomatic or not, don't shed the bacteria in their milk or feces consistently.

Goats can be infected with either a sheep or a cow variant of the bacteria. The disease is also present in wild ruminant populations of deer, elk, and bison. The cattle strain is more common in goats. It is more likely to affect large, commercial goat herds than small and backyard herds. A 2013 study showed a possible relationship between infection with *Map* and selenium deficiency.[9] Copper and other mineral imbalances are also likely to influence susceptibility.

Figure 10.8. This poor animal, in the end stages of Johne's, exhibits extreme emaciation and signs of diarrhea. *Photo courtesy of Michael T. Collins, DVM, Johne's Testing and Information Center*

Map is a particularly tough bacterium. It can survive freezing, boiling for two minutes, and exposure to UV light for 100 hours.[10] In a study in the UK, *Map* was found in bulk raw milk and in containers of pasteurized milk for retail sale. Because of the possible correlation between Johne's and the similar human disease called Crohn's, there is a good deal of urgency in limiting its presence in the milk supply.

In the United States a vaccine for calves is available, but it isn't 100 percent effective at preventing infection. Although maintaining a closed herd without any history of symptoms and with zero exposure to outside sheep, goats, cows, elk, bison, and deer will reduce risk, it is still plausible that the infection could become established through hay and feed contaminated with infected manure being brought onto the farm.

SIGNS AND SYMPTOMS

Gradual weight loss beginning after age one and usually before age three, rough coat, depressed and inactive behavior. Death.

TESTING

Necropsy of a symptomatic goat and testing of tissue sample is the only method for definitive result.
Fecal culture: This test shows positive only if the animal is actively shedding bacteria into its system; a symptomatic animal may or may not be shedding. Results take 7 to 13 weeks.
Blood or milk antibody—ELISA or AGID: If an animal is shedding bacteria and showing symptoms, this test detects antibodies to bacteria 85 to 100 percent of the time, but only 20 to 50 percent of the time if animal has no symptoms. Results take three to five days.
DNA PCR assay: This assay can be done via fecal or tissue culture. Reliability is similar to fecal culture, and results take 7 to 10 days.

TREATMENT

There is no form of treatment for infected animals. An aggressive testing and culling program is the best course of action to follow if you have an infected herd.

Not Getting Discouraged

Right about now, you might be feeling a bit hopeless about the possibility of keeping your goats healthy, much less buying healthy goats to start out with. We all have felt that way as we learn more about what can go wrong and even more so when something does go wrong. When focusing on illness it's easy to sink into despair, but there is hope! Your odds of maintaining a healthy herd are pretty good, especially if you focus on nutrition and environment. It's very important not to buy stock at auctions or through online ads. Instead, buy your stock from an honest breeder who also practices the same approach as you, and follow it all up with some laboratory testing.

When something does go wrong, try to keep your perspective. In 2005 our farm suffered from a couple of cases of mycoplasma infection (which is described later in this book). At that time, mycoplasma was a bit of a hush-hush problem: If you had it, you probably weren't going to talk about it. This made it seem like the end of the world for our farming venture. Not to mention how difficult it was to find information about what to do.

Thanks to a goat-raising friend who remained calm and reassured me that the problem was manageable, we moved forward. I shared our story broadly and am still contacted by other goat farmers across the country who are going through something similar. I also learned that buyers most often appreciate frank honesty and feel more trusting when you have aired all of your farm's "dirty" little secrets.

So expect some challenges along the way and know that goats are tough and goat farmers become even tougher.

PREVENTION

Test all new animals entering the herd, and quarantine until results are known. Most people choose to have the animal's blood tested, often at the same time (and from the same sample) as testing for other diseases, including CAE and CL. Buy from a herd with a testing program or from a closed herd. Maintain a closed herd if possible. Sanitation during kidding is important to limit the spread to other animals.

ELIMINATION PROGRAM

If disease is present, develop an aggressive eradication program. Increase barn and water trough cleanliness, don't spread manure on grazed fields, graze fields at no shorter than 6 inches (15 cm) of grass height, and address mineral deficiencies, especially selenium. Test feces (by PCR assay) from groups of up to 25 animals every six months—when positives result, each individual in the group should be tested if possible, or smaller groups within the positive group should be tested. Rear kids separately until the herd tests free of *Map*.

SALMONELLOSIS
CONTAGIOUS FROM GOATS TO HUMANS

Salmonellosis is caused by *Salmonella*—a large group of bacteria causing gastrointestinal illness in many species. You might think of it as a human health issue associated with poultry and eggs rather than with goats, but there are documented cases that feature goats as the origin of an outbreak, particularly from petting zoos. Three age groups are typically affected by salmonellosis: kids under a week of age, kids from two to eight weeks old, and adult goats. In each presentation, the outcome is iffy, with death rates higher the younger the animal.

SIGNS AND SYMPTOMS

Newborns: Depression, death. Occasionally with abdominal distension, pain, and diarrhea.

Use Antidiarrheals with Caution

It seems to be the instinct of our upbringing, no doubt thanks in part to persistent TV advertising, to run for a bottle of pink liquid whenever a "runny tummy" and too many trips to the bathroom are a problem. Bismuth subsalicylate (common brand Pepto-Bismol) and kaolin/pectin (common brand Kaopectate) are commonly administered by farmers to goats that are suffering from diarrhea. According to the *Merck Veterinary Manual*, however, "their only established effect is to increase fecal consistency; they do not reduce the loss of water and ions." In addition, in the alternative medicine circle (within which I was raised) there's a belief that diarrhea is the body's attempt to flush out the offending organism—in other words, diarrhea is the body's way of helping fight the underlying problem. However you choose to treat "the runs," remember that it is just a symptom of an underlying problem. Dehydration is the biggest threat to the patient, in most cases. I haven't used any "pink stuff" for my goats (or myself for that matter) in years.

Kids, unweaned: Depression, loss of appetite, watery yellow or greenish-brown diarrhea with a foul odor. Fever up to 107°F (41.7°C). Most die within 24 to 48 hours.

Adults: Depressed; no appetite; fever; watery yellow, gray, or greenish-brown diarrhea with a foul smell.

TREATMENT

Rehydration with electrolytes, either IV or forced orally. Probiotics; homeopathic remedies as advised by a practitioner, including pulsatile, arsenicum, phosphorus, mercurius, and sulfur. Aloe vera orally. Antibiotics are controversial. They can help save the animal's life but are associated with causing bacteria to be shed over an extended period of time in their feces, increasing the potential of spreading the disease. In essence, the infected animal that recovers becomes a carrier of the microbe.

PREVENTION

Avoid the introduction of new animals that could be shedding the bacteria. When new animals are purchased, the farm from which they came should be healthy. Avoid auctions, online classifieds, and the like. Stress, overcrowding, dirty pens, and unclean water troughs all contribute to the likelihood of succumbing to the illness. Poor colostrum quality or failure to ingest any colostrum can predispose newborns to the illness. Increase hygiene overall.

CHAPTER ELEVEN

The Respiratory System

A well-built goat has a wide chest and enough width at the elbows to accommodate its lungs and provide the ability to take deep breaths. The goat's respiratory system is quite similar to a human's and is affected by many of the same substances that trouble humans, including dust, smoke, and chemical fumes. Goats are also prone to similar illnesses, such as upper respiratory infections (a cold) and deeper lung infections (pneumonia).

The most common negative respiratory symptom in goats, rapid breathing, often isn't due to a disease of the respiratory system but is more likely due to something such as pain, a metabolic disorder, or a circulatory issue (see table 11.1). In addition to the signs listed in the table, you can listen to the goat's lung sounds if you have a stethoscope. They are difficult for an untrained person to assess—the goat's rumen sounds might be audible, too, and you may hear the goat's hair coat making crackling sounds against the stethoscope. If the goat moves, that can cause even more background noise.

PARASITIC PNEUMONIA (Lungworms)

Lungworms are a difficult parasite to deal with. The best news is that they aren't believed to be a widespread problem, even though many people jump to the conclusion that a goat with a chronic cough must have lungworms. Lungworms require wet conditions and a snail or slug as an alternate host to complete their life cycle. There are two varieties of lungworms in the United States: *Muellerius capillaris* is the more common, and *Protostrongylus rufescens* comes in second. Several other varieties are present in other parts of the world.

It's possible to detect all of these species in a fecal flotation, but not reliably enough for diagnoses. The only way these parasites end up in manure is if a goat happens to swallow some of the lungworms after coughing them up from its lungs. Those worms will pass through the digestive system and be excreted in feces.

There is also no anthelmintic that works reliably to eradicate them. After treatment, symptoms may recede for a while, but then typically return. If you suspect that lungworms are a problem in your herd, talk to your vet about a plan to eradicate them. In some parts of the world prolonged low-level treatments have been shown to decrease coughing, but the milk from such an animal isn't safe for human use, and the risks to a pregnant goat's fetus are not fully known. Many common dewormers are effective against *Dictyocaulus filaria*, but this species of lungworm is uncommon in goats. Anthelmintic labels do list the specific parasites that they are effective against, but only on the back. On the front

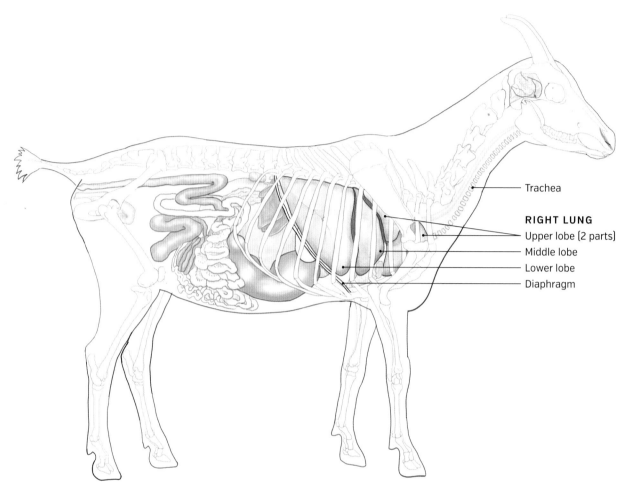

Figure 11.1. The respiratory system, right side.

of the packaging, the wording will be general, such as "effective against lungworms." This leads many producers to administer these drugs believing they are effective for treating a particular parasite when, in fact, they are not.

SIGNS AND SYMPTOMS

Shortness of breath, persistent cough, weight loss, reduced milk production. On necropsy, firm tan nodules present in lung. Larvae may or may not be seen in fecal flotation.

TREATMENT

The most common lungworm in goats, *M. capillaris*, is difficult to treat. It requires repeated deworming at intervals, but no currently available anthelmintic is 100 percent effective. Prevention is the best remedy.

PREVENTION

Avoid wet pastures and grass. Limit access to fields until morning dew is gone. Graze at heights above 6 inches (15 cm). Optimal nutrition, including access to browse, will help limit the goat's desire to access these areas.

PNEUMONIA

Many factors can lead to pneumonia, infection of the lung. It is often a secondary condition brought on by decreased immunity when a goat is fighting another

Table 11.1. Common Respiratory System Signs and Symptoms

Sign or Symptom	Possible Causes*	Sign or Symptom	Possible Causes*
Difficulty breathing or rapid breathing (dyspnea)	Acidosis Advanced pregnancy with multiple fetuses Anemia Bloat Heart murmur Heatstroke Lung CL abscesses (caseous lymphadenitis) Lungworms Obstruction in nose Pain, stress Poisoning or toxicity Pneumonia Toxemia or ketosis Urinary stones (urolithiasis) White muscle disease (nutritional muscular dystrophy)	Cough	Ammonia or other irritating fumes Congestive or other type of heart failure Difficulty swallowing due to neurological problem or white muscle disease (nutritional muscular dystrophy) Dusty or moldy feed Lungworms Pneumonia Obstruction in trachea Tight collar Upper respiratory infection
		Nasal discharge	Ammonia or other irritating fumes Dusty feed Nasal bots Nasal obstruction Upper or lower respiratory infection

* Causes listed include the more common possibilities, but the lists are not exhaustive or intended to replace the diagnostic services of a licensed veterinarian.

disorder, or is exposed to stress or poor environmental conditions. Inhaling of medications or foreign matter can also result in pneumonia. Pneumonia is a frequent cause of death in young and aged goats. Depending on the causative agent—bacterium, virus, foreign matter, abscess, lesion, or parasite (lungworms)—it may or may not be treatable other than with supportive care.

Bacterial pneumonia can result when common upper respiratory bacteria make their way into the lungs, usually as a secondary infection during another illness or even during a case of viral or other pneumonia. Pneumonia that comes on due to stress from shipping is still sometimes referred to as "shipping fever." Bacterial pneumonia can be treated with antibiotics, but if it

is secondary to viral pneumonia, the condition overall might not be cured. Sometimes antibiotics are given to an ill goat simply to reduce the likelihood of secondary bacterial pneumonia. Mycoplasma infection, which can cause pneumonia, is of great concern for goat producers because it is difficult to detect, is contagious to other goats, and can also cause mastitis (see the Mycoplasma entry on page 290).

Infection by the caprine arthritis encephalitis (CAE) virus can cause chronic, mild pneumonia, usually following a stressful event such as kidding. These cases of pneumonia often go undiagnosed.

Inhalation pneumonia occurs when a goat inhales something meant to be swallowed. This can happen during the dosing of medication using a balling gun or drenching gun or when attempting to force-feed using a stomach tube. The foodstuff then begins to deteriorate in the lungs, which causes infection. Careful technique on your part won't always prevent the goat from inhaling when you don't think it will. Saliva, feed, and vomit might also be the cause of inhalation pneumonia. Saliva and feed are usually inhaled only in situations where the goat is not able to swallow properly, such as white muscle disease, nutritional muscular dystrophy, cleft palate, listeriosis, and other neurological diseases that affect the ability to swallow.

Abscesses from caseous lymphadenitis (CL) in the lungs can cause chronic pneumonia in goats. Goats might have internal abscesses that are not visibly affecting the more superficial lymph nodes. On necropsy these may appear as yellow-green, round abscesses in the lung tissue. They might be detectable by an X-ray. (See the Caseous Lymphadenitis entry on page 275.)

Tuberculosis, caused by one of several *Mycobacterium*, can cause respiratory disease and other symptoms in goats, contrary to the common belief that goats don't contract tuberculosis. They can also be asymptomatic carriers capable of transmitting it to people and cattle. Fortunately a vet can reliably perform a TB skin test to rule out the disease or carrier status. Tuberculosis is a reportable disease in the United States.

SIGNS AND SYMPTOMS

Shortness of breath; weak, moist cough; weight loss; reduced milk production. Fever with bacterial infection. On necropsy, results will vary depending on cause (for example, CL will leave abscesses in lungs), but in general lung tissue with evidence of pneumonia may have fluid present and may be reddened or blotchy and show other color changes in one or more lobes.

TREATMENT

In cases of bacterial and inhalation pneumonia, antibiotics are often needed to prevent death. Supportive care, including rest, ventilation, fluids, probiotics, and nutritional support, is critical.

There is no treatment for pulmonary caseous lymphadenitis; euthanasia is the only humane option.

PREVENTION

Secondary pneumonia can sometimes be prevented or anticipated. Be aware that it is a risk during stressful events, such as transport, and give additional nutritional support, including probiotics and optimized vitamin E and selenium.[1] Also note when environmental conditions that stress the respiratory system—such as ammonia and methane fumes from unclean bedding—are present and take steps to address them, especially where animals that might be under stress or sick are housed. When ammonia fumes are detectable by most people, they are far beyond the level that is healthy for all animals.

UPPER RESPIRATORY INFECTION
(Rhinitis, Runny Nose)

Goats can suffer from a condition that has the same symptoms as the common cold in humans. They might have a runny nose with clear to whitish mucus, a mild cough, and runny eyes. Dust and fumes can contribute to a mild cough and runny nose. If the cause is an upper respiratory infection, the symptoms are likely to spread throughout the herd. Typically there is no fever and no loss of appetite (unless mucus interferes with swallowing). Breathing might sound obstructed when mucus blocks a goat's nose. As long as the animal is free from fever and eating well, no treatment is required. If the affected goat is penned with other goats where it can't compete for food or is otherwise stressed, though, it's wise to move it to a less stressful environment and allow it to recover before returning it to the group pen.

CHAPTER TWELVE

The Cardiovascular and Lymph Systems

The goat's heart and blood vessels make up the cardiovascular system. Goats are affected by many of the same cardiovascular problems that bother humans, but also by some conditions that are unique to caprines. Problems with the blood, especially anemia (low red blood cell count), can be due to a disease or may simply be a symptom of other, underlying problems.

The lymph system, with its network of nodes, is a part of the goat's immune system. The immune system also includes components such as gut bacteria, white blood cells, and much more. The support of this vast defense system depends greatly on nutrition and stress, as discussed in depth in other parts of this book. A goat's lymph system can be affected by the same problems as occur in human lymphatic systems, including cancer.

CASEOUS LYMPHADENITIS (CL)

CONTAGIOUS FROM GOATS TO HUMANS
CONTAGIOUS FROM GOAT TO GOAT

This highly contagious disease is one of the most prevalent health problems for the goat industry. Its occurrence causes lost productivity for milk goats and reduced value in goat hides. It is a concern for meat producers because it is contagious to humans. While the goal for every herd should be to be CL-free, it's a daunting prospect because of the infecting bacteria's persistence in the environment and ineradicable status in the animals. Once a herd is infected, it's essentially permanently infected.

The disease is caused by the bacterium *Corynebacterium pseudotuberculosis*, which infects lymph glands close to the surface of the skin through small wounds; it also infects internal lymph nodes, especially in the lungs, through inhalation. Once infection occurs it can take about two to six months for superficial abscesses to become visible or palpable. Additional abscesses may develop when the bacteria invade nearly invisible injuries to the goat's skin. When a CL abscess ruptures, the bacteria are spread into the environment and can then be picked up by other goats, through contact with fencing, head gates, and places where goats rub and scratch. When internal abscesses rupture, the bacteria are also spread through inhalation and direct contact. The bacteria can survive in the environment for months. They can be present in milk through contamination during milk collection. Because of this, milk from CL-positive herds should always be pasteurized.

Caseous lymphadenitis is so named because of the curdy, cheese-like texture and color of the gland when the abscess ruptures. It is sometimes called cheesy gland. Superficial CL abscesses are not contagious until they rupture. A blood test can fairly

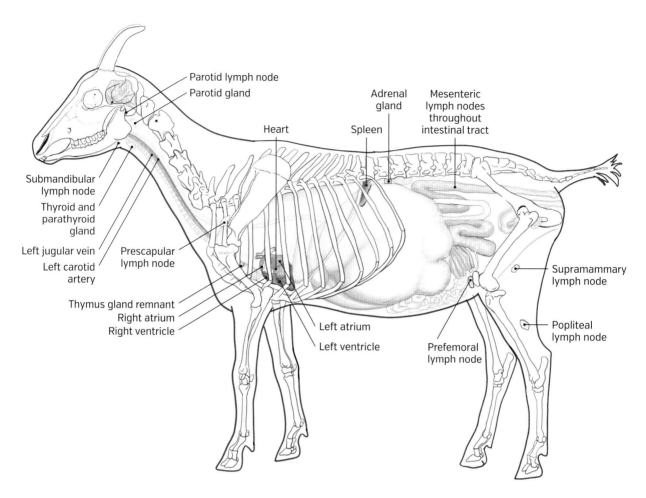

Figure 12.1. The goat's heart and lymph system as well as the spleen, and the thyroid, thymus, and parotid glands.

reliably determine if a goat has an undetected or internal case of CL. Where vaccination has occurred, though, a false positive will result, which can negatively influence sales, reputation, and more. Because of this, the vaccine, which is labeled for sheep but often used in goats, is not as widely used as it could be. Also, there is a risk of adverse reactions to the vaccine[1] including, according to the manufacturer, large areas of swelling at the site, lameness, and fever and lethargy.[2]

SIGNS AND SYMPTOMS

Swelling at any of the lymph node points (see figure 12.1 above). Open abscess with yellow, white, or greenish pasty or curdy pus. Coughing, weight loss (due to internal lung abscesses).

TREATMENT

If eradication is not the goal, goats with an undiagnosed abscess should be isolated in pens that are completely cleanable—with a concrete floor and smooth walls, for example. Because the disease can spread to humans, wear gloves when handling an infected goat. If possible, abscesses should be lanced and drained in controlled conditions, such as a separate "hospital" pen with smooth walls and flooring that can be sanitized, to prevent the bacteria from being spread in the environment. Superficial abscesses can be removed surgically

Table 12.1. Common Cardiovascular and Lymph System Signs and Symptoms

Sign or Symptom	Possible Causes*
Sudden death/ heart attack	Foot and mouth disease Poisoning Trauma White muscle disease (nutritional muscular dystrophy)
Anemia	Chronic wasting diseases Copper, cobalt, iron, or phosphorus deficiency Copper toxicity Enterotoxemia External parasites Internal parasites Plant toxicity Trauma
Heart murmur	Heart murmur
Chronic cough	Heart failure Pneumonia White muscle disease
Enlarged lymph glands	Caseous lymphadenitis Temporary local infection Tumor

* Causes listed include the more common possibilities, but the lists are not exhaustive or intended to replace the diagnostic services of a licensed veterinarian.

Figure 12.2. This doe has a sizable caseous lymphadenitis abscess that hasn't yet ruptured.

Figure 12.3. The bulge on the side of this doe's mouth, which at first glance looks like an abscess, would periodically appear, disappear, and then reappear. By propping her mouth open with a smooth piece of wood, I was able to examine her and determine that the cause was feed retention—a wad of feed lodged in a gap in her teeth and extended out to her cheek.

by a vet. All discharge and any materials that contacted it—such as bedding and gauze—should be burned.

PREVENTION

Sample and test the contents of any abscess to confirm whether it is CL (unless the presence of CL in the herd is already documented). Blood testing should be done before new animals enter the herd and again 15 days after purchase (when exposure will be detectable), which should be before quarantine is over. Maintain a closed herd.

ELIMINATION PROGRAM

If the herd is already infected a control program can be successful, but it will require heavy culling. All individuals that test positive should be culled, with the *possible* exception of highly valuable animals, which should be kept separate from the herd in a dedicated quarantine pen that is completely cleanable. No milk from positive animals should be fed to kids unless carefully and properly pasteurized. New animals should be quarantined and tested as above.

HEART FAILURE

OCCURS MOSTLY IN AGED GOATS

This term applies to many conditions that cause the heart to not work at its optimal level. In most herds the symptoms are associated with old age, and the producer usually retires or culls the animal, rather than seeking a medical diagnosis and care. Some symptoms, such as weight loss and chronic cough, that are often associated with other treatable or contagious disease may actually be the result of heart failure. If this can be determined, then you can save time, money, and medications.

SIGNS AND SYMPTOMS

Chronic, non-productive, deep and moist cough, especially when exercising. Jugular vein distension and visible pulsation of the jugular (also best observed when the goat has exercised). Reluctance to exercise. Weight loss.

TREATMENT

If the animal is a beloved pet, a veterinarian can prescribe medications similar to those used in humans with heart failure. Homeopaths and herbalists can offer remedies that might be helpful. For a retired goat that remains part of a working herd, minimize undue stress, both physical and emotional, and provide adequate nutrition. This will help keep the animal comfortable. Monitor the goat's diet to attempt to keep its weight at an ideal level so that its heart doesn't have to work any harder than necessary. Watch the animal for signs of stress and a reduction

Figure 12.4. Listen to heart sounds just behind and slightly below a goat's left elbow.

in quality of life that indicates when euthanasia becomes the most humane treatment.

PREVENTION

Sometimes unavoidable after a long, hardworking life. Good nutrition and consistent fitness are the best bet for prevention and management.

HEART MURMUR

Heart sounds in the goat are difficult to hear without some practice and facility with a stethoscope. A heart murmur, sometimes present in a newborn kid, is particularly hard to hear, because the kid's heart rate is so rapid. The heart is located to the left in the goat's chest, so place the stethoscope disc on the left side. Undersized kids with a heart murmur will pant and exhibit shortness of breath. They will also have trouble nursing or bottle feeding, simply because they need to stop frequently and catch their breath. If the kid is otherwise healthy, a murmur will usually resolve itself as the kid grows.

CHAPTER THIRTEEN

The Musculoskeletal System and Skin

A well-functioning musculoskeletal system allows a goat to be an efficient browser and compete successfully with her herdmates. Any impairment in this system is likely to negatively affect her productivity and happiness. That being said, I have had many goats that were physically handicapped to one degree or another over the years. Some of them did quite well; others did not. If an injury leaves a goat permanently damaged, but it can still keep up with the herd, looks healthy, and seems content, it might enjoy a long life.

The skin is the largest organ for all mammals. I think of the skin and coat of the goat as a window into the overall health of the inside of the animal—any problem with nutrition and all chronic health issues almost always affect the skin and hair. Because many skin conditions are symptoms of another underlying problem, I am not a believer in topical treatments, especially without addressing the root cause. Refer back to chapter 4 for details on nutritional balance and how it relates to symptoms appearing in the skin and coat.

BACTERIAL FOOT INFECTIONS
(Foot Scald, Foot Rot, Foot Abscess)

Foot problems in goats can lead to loss of productivity both because the animals can't move without discomfort, so they don't compete well for feed, and also because they suffer chronic pain. Foot scald, rot, and abscesses are caused by several bacteria that in some cases work together to damage the hoof. Although these problems are more common in sheep, goat herds can easily experience them. Moist warm conditions favor all three of these problems. Spring and early summer, when the ground is still wet and the weather growing warmer, are the most likely times for problems to occur. Hooves that have grown long are more likely to become infected. Some goats are more resistant to foot infections than others. This should be considered when you're culling and selecting for hardy stock. Nutritional deficiencies and imbalances are also related to all types of hoof problems.

Foot scald is the mildest and least painful for the goat; foot rot, as the name might tell you, is the worst. Foot scald is caused by a bacterium commonly present in the environment. Foot rot is caused by a bacterial species that is uncommon until it is brought into an area on contaminated feet. Fortunately, foot rot bacteria survive in the environment for only a maximum of fourteen days. When present, foot rot causing microbes require the right conditions and a route of entry to infect new animals. So when conditions are right for foot scald, foot rot is a greater risk as well.

Abscesses can occur along with either of the other two conditions, or independently. Abscesses are incredibly painful but can be quickly alleviated by a

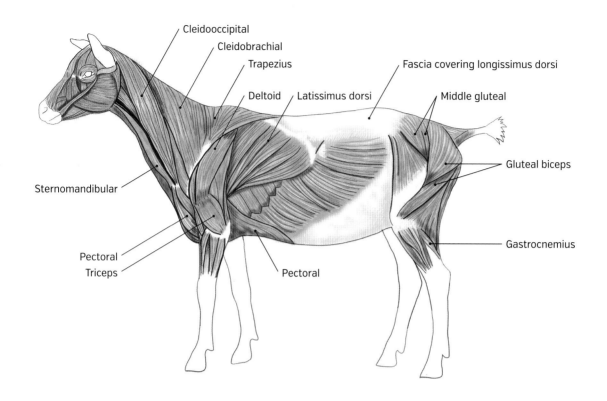

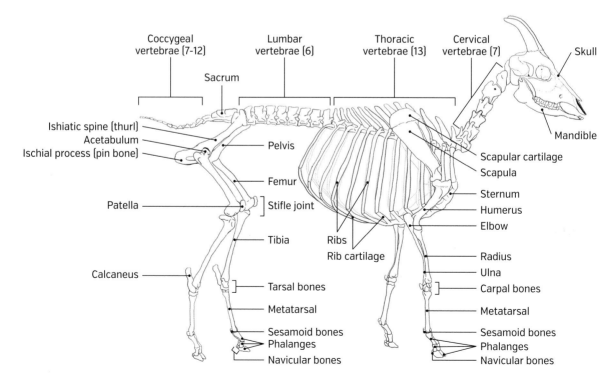

Figure 13.1. The goat's muscular (*top*) and skeletal (*bottom*) systems. Only a few of the major superficial muscles are shown here.

Table 13.1. Common Musculoskeletal System Signs and Symptoms

Sign or Symptom	Possible Causes*	Sign or Symptom	Possible Causes*
Sore feet	Foot scald Foot rot Hoof abscess Laminitis	Bowed limbs	Epiphysitis Rickets
Swollen joints	CAE arthritis Individual joint injury Mycoplasma arthritis	Scabby or mildly itchy skin	Dermatitis (multiple causes) Fungal, bacterial, or parasitic skin problem Nutritional/mineral deficiency Soremouth (orf, contagious ecthyma) Urine scald (on face and legs of bucks)
Stiff movement	Arthritis CAE or MVV Injury Rickets White muscle disease	Swelling of face, neck (not lymph node area)	Blue tongue Bottle jaw Goiter (iodine deficiency) Milk "goiter" Phosphorus excess Wattle cyst
Limb held in a bent position	CAE arthritis Injury		

* Causes listed include the more common possibilities, but the lists are not exhaustive or intended to replace the diagnostic services of a licensed veterinarian.

vet, who will locate the infection and open it up to drain. Or an abscess may burst on its own, and immediate relief follows.

SIGNS AND SYMPTOMS

Scald: Mild lameness, swelling and redness of the skin between the toes, no odor.

Rot: Severe lameness, to the point that the animal refuses to walk; swelling of the skin between the toes and around the top of the hoof (the coronary band or coronet), black color to the sole, foul odor, separation of the hoof wall (outside layer) from the interior.

Abscess: Severe lameness, usually only one foot. Heat in the hoof and foot—one foot hotter than the other feet. After an abscess ruptures on its own or has been opened up, there may be drainage at the coronary band.

TREATMENT

Scald and rot: Examine and treat the feet of all goats in the herd. Trim all hooves, and follow up with an antibacterial (such as a 10 percent zinc sulfate solution or 5 percent copper sulfate) walk-through

foot bath. Trimming tools should be disinfected between animals and all hoof trimmings discarded or, preferably, burned. In severe cases soaking in the antimicrobial foot bath should last for 30 to 60 minutes and occur weekly for two to four weeks. After soaking, the goats should not be allowed to return to wet, moist pastures until all signs of scald are gone and a minimum of two weeks has passed. Producers that aren't certified organic might be advised by their vet to administer antibiotics. Address nutritional deficiencies.

Abscess: Once the abscess has opened up, soaking the hoof in an Epsom salts solution is helpful to flush out as much pus as possible and prevent the opening from resealing until after it has been cleaned.

PREVENTION

Maintaining a closed herd and never taking any of your goats off the farm (even for shows or breeding) is a good way to avoid problems with rot or scald. Strictly quarantining and assessing new animals is also important.

Figure 13.2. According to the owner, this buck had had a hoof abscess that resolved itself. But when the goat started limping again, the owner inspected the hoof and found this large stone buried in it [*left*], presumably where the abscess had opened up and left a cavity [*right*] in which the stone then lodged itself. I recommended wrapping the hoof with gauze, then Vetrap, then duct tape to prevent more debris from packing the hole before it had a chance to grow out.

Herds that have had a problem may be able to eliminate it, but only with a strict program that includes culling of chronic carriers. Ten percent of all animals infected will remain so for life and can continue to reinfect the entire herd when conditions are right.[1]

Where control rather than eradication is the choice, treat animals with a foot bath as described above once a week when outbreaks are likely to occur because of weather conditions. Diligent, regular hoof care is also a necessity.

BACTERIAL POLYARTHRITIS (Joint Ill, Navel Ill)

OCCURS MOSTLY IN YOUNG GOATS

The umbilical cord can be a quick route for bacteria to enter a newborn kid's internal body systems. When this happens the bacteria may cause a localized infection and swelling in the umbilical veins. This can move through the goat's bloodstream and settle in the joints, causing pain and swelling. Polyarthritis means inflammation in many (poly) joints, rather than in just one, as can occur in other types of arthritis. Mycoplasma arthritis (see the entry on page 290) also causes polyarthritis. When the source of the bacteria is the umbilical cord, this condition is called navel ill or joint ill. It is rare, but bacterial polyarthritis (not mycoplasmal) can also occur in adult goats.

SIGNS AND SYMPTOMS

Fever between 103 and 105°F (39.4–40.6°C); painful, swollen, hot joints. Unwillingness to stand up or walk. Often preceded by diarrhea, pneumonia, or navel abscess.

TREATMENT

Joint fluid can be drawn out of a joint by a vet and sent for testing to rule out mycoplasma infection. Antibiotic treatment and anti-inflammatory drugs may help, but recovery rates aren't high. If a kid dies or is euthanized, consider having joint fluid drawn and tested (the vet will still have to draw the sample) to rule out mycoplasma, which is contagious, raising the concern that other animals in the herd might also be at risk or be carriers.

PREVENTION

Goats on range and those that might kid in pastures must have dry, clean areas that they can access for kidding. If maternity pens or paddocks are used, they must be clean with dry bedding that is changed between animals. Traditional wisdom advises that umbilical cords should be cut to ½ to 1 inch (1.25–2.5 cm), but holistic advice is that they be allowed to break and then left long. Good management suggests the stump should be dipped in strong (7 percent) iodine as soon as possible after birth and again the following day if navel ill has been a problem in other kids. Kids that haven't received adequate colostrum are at a higher risk, even when pens are kept clean and umbilical dipping has occurred.

BACTERIAL SKIN PROBLEMS

CONTAGIOUS FROM GOATS TO HUMANS

Goats can be plagued by bacterial skin infections that cause a variety of scabby or bumpy lesions. The most common culprits, which are from the *Staphylococcus* family, cause staphylococcal dermatitis. *Staphylococcus aureus* is one of the staph bacteria that can cause dermatitis. Because it is also of concern as a human health pathogen, as well as a cause of severe mastitis in goats, its presence might need to be ruled out by having the lesion sampled and tested by your vet. Staphylococcal dermatitis is sometimes confused with contagious ecthyma (soremouth); see the Contagious Ecthyma entry on page 286. When the infection occurs on the udder, bacteria may enter the udder and cause gangrenous mastitis.

The second most likely cause of skin infection is *Dermatophilus congolensis*, which causes dermatophilosis (which can also affect humans and other livestock). This infection often occurs in humid and wet conditions (in horses it's often called rain scald or rain rot). It usually resolves itself after a few months, with no treatment required, especially when the weather becomes drier. Dermatophilosis is easily confused with contagious ecthyma, especially when it appears on the face. But unlike ecthyma, dermatophilosis is usually isolated to one or a few animals. I have had several goats that suffered from dermatophilosis on their lips and feet.

SIGNS AND SYMPTOMS

Staphylococcal dermatitis: Pimple-like bumps on skin and hair follicles; bumps may enlarge and form scabby crusts. Commonly found on teats, udder, and around tail; in some cases on inner thighs, belly, neck, and back.

Dermatophilosis: Starts as small, round scabs usually on and in the ear. Can expand and become raised in thick matted crusts. Also found on the nose, muzzle, feet, scrotum.

TREATMENT

A milk goat with an infected udder should be milked separately, after other goats. Do not allow clothes or towels that humans use to touch the infected goat, and do not allow her to nurse kids. Wear gloves while milking. Clean the lesions with tea tree oil, iodine wash, or another antimicrobial, and then let them air-dry, or use a hair dryer on medium to low heat. Don't administer antibiotics without first taking a culture to determine which type of bacterium is the cause; have milk tested with a "culture and sensitivity" test (which is meant to not only identify the bacteria present, but also determine what antibiotics they are sensitive to). Methicillin-resistant staph can be present on the farm; if so, consider culling the animals that test positive. Lesions might also clear without any treatment.

PREVENTION

Provide nutritional support for skin and immune function. Remember that all the essential minerals and vitamins work together, but in particular copper, vitamin E and selenium, sulfur, zinc, iodine, and the B vitamins are associated with skin and immune health. (Review chapter 4 for more details if necessary.) Dry shelter is also an important part of preventing some bacterial skin problems.

CAE ARTHRITIS

CONTAGIOUS FROM GOAT TO GOAT
OCCURS MOSTLY IN ADULT GOATS

Caprine arthritic encephalitis (CAE for short) is the third important contagious disease, along with paratuberculosis (Johne's disease) and caseous lymphadenitis, that commonly plagues the goat industry. This reportable disease is caused by a virus from the same family as the HIV virus and the maedi-visna virus (MVV), which affects primarily sheep but also goats. Several other related viruses affect horses, cattle, cats, and apes. The neurological disease caused by the CAE virus was described for the first time in 1974 and the virus first isolated in 1980. Within a short time, it was found in a high percentage of dairy goat herds from many countries, especially those with larger-scale goat dairy operations. Although MVV used to be thought of as only affecting sheep and the CAE virus active only in goats, there is evidence that the two viruses can cross species. In fact, as part of a CAE prevention program, it is recommended that you avoid allowing a CAE-free goat herd any contact with sheep.

The primary symptom of CAE is chronic or progressive, debilitating arthritis. A goat infected with the CAE virus can experience symptoms in several body systems. In the lungs it manifests as chronic pneumonia; in the central nervous system as encephalitis (see the Caprine Arthritis Encephalitis entry on page 297), an infection of the brain. Other CAE-related problems are mastitis and progressive weight loss (not only due to the loss of mobility from CAE-related arthritis). Increasingly, though, a large number of goats may carry the virus and remain without any symptoms for all or most of their lives.

The most common way that the virus is spread is through the mother's colostrum and milk to newborn

Biosecurity at Goat Shows

For those of us who enjoy the competition, learning opportunities, and camaraderie of goat shows, it is always a bit of a conundrum how to maintain herd health and biosecurity. If you want to be a part of the goat show world, you do have to accept an additional level of risk, but there are many things you can do to decrease your odds of bringing anything back from the show other than ribbons, trophies, and dirty clothes.

Here are some things that you can consider when deciding if a show is up to your biosecurity standards:

- Ask whether there will be a health or vet check for all goats upon arrival and before entering the facility.
- Avoid shows in which the public might be a strong presence and be tempted to go pen-to-pen petting goats.
- Ask whether the show allows for buffer zones between pens. If not, ask if they will allow you to hang tarps to block contact between your goats and other goats.
- Consider buying extra pens that you can keep empty to provide a buffer between your herd and others.
- Understand that one or more goats will usually get loose at a show, and they could have contact with your goats—hanging tarps can help prevent this, but might be impractical. Hang feeders and place stored feed where they won't be reachable by goats outside your pens. This will encourage roamers to bypass your pens in favor of easier targets.
- When off-loading your goats upon returning from a goat show, send them through a foot bath of sanitizer.
- Change your own shoes upon return to the farm. Sanitize before using again on your own property.
- Don't let goats socialize with other goats while waiting to enter the show ring—"No nose-to-nose with does you don't know."

The ideal practice is to quarantine goats for at least 14 days upon returning from a show, but in most cases this is impractical. You should, though, closely observe all goats in your herd after returning from a show. The goats that have never attended a show before are the ones most likely to have contracted an upper respiratory virus that might then make its way through the entire herd. The same is true for a goat that is part of a herd that hasn't been to shows. In a herd where some animals have been exposed to outside viruses at shows, even those animals that haven't attended will have had some exposure indirectly through their herdmates. Some goat owners intentionally start exposing young stock to shows at an early age to "toughen" them up.

kids. Birth fluids and licking of kids by the mother goat are also routes for transmission. Infection is considered possible within the uterus; a percentage of kids who are raised completely separate from any contact with their mothers or other positive goats will begin showing positive later in life. When the disease passes to the kid from the mother, it is called vertical transmission.

Horizontal transmission, from adult goats to other goats, also occurs, through both direct and indirect contact. Indirect routes include feed troughs, water troughs, and milking equipment. The virus can be found in the semen of infected bucks, but transmission through breeding or artificial insemination is thought to be unlikely. The virus can also be spread on equipment that contacts blood, such as injection needles and tattoo equipment.

Prevention and control of CAE are all possible through three different approaches: avoidance, elimination, and coexistence. Producers who establish a new herd have the opportunity to create a CAE-free herd by purchasing from other CAE-free herds and instigating their own testing program. Owners of herds that are already infected can work toward the elimination of all positive animals, or they can choose to manage and control a positive herd, never seeking to eliminate the disease. It may seem as though the only correct choice is to work toward elimination of the virus, but consider the cost in genetics and production that many goat producers face. For years the disease was thought to only be transmitted vertically—from mother to offspring. Most producers focused only on CAE prevention through pulling kids from mothers, only to have generation after generation convert to positive later. Early in my goat career, we were penned nearby (separated by barriers) a gorgeous herd of Toggenburgs at a show. The owners were very kind to honestly share with my greenhorn daughter and me the fact that the herd was positive for CAE. Even the oldest positive does looked healthy and vigorous. The owners raised the kids on a prevention program so that buyers avoiding CAE could still purchase stock from them, but any retained offspring that reentered the main herd almost always converted to positive.

Figure 13.3. These kids exhibit very bent limbs at a young age. Unfortunately the cause wasn't being investigated by the farmer at whose dairy this image was taken, but I suspect it was nutritional in nature, given a few other indicators I observed.

SIGNS AND SYMPTOMS

Onset of symptoms usually begins between ages one and two, with most cases manifesting at about age two.
Early: Discomfort when walking or getting up; weight loss due to not being able to compete as well at the feeder. Possible swelling in the knee (carpal) joints of the front legs.
Advanced: Holding front leg flexed with swelling. Walking on knees to avoid standing. Weight loss and rough coat (due to decreased nutrition).

TREATMENT

There is no treatment other than supportive care. Euthanasia should be considered when pain limits an animal's quality of life.

PREVENTION

In positive herds, kids should be separated from their mothers immediately after birth. They should be washed and dried and fed heat-treated colostrum. Colostrum from CAE-free goats can be fed, but other illnesses that you can't test for, such as mycoplasma infection, could be transmitted. The kids should be raised completely separate from positive adults.

(Review chapter 9 for more details on a CAE prevention programs for kids.)

ELIMINATION PROGRAM

Aggressive: Kids are raised as above. All positive adults are culled. Testing is done every six months on all age groups.

Moderate: Kids are raised as above. All positive adults are kept separate from the negative animals—a physical barrier of at least 4 feet (1.2 m) of open space must separate them. No feeders, water troughs, or equipment should be shared without sanitation. If you have milk goats, milk the negative herd first, then the positive herd. Testing occurs every six months.

For new goats: Even when purchased from a CAE-free herd, testing before entering the herd and again in six months is advised. Until an animal is verified CAE-free, raise any kids borne by or sired by that animal on the CAE prevention program. If the herd the goat originates from is not tested, it's prudent to keep the new goat separate for the full six months. Sheep should not be kept with the herd unless they come from a herd with a rigorous prevention program, due to concerns of cross-species transfer.

TESTING

Testing for antibodies, using the ELISA or AGID test (described in the Paratuberculosis entry on page 267), is the most common form of detection. False negatives can occur, and goats might test negative between the time of exposure and when antibodies are produced (six months). Kids fed heat-treated colostrum from positive does will have false positive test results until several months after weaning.

CONTAGIOUS ECTHYMA
(Contagious Pustular Dermatitis, Soremouth, Orf)
CONTAGIOUS FROM GOAT TO GOAT
CONTAGIOUS FROM GOATS TO HUMANS

This painful condition is fortunately self-limiting. During its course, though, it can cause fatalities in kids—due to secondary infections in the sores and from starvation when kids stop eating because of pain. Soremouth is caused by the parapox virus, which enters through tiny breaks in the skin. The mouth is often the first area affected, because abrasions from browsing and eating are common. The disease is contagious to humans, so gloves must be worn when handling infected goats. After the disease has left the herd, to prevent human infection it is a good idea to clean and disinfect all equipment, head gates, fencing, et cetera that the goats have made contact with. A few other diseases, rare in goats, can look similar to soremouth. Some are reportable, so be sure to consult your veterinarian if there is any doubt as to the diagnosis.

It's believed that parapox is spread through contact with scabs on other animals, or spots on the skin of an infected animal where a scab has fallen off or been rubbed off. The disease has also now been documented to spread through asymptomatic carrier sheep; the same is likely to be true in goats.[2] Once an animal is exposed, it takes three to eight days for symptoms to begin and three to four weeks before they disappear.

Sometimes sores will occur on many other parts of the body, including the udder. Sores on the mouth can interfere with eating, especially for the nursing kid. When sores are present on a doe's udder, she may not allow kids to nurse adequately. She may also resist being milked, due to discomfort. Sores can plug teat orifices, making milking even more difficult.

SIGNS AND SYMPTOMS

Small sores that rapidly progress to large scabs, usually on the lips but also anywhere on the head, udder, vulva, and scrotum. Appearing in more than one goat on the farm. Scabs are easily dislodged with fingers. (Be safe—wear gloves if you try this.)

TREATMENT

Supportive measures are the usual form of treatment: Ensure kids are eating (by bottle-feeding if necessary or checking their belly for fill). Apply an udder cream or ointment to keep the teats supple so that udder sores don't obstruct the teat orifices. Apply tea tree oil or calendula ointment or tincture.

PREVENTION

Avoid contact between the herd and new goats that have signs of soremouth. Avoid shows or contact with fencing panels and equipment used by other goats. Vaccination is possible, but not always recommended unless exposure is likely.

EPIPHYSITIS

OCCURS MOSTLY IN YOUNG GOATS

This uncommon problem results from a high intake of calcium in feed, above the recommended 1.5:1 or 2:1 calcium-to-phosphorus levels. The end of the long bone in the front leg, called the epiphysis (*e-PIFF-i-sis*), experiences uneven growth, causing the goat pain on walking and rising. Joints may be swollen, and limbs can appear bent or bowlegged. The issue is easily preventable when proper mineral balance is provided. It will improve and gradually vanish if improvements are made in the mineral balance, but it takes time.

EXTERNAL PARASITES

Depending on where you live in the world, your goats might be bothered by any number of external parasites and pests. Fortunately most of these problems are more of a nuisance to the animal than they are a health concern.

Lice

Two types of lice commonly plague goats, especially during cool, cloudy weather: biting lice and bloodsucking lice. Biting lice cause severe itching, sometimes to the point of hair loss and damage to fleeces on fiber goats. Bloodsucking lice can cause anemia and even death in heavily infested kids. You don't have to determine which type of lice is the problem; the treatment is the same for both.

SIGNS AND SYMPTOMS

Nits and eggs, present at the base of hair shafts, particularly on the top of the head and on the withers. Scratching, anemia. You can shine a flashlight (not an

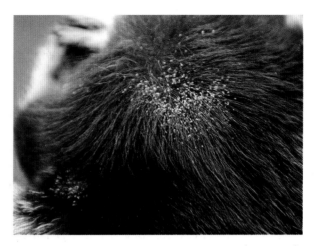

Figure 13.4. Inspecting a kid's head for nits (lice eggs) is easy to do, especially when the animal has a dark-colored coat.

LED type which is too cool) close to the skin and look for lice crawling toward the heat of the light.

TREATMENT

There are many synthetic and natural chemical treatments for lice, but many are not allowed in organic production. A simple 50–50 mixture of sulfur powder to diatomaceous earth or 100 percent of either rubbed into the animal's coat and repeated 10 to 14 days later—when eggs hatch—is an effective and organically approved treatment.

PREVENTION

Inspect new animals coming into the herd and treat if nits or lice are present. Provide sunshine when possible. Provide proper mineral balance for healthy skin and immune system. Some natural resistance to biting lice has been documented; if you observe this in any animals in your herd, you can continue to select for it.

Mites

It's easy to get confused about the difference between mites and lice. Mites are much smaller than lice—microscopic, in fact—and they are less common in goats than lice are. Their feeding action creates a

Figure 13.5. Kids should be checked for lice at several stages. Here a kid gets rubbed down with a nylon stocking filled with diatomaceous earth and dusting sulfur.

condition called mange. There are four types of mites and mange that can be a problem for goats. Sarcoptic mange, or scabies, is the most common, and most damaging, form of mange. This form of mange is reportable if detected in the United States. The mites burrow under the skin, and their feeding can damage hides for market purposes; in severe instances it can cause death. Chorioptic mites do not burrow as deeply into the skin as does the scabies mite. Psoroptic mange affects a goat's ears, and demodectic mites affect the oil glands of the goat's skin. Kids are more likely to suffer from this type of mite problem than are adult goats.

SIGNS AND SYMPTOMS

Sarcoptic mange (scabies): Raised, thickened, hairless tissue and severe itching, typically around the eyes and head and sometimes spreading to the udder, scrotum, inner thighs, and chest. Weight loss and even death possible.
Chorioptic mange: Scabs, usually on the legs and feet. Scabs less thick and less irritating than sarcoptic mange.
Psoroptic mange: Shaking the head, scratching at the ears (much like a dog with ear mites).
Demodectic mange: Small to pea-sized nodules on skin; generally no itching.

TREATMENT

Often cases of mange will resolve on their own, and treatment is often ineffective. For severe sarcoptic mange, organic treatment of a lime-sulfur dip applied in six treatments, one every 7 to 10 days, can be effective. This product is usually packaged for use on dogs or for agricultural use (usually for spraying on plants). Follow the directions for mange in sheep and cattle usually included on the product label.

PREVENTION

Provide nutritional support for strong skin and immune function. Breed and select for animals that are resistant or recover quickly.

Ticks

Ticks can be a problem for goats but are usually fairly limited in their numbers and effect. Goats may pick up ticks during a browsing walk in the forest, and afterward some may bite into the goats' udders. They should be removed by whatever method you prefer.

Lyme disease, the tick-borne illness caused by the bacterium *Borrelia burgdorferi*, has not been conclusively diagnosed in goats, but it's possible that they serve as a carrier. Blood work often shows evidence of infection, but clinical symptoms of the disease are absent. Smith and Sherman comment that Lyme disease may indeed occur in goats, but that definitive proof is at this time lacking. (Symptoms in cattle include arthritis, fever, swollen joints, and decreased milk production.)

Fleas and Flies

Fleas can bother goats but are uncommon unless widely present on the farm or on farm dogs and cats. Prevention and treatment should focus on reducing their presence on these other animals. Flies of many kinds are present on farms. Most don't bite goats but are annoying to both the animals and the farmer. They are also a main cause of the spread of infectious conjunctivitis (pink eye). Biting flies can also be present and a problem for all mammals on the farm.

Reduction of fly populations should focus on manure management methods that prevent flies from reproducing, as well as control of food matter that attracts flies, such as milk, compost, and wet grains. If other livestock animals that attract flies are present on the farm (those with wet manure, such as cows and pigs, are the most appealing to the pests), then extra work will be involved, and likely a bigger fly problem will be present. Some people have luck with the use of predatory wasps (which you must purchase and release periodically on your farm), blacklight sticky traps (you can purchase these online), and other organic means of control. In the summer fans in loafing areas can help alleviate annoyance. A note about the blacklight traps: I recommend turning them off at night, as moths are quite drawn to them when it is dark.

FOOT AND MOUTH

Foot and mouth disease (FMD) is a highly contagious reportable illness caused by a virus. Fortunately it is not present in North and Central America, Australia, New Zealand, and Japan, nor on many islands in the Pacific Ocean. Many countries have implemented harsh controls that have, for the time being, eradicated the virus. For example, in 2001 an outbreak in the United Kingdom was controlled by the slaughter of over 10 million animals. Not all countries seek to control the disease in this fashion.

Goats are considered likely carriers of the virus, usually not exhibiting any signs of the disease.[3] Human travel to countries that host FMD is a source of the spread of the illness—it is exceedingly rare for the virus to infect humans, but it does hitch a ride in our lungs and can remain there for 24 hours. Contact with animals during that time can be the start of an outbreak. When reentering a country free of the disease, travelers are asked to complete a form that includes questions about visits to farms. People who have visited a farm and don't inform authorities might be the unwitting source of a disaster.

SIGNS AND SYMPTOMS

Usually asymptomatic. Listlessness, loss of appetite, fever, increased breathing and heart rate possible. Lameness due to sores between the toes

How to Get Your Goat Through a Foot Bath

Goats are naturally averse to water. They can be trained to overcome this distaste, which is important if you're training them as cart or pack animals. But for most farms, when a foot bath is needed, you'll need clever tactics or hands-on guidance, or goats will simply leap over it.

In areas where scald and foot rot are common problems, foot baths can be integrated in a way that makes them unavoidable—such as placing them in a long chute that takes the animal to a feeding station (which provides the incentive). Even in a chute, the bath will have to be about 8 to 10 feet (2.4–3 m) long, or some goats will attempt to jump over it, and perhaps get only a couple of their feet wet. If you're using a foot bath for treatment, rather than just prevention, soaking must last for about 30 minutes, so you will need a way to keep the goats in the chute in groups. To avoid splashing, foam rubber (an old or new egg-crate-type mattress from a bed works great) or heavy wood chips can be put in the bath.

For temporary use, goats can be hand-walked through a bath, if the herd is small enough, or you can dip their feet manually one at a time.

If a foot bath will become a permanent part of a facility, it should be designed so that it can be drained and refilled easily. Suppliers of sheep and goat foot bath supplies are listed on page 337 in appendix D.

and on the goat's heels. Sores in the mouth, drooling, and lip smacking.

TREATMENT

In most countries, animals with FMD are destroyed. In those that don't follow this aggressive approach, supportive care can be given and adult animals are likely to recover. In young animals, death rates can be quite high.

PREVENTION

In the United States and other countries where food and mouth disease is not present, do not allow people who have traveled outside your country to a high-risk area within the last five days to visit your farm.

LAMINITIS (Founder)

OCCURS MOSTLY IN ADULT GOATS

Laminae are delicate living structures that connect the hard outer portion of a hoof to the internal soft tissue. During times of fever, acidosis, or the release of bacterial toxins in the goat's bloodstream, the laminae can become inflamed, a condition called laminitis. When the condition worsens to the point that the interior structures die and the bone rotates and drops, it's known as founder. Depending on the degree of inflammation, lamina damage varies. There can be so much destruction that the laminae die and lose their grip on the interior structures of the hoof. This allows the bone inside the hoof to move and begin to point downward, causing severe pain. The condition can occur in any hooved animal and is quite well known in the horse world, where it's often a death sentence for the animal. In goats it can occur suddenly and severely but more often, it's a chronic condition. Founder is more often seen in high-production dairy goats in more intensively managed situations. The acute form often follows a sudden change in diet, overeating of grain, kidding, or an episode of mastitis or other bacterial infection.

SIGNS AND SYMPTOMS

Acute: Severe pain, grinding of teeth, holding front feet out in front of shoulder line or back feet under body (to keep weight off soles of feet). Hooves hotter than normal (compare with another goat without symptoms). Unwillingness to walk or stand. Walking on knees. Fever possible.

Chronic: Abnormal posture as described for acute laminitis. "Slipper" or "sleigh" appearance to hoof, with the heel long, but curved forward; see figure 13.6.

TREATMENT

Treat acute cases with pain medication and available or prescribed non-steroidal anti-inflammatories. Treat for acidosis (see the Acidosis entry on page 247). Feed only grass hay until the goat's condition has improved.

The management measures described for prevention of laminitis (see below) should help with chronic laminitis. If a chronic case does not improve over time, consider culling or sending the goat to a home where mobility can be managed without stress and grain can be removed from the diet. Long-term use of aspirin or alternative pain management is acceptable. Your vet may reccomend aspirin at the rate of 100 mg/kg body weight twice daily and reduce over time to determine maintenance dose.[4] Frequent trimming of deformed hooves is necessary to prevent additional pain and strain on the legs.

PREVENTION

Avoid sudden changes in feed. Limit grain volume in the diet. Provide free-choice baking soda or other buffers to help prevent acidosis.

MYCOPLASMA ARTHRITIS

OCCURS MOSTLY IN YOUNG GOATS

Mycoplasmas are bacteria in which the cells are enclosed only by a thin cell membrane, instead of a cell wall. Their presence in goats is widespread, but disease is not always present—goats can remain asymptomatic carriers. There are many types of mycoplasmas, and their study and classification is ongoing and evolving.

Mastitis caused by mycoplasmas is becoming an increasing problem (see the Mastitis entry on page

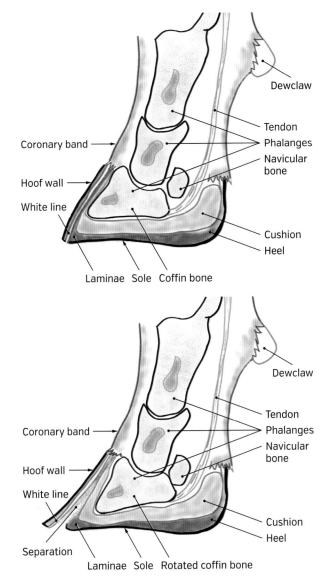

Figure 13.6. The hoof, normal (*top*) and showing rotation of the bone in severe laminitis (*bottom*). The foot is not erect and in line with the pastern.

305 for more about this). Carrier does, whether sick or not, can shed mycoplasmas in their milk and infect kids. Mycoplasmas are more likely to be shed during conditions of stress, including giving birth. Birthing fluids, colostrum, and the first milk are more likely to harbor the microbes than milk produced later in the lactation. Kids that have been infected may become carriers without illness. When an animal is affected,

a body-wide infection occurs, septicemia, which also settles in the joints causing bacterial polyarthritis (see the entry on page 282). Death rates are high in symptomatic kids.

The disease caused widespread problems in the United States in and around the 1980s, resulting in the death or culling of many dairy goats. It has largely been controlled in a similar fashion as CAE—milking hygiene and feeding of heat-treated colostrum and milk to kids. As herds have eliminated CAE, however, and returned to previous kid-rearing practices—namely, the feeding of raw milk from multiple mothers to kids—there are more reports of mycoplasma arthritis.

SIGNS AND SYMPTOMS

Fever of 104 to 108.5°F (40–42.5°C), loss of appetite, depression, unwillingness to walk. Swollen, hot, and painful joint or joints possible.

TREATMENT

The right antibiotic, such as one from the tetracycline family, may alleviate the symptoms, but the animal will still be a carrier. Have a joint fluid sample taken by a vet and sent for testing.

PREVENTION

Strict milking hygiene can prevent the transmission between adult goats. Follow CAE prevention protocols for kids as described in the CAE Arthritis entry on page 283.

NUTRITIONAL MUSCULAR DYSTROPHY (White Muscle Disease)

OCCURS MOSTLY IN YOUNG GOATS

Nutritional muscular dystrophy (NMD) is caused by a deficiency in selenium or vitamin E. (It is unrelated to muscular dystrophy in humans, a group of genetic diseases.) Goat producers generally call it white muscle disease because of the visible changes (seen on necropsy) to the muscles that are affected—usually the hind legs, heart, and diaphragm. NMD is a

Pholia's Mycoplasma Story

In 2005 we moved our herd of goats from Southern California to my family farm in Oregon. Several does had kidded in California just before the move. The kids received colostrum from their own mothers, but following that we bottle-raised them on comingled (pooled) raw milk from all of the dams. About a week after the stress of the move, one young, pre-sold buckling from our best doe became ill. He had a high fever, wouldn't eat, and didn't want to move. I tried to make an appointment with a veterinarian, but being new to the area we hadn't yet established ourselves as clients of any veterinarian, so had no luck. We called our vet from California and he overnighted what was then considered the best broad-spectrum antibiotic available to treat what sounded to him like joint ill (bacterial polyarthritis). It had no effect. The kid, named Cupid, died a painful death in my arms.

Another breeder suggested that mycoplasma infection might be the cause. I thought, "How could that be? We haven't had any mastitis." At that time, most literature on the subject of mycoplasma led one to believe that prior to arthritis symptoms, some adult milking animals would have developed mycoplasma mastitis. I really didn't want to accept the possibility that my herd could be infected. I feared it could mean the end of our reputation and herd, according to what I had read about what happened in the 1980s to herds in California.

Then another kid started to show the same symptoms. This time I was able to get an appointment with a vet, and the vet drew some joint fluid drawn from the kid's swollen front knee for testing. The test for mycoplasma takes a long time, so, after hearing that we had already lost a kid to the same symptoms even though he had been on "the best" antibiotic, the vet treated the infection as he would mastitis. He instructed us to administer over-the-counter oxytetracycline, known as Bio-Mycin. This antibiotic, unlike many others, doesn't kill by attacking the cell wall. (Because mycoplasmas have no cell walls, many common antibiotics are ineffective against them.) The Bio-Mycin worked almost immediately. Another kid, Cupid's sister Puzzle, was the next one to show symptoms. I treated her immediately, and she recovered. A few weeks later the joint fluid sample results came back positive for *Mycoplasma mycoides* subsp. *mycoides* Large Colony.

I was devastated. I thought it was all over for us. I sent milk samples out on each individual doe, but they were all negative. Whoever it was, she was not shedding anymore. I called everyone who had already received a kid from that herd, expecting to have to buy them all back. People were understanding and helped calm me.

From that time on, I fed pasteurized milk to all of the kids but had to treat every kid in that batch as if it were a carrier for life. I didn't cull anyone or segregate them, but after they matured and had kids of their own, I did not let any of those does nurse their kids. Once the last doe from my group of affected kids died or left the herd, I went back to letting kids have raw colostrum from their own moms. But still, whenever I feed pooled milk to kids, I first pasteurize it. This way, if a kid ever does come down with mycoplasma infection again, I will know for sure that the organism was passed through the kid's mother via the colostrum.

Puzzle, the doe kid that had been symptomatic, never had any recurrence of symptoms; she died years later of old age. Nor have I seen mycoplasma again in any goat in our herd. I share this story when I give presentations, and I have written about it in magazine articles. In response, I've heard from many people across the United States who own goats that have had the disease. I am of the opinion that many goats are carriers, but in a healthy, unstressed herd, they may never shed, and in the absence of symptoms, the owners never realize the organism is present in their herd.

common problem in many parts of the world from selenium deficiencies in the soil. Legumes, such as alfalfa, take up less selenium than do grasses and most other plants, so goats might suffer deficiencies if fed a high-legume diet, even if those legumes were grown in soils containing plenty of selenium. Fortunately NMD is fairly easy to avoid by supplementation, either in the feed or through routine injection of vitamin E and selenium mixtures. Treatment is also very effective when given in time. Symptoms will start to recede within 24 hours of treatment and will continue to improve over time—how long until full recovery will depend on how much damage exists.

Vitamin E and selenium work together to prevent damage to cell tissues. When either or both are low in a goat's diet, damage to muscle tissue occurs, and there is destruction of muscle fibers. The destroyed fibers are replaced by pale connective tissue that looks a bit like partly boiled chicken meat—gray to whitish (again, this can be seen during a necropsy). Such muscles don't work properly. In the case of the heart, sudden death can occur. Damage to the diaphragm muscle results in difficulty breathing. If leg muscle is damaged, the animal has difficulty walking and standing; if the tongue and throat, then difficulty swallowing. Does that are deficient may deliver stillborn kids or kids that suddenly develop symptoms, usually anytime from three days to six months of age.

SIGNS AND SYMPTOMS

Kids: Sudden death. Previously active kid found lying down, unable to stand, or walking with a stiff gait or muscle tremors in hind legs and alert otherwise. Often symptoms occur following vigorous, unusual exercise. Difficulty swallowing and nursing. Rapid and labored breathing.
Adults: Delivery of stillborn or weak kids. Fertility issues—low conception rates and poor semen quality. Difficult deliveries, retained placenta, kids fail to thrive.

TREATMENT

Administer injectable vitamin E and selenium (a common brand is Bo-Se; this must be purchased with a vet's prescription or directly from the vet). Follow instructions on the label.

PREVENTION

Determine whether you live in a selenium-deficient area and whether the area where your feed is grown is deficient in either selenium or vitamin E. Anticipate the problem by adding supplements to feed or providing a mineral mixture (most of the commonly available brand-name mineral mixes are not specific to regional deficiencies) or by giving regular (usually one or two times per year) injections of vitamin E and selenium. Common protocol is to give one before breeding season and then again four to six weeks before the doe kids. For bucks, give before breeding season and again six months later. Monitor liver levels of selenium whenever possible (harvest a liver sample from a butchered or necropsied animal as described in chapter 7).

NUTRITION-RELATED SKIN PROBLEMS

When a nutritional deficiency causes a skin problem, the most common culprit is zinc deficiency. An excess of calcium, usually from too much alfalfa in the diet, can be the underlying cause. Other nutrition-related skin problems include iodine deficiency, vitamin A deficiency, selenium toxicity, vitamin E and selenium deficiency, and sulfur deficiency. All of these possibilities should be considered. Here I describe signs and symptoms, treatment, and prevention of zinc deficiency.

SIGNS AND SYMPTOMS

Redness and itching, loss of hair. Thick cracked crusts on the skin of the back legs, escutcheon, face, and ears; scales on the skin over the rest of the body.

TREATMENT

Administer a daily dose of 1 g zinc sulfate orally for two weeks. You can use tablets intended for humans. If there's no improvement after two weeks, reassess the problem, because zinc deficiency may not be the cause.

PREVENTION

Address deficiencies of zinc in mineral rations and excess calcium in diet and mineral mixes (for specific measures, see the Calcium section on page 81).

RICKETS

OCCURS MOSTLY IN YOUNG GOATS

Rickets is caused by a nutritional imbalance, and the resulting condition includes deformities in the growing bones of the young animal. It is similar to epiphysitis, which is described on page 287, but affects the epiphyseal growth plate rather than the entire end of the long bone. The disease, which also occurs in humans and other mammals, was originally believed to be due to a deficiency in vitamin D resulting from insufficient exposure to sunlight (the skin can manufacture vitamin D when exposed to the sun). It's now understood that vitamin D levels can be adequate, but if there is a deficiency of calcium or phosphorus, rickets will also result. The growth plates of the long bones in the limbs are weak and become deformed due to the weight they are bearing. The classic sign is bowleggedness, but it can instead result in a dropping of the spine between the shoulder blades as the goat's weight settles. Goats may also simply appear stunted. As with epiphysitis, rickets is easily prevented by providing the proper mineral balance in the young goat's diet.

RINGWORM

CONTAGIOUS FROM GOATS TO HUMANS
CONTAGIOUS FROM GOAT TO GOAT
OCCURS MOSTLY IN YOUNG GOATS

Ringworm, round or roundish scaly patches that usually occur on the head and neck of young animals, is not caused by a worm but by various fungi that use the keratin in skin to grow. The patches start small and then spread outward from the center. These fungi can be found in the environment of dark, damp, and unclean barns and may infect every crop of kids housed there. Adult animals can also contract the fungus. Fortunately it is easily treated and temporary. Because ringworm is contagious, though, it's important to isolate infected animals and not bring an animal that has symptoms onto your farm.

SIGNS AND SYMPTOMS

Scaly, dry circular patches on the face and neck, occasionally other areas.

TREATMENT

Wearing gloves, scrub the affected areas with a Betadine or an equivalent povidone-iodine 7.5 percent concentration solution daily until improved. Tea tree oil can be tried rather than an iodine product.

PREVENTION

If ringworm has occurred in your facility, clean the facility thoroughly, paying special attention to dark areas. Remove all bedding; spray the walls with bleach water (0.5 percent sodium hypochlorite), and let them dry. If it's possible to expose the area to sunlight, that will help, too.

WATTLE CYST

A wattle cyst is a benign swelling at the base of a wattle. Such cysts are usually present at birth but often go unnoticed. They can be removed surgically along with the wattle.

CHAPTER FOURTEEN

The Vision and Nervous Systems

People new to goats often find goat eyes unsettling, especially the rectangular pupil, which is also found in sheep, cows, and horses. The goat is hardwired to keep its focus and visual awareness on the ground as much as possible. This is one reason a goat grows fearful when you try to lift its head, such as when you're training it to use a head gate on a milking stanchion. Losing track of the ground is, by their nature, not a good thing.

While some eye problems, such as pink eye (conjunctivitis), are quite benign, other problems that affect the vision and nervous system of the goat are often related to something very severe, of an emergency nature, and possibly lethal. They should be addressed immediately and with an awareness of the odds for the goat to heal.

ANENCEPHALY

OCCURS MOSTLY IN YOUNG GOATS

Anencephaly is a birth defect in which the baby is born missing a part of its brain and skull. Skin and tissue cover the soft spot. The kid might be dead or alive at birth, but death will result soon after delivery. I have seen this defect only one time. The kid was able to breathe and cry out, but during delivery the skin over the defective spot broke open. It was a traumatic experience for everyone. I immediately euthanized the kid.

Figure 14.1. A goat's head and the rectangular, horizontally oriented pupils of its eyes continuously shift to orient the animal in relationship to where the ground surface is—one reason goats are so sure-footed.

BOTULISM

Documented cases of botulism in goats are rare, but this disease can lead to death without symptoms ever appearing. Botulism is caused by the bacterium *Clostridium botulinum*. This type of bacterium forms spores that can survive severe conditions, including no oxygen and the heat of boiling. It's found commonly in soil, on plants, and in the intestines of animals. When eaten, it produces strong toxins that affect the nervous system and often cause death. Livestock are usually exposed through the accidental ingestion of decaying animal parts. Several decades

Table 14.1. Common Vision and Nervous System Signs and Symptoms

Sign or Symptom	Possible Causes*	Sign or Symptom	Possible Causes*
Facial nerve paralysis Head tilt	Brain abscess CAE Inner ear infection Listeriosis Ruptured eardrum Trauma	Incoordination (ataxia)	Brain parasites CAE Listeriosis Milk fever, polioencephalomalacia, grass tetany Poisoning Rabies Scrapie Spinal cord lesion Swayback (ezootic ataxia)
Muscle tremors	CAE Fear Low blood sugar Plant poisoning Polioencephalomalacia, grass tetany (staggers) Rabies Scrapie	Weakness	Botulism CAE Floppy kid syndrome Listeriosis Milk fever Plant poisoning Pregnancy toxemia Rabies Tick paralysis
Seizures	Brain parasites Enterotoxemia Low blood sugar Plant poisoning Polioencephalomalacia, grass tetany Pregnancy toxemia Pseudorabies Tetanus	Paralysis	Botulism Brain abscess CAE Milk fever Mycotoxin poisoning Pseudorabies Rabies Trauma
Severe localized itching	Pseudorabies Rabies Scrapie	Eye clouding, weeping	Contagious conjunctivitis (pink eye) Non-infectious pink eye
Circling	Brain abscess CAE Inner ear infection Listeriosis Polioencephalomalacia Rabies	Blindness	Enterotoxemia Hydrocephalus Polioencephalomalacia Vitamin A deficiency

* Causes listed include the more common possibilities, but the lists are not exhaustive or intended to replace the diagnostic services of a licensed veterinarian.

ago, I boarded horses at a stable near, and frequently rode at, another horse facility that experienced the sudden death of 15 horses from botulism. The source turned out to be hay cubes. During the baling process, the remains of a dead rabbit had been incorporated into some of the bales. Goats could be similarly exposed through contaminated processed feeds or a dead rodent decaying in a water trough.

SIGNS AND SYMPTOMS

Trembling of muscles, stiffness, unwillingness to walk, difficulty chewing and swallowing. Sudden death.

TREATMENT

The only form of treatment is supportive care, including keeping the animal well hydrated. Give charcoal and probiotics. Treatment can be successful if only a small amount of toxin-producing bacteria have been consumed.

PREVENTION

Inspect water troughs daily for drowned rodents and birds. Clean feeders of rotting plant material.

CAPRINE ARTHRITIS ENCEPHALITIS (CAE)

OCCURS MOSTLY IN YOUNG GOATS

As its name suggests, caprine arthritis encephalitis can cause both encephalitis and arthritis. The nervous system form of the disease is encephalitis. This form of CAE can be confused with other diseases. It typically affects goat kids between the ages of one and six months. Other diseases that occur about the same age, such as copper deficiency in kids and nutritional muscular dystrophy, might produce similar signs and symptoms. It can also be confused with listeriosis, another neurological condition described later in this chapter. If CAE is present in the adult herd in its arthritis form (this form is described in the CAE Arthritis entry on page 283), that is often a good indicator that neurological signs are due to CAE as well, manifesting as encephalitis.

If there is uncertainty whether goats are suffering from CAE or nutritional muscular dystrophy (see the entry for this disease on page 291), one option is to treat the animal with vitamin E and selenium. If there is no improvement, that's a good indicator that the underlying cause is CAE.

SIGNS AND SYMPTOMS

In goat kids aged one to six months, incoordination and inability to lift hind feet properly. Progressive signs include loss of ability to use the limbs (usually the hind legs are first), blindness, head tilt, difficulty swallowing, and loss of facial nerve use.

TREATMENT

There is no effective treatment for the encephalitic form of CAE. Euthanasia is the humane choice and should be implemented before a secondary case of pneumonia or starvation causes death.

PREVENTION

For preventive measures for this disease, refer to the CAE Arthritis entry on page 283.

CONJUNCTIVITIS (Pink Eye)

CONTAGIOUS FROM GOAT TO GOAT

Bacteria or an injury to the eye can cause conjunctivitis, commonly known as pink eye. When contagious bacteria are involved, the disease can spread through the herd, usually transferred by flies or direct contact. Although the traditional choice is to treat conjunctivitis with easily accessible antibiotic eye ointments, cases are just as likely to heal on their own, even cases where a goat appears to have become blind, with the eye appearing cloudy or opaque. The only cases of conjunctivitis that have occurred in my herd

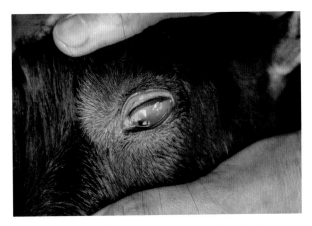

Figure 14.2. The classic symptoms of vitamin-A-responsive conjunctivitis are visible here: redness and clouding. This goat of ours responded well to treatment.

have been during the winter months when vitamin A has been at its lowest supply in the stored feed. Supplementing my herd's diet with leafy green crops, stored pumpkins, carrots, or other supplements has seemed to prevent further cases.

SIGNS AND SYMPTOMS

Tearing and redness of the white part of the goat's eye (visible only when the eyelid is pulled back) progressing to clouding of the entire eye and ulceration of the cornea. When ulceration occurs, the animal appears to be permanently blind, but when the ulceration heals, the eyes will be clear. You may still be able to see a small indentation or amount of scar tissue on the surface of the eyeball, but it doesn't seem to bother the goat.

TREATMENT

Rinse the eye with sterile saline solution; over-the-counter products used by contact lens wearers work well, or use calendula tincture. Administer vitamin A sub-Q and feed cut carrots or other foods high in vitamin A. If possible, keep the goat out of bright light and in a low-stress environment. Antibiotic ointments might be suggested to treat affected animals in a conventionally managed herd, but their effectiveness is often questioned.

PREVENTION

Closed herds are unlikely to experience contagious versions. Keeping fly populations and stress levels low, but nutritional status high (particularly vitamin A), helps prevent infections.

COPPER DEFICIENCY IN KIDS
(Swayback, Enzootic Ataxia)
OCCURS MOSTLY IN YOUNG GOATS

Two terms, *swayback* and *enzootic ataxia*, are used interchangeably to describe the symptoms of copper deficiency in kids. Swayback is the term applied to kids born with copper deficiency. Those who become copper deficient a few weeks after birth are said to suffer from enzootic ataxia. *Enzootic* means "widespread" (similar to the term *endemic*, which is used in relation to human health problems). *Ataxia* means "incoordination" and "stumbling." For clarity, it would be better to simply call these conditions early copper deficiency and delayed copper deficiency.

Both swayback and enzootic ataxia result when the mother didn't have enough copper in her diet to allow proper formation of the spinal cord in the developing fetus. The early form indicates that the dam was deficient for the entire pregnancy; the late form indicates deficiency only for the last stages of pregnancy. Since goats store copper in the liver, typically a mother will pass her stores to the fetuses and then may end up deficient before the end of her pregnancy. For that reason the early form is uncommon and the delayed form more common. Appearance of the problem in one kid may well indicate a herd-wide problem related to insufficient copper in the diet.

SIGNS AND SYMPTOMS

Swayback: Weakness of kids at birth. Inability to stand or remain standing. Muscle tremors and shaking of the head.
Enzootic ataxia: In kids from ages 1 to 28 weeks, weakness, tiredness, difficulty standing, equal degree of muscle weakness on both sides of the body, muscle tremors, stumbling. Most common for symptoms to appear around 13 weeks.

TREATMENT

Early: The kid is not likely to survive. If it dies, and copper deficiency is suspected, a liver biopsy should be done to determine whether swayback is the problem so that other pregnant does can be treated.
Delayed: Early treatment by a vet with intravenous copper glycinate might be effective.[1] As with the early form, it is very important to determine (by liver tissue sample) if enzootic ataxia is present so that the problem can be addressed herd-wide.

PREVENTION

The copper needs of goats are not fully known. There is much speculation about the required minimum.

Copper is easily blocked by excesses of several other minerals including sulfur, molybdenum, and iron. Even if you are supplying your herd with the current recommended copper levels in the diet or mineral mix, if there are inhibitors present in excess, you may have to adjust the amount of copper. I suggest taking liver samples of any goats whose death or euthanasia allows for collection. (Refer back to chapter 7 for instructions for collecting a liver sample).

LISTERIOSIS ENCEPHALITIS

CONTAGIOUS FROM GOATS TO HUMANS
CONTAGIOUS FROM GOAT TO GOAT

Listeriosis is caused by the bacterium *Listeria monocytogenes*. It can take several forms of illness in goats, but the most common form is encephalitis, inflammation of the brain. Another form causes septicemia—a generalized blood infection—and abortion. Rarely, listeria infection can also manifest as a case of mastitis.

When *Listeria* causes encephalitis, it enters the goat's system through the mouth and tiny scratches or wounds and moves to the brain. In the septicemic form, it enters the bloodstream through the intestinal wall.

Many types of *Listeria* bacteria are naturally occurring and very common in the environment. Most are harmless, but several strains of *L. monocytogenes* can be deadly to both goats and humans. Healthy goats and humans can come into contact with an unknown number of these bacteria on a regular basis, but for those in a compromised state, illness and death can result. We have seen one case of listeriosis at Pholia. An extreme weather change from freezing one weekend to over 100°F (38°C) the next was very stressful to a pregnant doe in her fourth month. By the following weekend she was showing the classic signs and symptoms. The poor odds of recovery, in addition to the negative effect that antibiotic treatment would have on the fetuses, led me to choose to euthanize the doe. Necropsy of her brain was positive for listeriosis.

L. monocytogenes is a major pathogen of concern for human health. Because of its regular presence on farms, the contamination of milk and meat through contact is always of concern for food producers. Consumption of milk that carries the bacteria can lead to illness. Goat products such as milk and meat can also become contaminated after harvest by *Listeria* in the farm environment, again leading to illness. The microbe can take up residence in the intestinal tract and be shed in the feces. There have also been rare cases of listeria mastitis and food-borne outbreaks from contact with the raw milk from such an animal.

SIGNS AND SYMPTOMS

Depression, loss of appetite, fever up to 107.6°F (42°C), loss of blink and reflex to stimuli on one side of the face, ear drop on the same side, slobbering or drooling, difficulty swallowing, and circling in the direction opposite the facial symptoms. Subsequently, inability to rise, and death.

TREATMENT

Recovery is possible only with aggressive antibiotic treatment early in the disease. Even then, the chances are not good. Administer IV penicillin as recommended by a vet every six hours until you see improvement and then IM for a week. Intensive supportive treatment during the antibiotic therapy is a must. Consult your veterinarian for directions and perspective.

PREVENTION

L. monocytogenes prefers cool, moist environments and is found in and around water troughs, in wet soil, and in wet feeds such as silage that is not kept at a safe acid (pH) level. Regular cleaning and sanitation of water troughs, ensuring that soil does not contaminate feeds, and proper fermentation and maintenance of silage at a pH of less than 5.0 will reduce the likelihood of listeriosis. Feeding, nutrition, and environment that support immune function also reduce the risk that animals will contract the disease.

POLIOENCEPHALOMALACIA
(PEM, Goat Polio, Stargazing)

The word *polio* means "gray"; *encephalo* refers to the brain, and *malacia* to a softening. In polioencephalomalacia (PEM), goat polio, the gray matter of the brain experiences abnormal softening. The condition is also called stargazing—a reference to one of the common signs observed in animals suffering from PEM.

It was previously believed that PEM was synonymous with thiamine (vitamin B_1) deficiency, but new research has brought that into question. PEM also results from sulfur excess, lead poisoning, and salt poisoning. Interestingly, however, no matter the cause, thiamine is an effective treatment. It's known that thiamine plays a role in keeping the brain cells and nerve cells healthy. So even if an animal is producing enough thiamine in the rumen, additional thiamine can help alleviate PEM.

Rumen bacteria produce thiamine, so feed changes that shift the rumen's microbial balance can short-change the goat on vitamin B_1 (as explained in more detail in chapter 4). PEM occurs more frequently in situations where grain is a major part of the animal's diet. It also occurs more often in weanlings and young adult goats, perhaps due to changing diets. One case we have had at Pholia followed a mild bloat, which no doubt changed the rumen microbial balance. I caught it early, and the goat responded immediately to intramuscular injections of thiamine. Early symptoms of PEM mimic other diseases, but the administration of thiamine will never harm the goat, so its use is a way to cure, improve, or rule out PEM.

SIGNS AND SYMPTOMS

Early: Depression, loss of appetite, nervousness, holding the head high, staring into space forward or upward (stargazing). Diarrhea possible.
Progressive: Wandering, circling, staggering, muscle tremors, acting blind.
Final: Cross-eyed appearance, side-to-side eye movement, lack of blinking, convulsions, death.

Figure 14.3. This young doe, Greta, exhibited stargazing symptoms for over a week, during which time I treated her aggressively for both listeriosis and polioencephalomalacia. She also circled and acted blind on one side. Initially she was unable to swallow. She recovered fully.

TREATMENT

Administer 10 mg/kg (5 mg/pound) thiamine every 6 hours for 24 hours. Give the first dose IV if possible. The others can be IM or sub-Q. If only B complex is available, check the label to see how many mg/ml of thiamine it contains, and do the math to make sure you are administering the correct amount of thiamine.

PREVENTION

Avoid sudden feed changes, reduce amount of grain in the diet, avoid moldy feeds, avoid feeds high in molasses (these contain too much sulfur), make sure that kids are eating plenty of fiber during weaning.

RABIES
CONTAGIOUS FROM GOAT TO GOAT
CONTAGIOUS FROM GOATS TO HUMANS

While rare in goats, cases of rabies can easily occur if a goat is exposed to a carrier or diseased animal.

Goats aren't less susceptible than other animals, just less likely to interact with common carrier animals such as raccoons, skunks, or foxes. This reportable disease is caused by a virus that is transferred through contact with the infected animal's saliva, usually through a bite. There may be an extended incubation time before symptoms appear.

SIGNS AND SYMPTOMS

Extensive itching—to the point of rubbing off the skin. Excitement, aggression, depression, excessive salivation.

TREATMENT

There is no effective treatment. If a goat is suspected of having rabies, it should be isolated and euthanized (not by shooting, however, which destroys the brain—which must be necropsied) and then sent for necropsy. Anyone handling the animal should wear gloves and a mask and avoid contact with saliva.

PREVENTION

Although there is no rabies vaccine approved for goats in the United States (one does exist in many other countries), if your area is one of high risk and cases of rabies in goats have occurred, you should consult with your veterinarian to see whether vaccination of your herd could be done.

SCRAPIE

CONTAGIOUS FROM GOAT TO GOAT

Scrapie refers to two diseases: transmissible spongiform encephalopathies (TSE) and bovine spongiform encephalopathy (BSE). Both are reportable diseases in the United States. It is unclear how closely related they are—more research is needed. Indeed, much is unknown about BSE—also known as mad cow disease—and its small-ruminant cousin (affecting goats and sheep) TSE, more commonly called scrapie. The name *scrapie* describes the itching behavior of infected animals. BSE was first documented in cattle in 1986; scrapie has been known for over 250 years. Both diseases are caused by abnormal proteins known as prions. BSE can be transmitted to goats; there has been one documented naturally occurring case.[2] When contracted by humans, BSE causes a similar disease called Creutzfeldt-Jakob disease (CJD).

Although active cases of scrapie are believed to be extremely rare in goats, it is of increasing concern to regulators whose focus is protecting the human food supply. The laws focusing on it are becoming more stringent in the United States and other countries as more is learned about its method of transmission. Regulations include having a registered scrapie premises ID number and having animals inspected before transporting them across state lines. Exporting them requires even more oversight.

BSE can be transferred to humans through contaminated cow products, especially meats and parts of the central nervous system. In fact, that is the way that cows became infected—they were fed feeds containing ruminant-sourced ingredients. It is unknown if TSE has the same potential. There is some speculation that the first BSE case may have come from contact with scrapie through ingesting of parts of an infected sheep or goat. Laws in most countries now prohibit the feeding of animal parts, including blood meal and bonemeal, especially from ruminants, to other ruminants.

Signs and symptoms of these diseases come on slowly over the course of several months. Because so much remains unknown about these diseases, regulators in countries that have elimination programs are likely to react to a suspected case or to a single positive case by calling for destruction of an entire herd. Unfortunately the disease is almost impossible to conclusively diagnose unless the animal is destroyed and then necropsied. Perhaps new testing methods will be developed and existing tests will improve so that needless destruction of healthy animals and loss of genetics can be avoided.

SIGNS AND SYMPTOMS

Loss of interest in surroundings, gradual weight loss, severe itching, irritability, posture with rear legs held under belly and rump angle steep. Signs may be limited to progressive weight loss only.

TREATMENT

There is no effective treatment. Euthanasia with necropsy is essential if either condition is suspected.

PREVENTION

Avoid comingling your goat herd with sheep or cows from a herd whose health status for this disease is unknown.

TETANUS (Lockjaw)

Tetanus is caused by *Clostridium tetani*, a bacterium from the same genus as the organisms that cause botulism and enterotoxemia. All *Clostridium* species are spore-forming anaerobic bacteria (*anaerobic* meaning "able to survive and grow without the presence of oxygen"). *C. tetani* invades a body through a wound or by entrance into a body cavity, where it grows and produces powerful toxins that cause muscles throughout the body to spasm—tetany.

Common routes of entry are through puncture wounds; into the uterus during an assisted delivery or procedure done using obstetrical tools; and through wounds or skin openings after disbudding, dehorning, tattooing, and castration (primarily by banding). Bites, hoof trimming cuts that make contact with blood vessels, and other wounds can also serve as entry points. Once *C. tetani* enters the body, spores increase and release toxins. The typical time interval from infection until symptoms appear is about 10 to 20 days, but it can be shorter or longer.

Horses commonly harbor the bacteria in their intestines and spread them in the environment.[3] A farm with horses is evidently more likely to have the bacteria present in the environment. Farms where tetanus has never occurred may be at a lower risk.

Tetanus is usually a death sentence for the animal. It is rare for recovery to occur even with aggressive treatment, but it can happen. If an animal is of high genetic value and symptoms are discovered early, treatment might be considered. When I was young we had a steer come down with tetanus after castration by banding. We were able to pull him through with medication and intensive supportive care.

SIGNS AND SYMPTOMS

Early: Anxiety, slightly stiff movements, and mild bloat. Classic "sawhorse stance," a pose with front feet stretched forward and back feet stretched behind. Unwillingness to move. Difficulty opening the mouth ("lockjaw").

Advanced: Hypersensitivity to sounds and touch, even to the point of having seizures. Inability to rise; all limbs extended stiffly. At this stage death usually comes within a day or two.

TREATMENT

If treatment is considered, it will consist of aggressive antibiotic therapy along with tranquilizers or muscle relaxants and tetanus antitoxin. Supportive care must be intense: Place the animal in a quiet area in comfortable conditions and provide fluids either IV or through a stomach tube.

When death occurs, the momentary cause is usually inability to breathe due to spasms in the diaphragm. Euthanasia should be done before this stage is reached.

PREVENTION

Vaccination with tetanus toxoid is the usual method of prevention. Vaccination protocol: The first vaccination is followed in three to four weeks by a booster and then annually. Does can be revaccinated with their annual dose one month prior to kidding to also provide initial protection for their kids. Kids should receive a booster on this initial protection at age three to four weeks.

At Pholia, as well as other farms that I know, we don't regularly vaccinate with toxoid. We do give a dose of antitoxin—which only works against any toxin present and will not provide lasting immunity—prior to procedures such as disbudding and tattooing or if a wound occurs during hoof trimming or any other time. Wounds of any kind are cleaned immediately. A solution of mild soap and warm water is a good choice for wound cleaning. When I did do castrations (all of our pet-quality boy goats are now shipped off the farm before reaching castration age), I administered toxoid for longer-term protection. We haven't yet experienced a case of tetanus.

CHAPTER FIFTEEN

The Mammary System

Most of the health concerns related to a goat's udder are some form of mastitis. And rightly so, because with mastitis comes loss of quality milk and loss of production, whether that milk is destined for a dairy or for raising kids. As with most other health problems, the goat's immune system function is the key to reducing the likelihood of mastitis occurring and increasing the animal's ability to recover if her udder does become infected.

The udder is a gland, and the function of glands is to secrete a hormone or other substance. The job of mammary glands is secretion of milk. When the milk production ducts in the goat's udder excrete milk, they also shed a good portion of the cells lining the duct. This sort of glandular secretion is called apocrine. The way goat udders make milk is the way that the human mammary gland makes milk as well. Cows and sheep, on the other hand, don't shed cell wall material. Instead their glands produce milk by what is called the merocrine system.

The term *somatic cells* (somatic means "from the body") is used to describe both white blood cells and cell that make up lining (epithelial) material. White blood cells are the bodyguards of udder health and are on patrol at all times. When an invader is detected, more rush into the udder to attempt to prevent an infection. Somatic cell counts (SCC) are done to monitor the presence of white blood cells in milk as a way to determine whether an infection or inflammation is present. Because goats also shed harmless cell material, their somatic cell count is naturally higher than that of cows and sheep. Monitoring SCC is discussed in detail in the Mastitis entry on page 305.

BLOCKED TEAT

When one half of a doe's udder is unable to be milked or nursed, the problem may be that the orifice or teat cistern is blocked by scar tissue or as a birth defect. If that teat has given milk in the past, though, then trauma is the likely culprit. In cows, this type of damage is colloquially referred to as a "spider"—a web of tissue that develops somewhere in the teat and traps the milk inside the udder. A Jersey cow of mine developed one. I could insert a teat cannula all the way into the teat, but a mass of tissue prevented it from passing into the cistern area. A vet may be able to use a needle to open a blocked teat. If this is tried, then the animal will need to be milked frequently afterward while the tissue heals. The doe will be at risk of mastitis if the blockage was in the teat canal, because the needle will create an opening that is not properly protected by skin.

Another option is to leave a blocked teat untreated, and the udder half should dry itself off over time. Often animals will develop increased production in the other half. But every freshening will bring repeated discomfort on the blocked side, and thus culling is another option to consider.

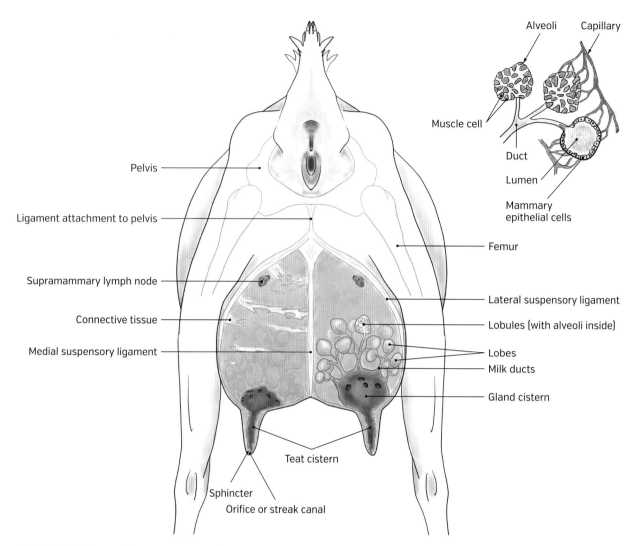

Figure 15.1. The goat's mammary system.

EDEMA

Before freshening, some animals' udders will become unusually swollen, due to fluid accumulating in the tissues—not just milk in the milk-producing tissue. When you press your finger to the skin of an udder with edema, it may feel a little like a memory foam mattress. It is important to determine whether or not the doe has mastitis. Plan on observing the milk and checking the somatic cell count as soon as the doe freshens. If it is a case of simple edema and the doe is uncomfortable, try gently massaging the udder with circulation-stimulating ointments, such as peppermint oil; applying a hot pack may help, too. If the animal isn't in too much discomfort, you can ignore the edema and it will resolve on its own. If the animal has mastitis, treatment is important. See the Mastitis entry on page 305.

GYNECOMASTIA

On rare occasion a buck's mammary tissue will produce milk. This is known as gynecomastia. (This syndrome can happen with all mammals, even

Table 15.1. Common Mammary System Signs and Symptoms

Sign or Symptom	Possible Causes*	Sign or Symptom	Possible Causes*
Bumps on udder skin	Contagious ecthyma (soremouth) Staphylococcus dermatitis (see bacterial skin infections) Warts	Nodule inside udder	Blocked milk duct
Lopsided udder	Clinical or subclinical mastitis Partial dry-off from uneven nursing Self-sucking or letting others suckle	Crystals in milk or blocking teat	Milkstones
		Blocked teat with full udder	Congenital problem or previous trauma
Hard, swollen udder	Blocked teat (on doe being nursed, not milked) Edema Mastitis	Hot udder	Mastitis
		Cold, blue udder	Gangrenous mastitis
		Stringy, thickened milk	Mastitis
Swollen bump on or near teat that leaks milk	Weeping teat, teat wall cyst	Milk in udder of young goat or buck	Gynecomastia Precocious udder Witch's milk

* Causes listed include the more common possibilities, but the lists are not exhaustive or intended to replace the diagnostic services of a licensed veterinarian.

humans.) In some cultures, milk from a buck is considered an aphrodisiac. There are several possible causes, including chromosomal imbalances, adrenal tumors, or hormonal imbalance. Bucks with this condition may or may not be fertile. It's thought that high-producing dairy goat lines are more likely to produce bucks with this condition due to selection for high-production genetics.

MASTITIS

Mastitis is an inflammation or infection in the goat's udder. As with many other mammary disorders, all mammals can get mastitis, including humans. Mastitis can be so mild that it isn't noticeable or painful or so severe that the affected part of the udder literally dies—gangrenous mastitis. Mild mastitis can appear in two forms, subclinical and chronic, and both are a problem for the dairy farmer, because both affect milk quality as well as production volume and overall animal vitality. For meat and fiber goat producers, mild mastitis can mean loss of milk

needed to help kids grow and strain on the immune systems of the mothers.

There are many causes of mastitis. The virus that causes caprine arthritis encephalitis (CAE) can cause one type. Eating too many avocado leaves from particular types of avocado trees has been shown to cause udder inflammation. And many species of bacteria can cause an infection in the udder.

Keeping mastitis at bay requires a multifaceted approach that includes a high level of nutrition to support the immune system, a clean environment, herd dynamics that don't cause undue stress, and a planned approach to monitoring udder health through milk testing.

Goats have a somatic cell count, SCC, that is naturally higher than that of cows and sheep, as I mentioned earlier. Fortunately the laboratory equipment (the most common machine in use is called the Fossomatic) used for cell counts today is much more accurate when used on goat's milk and even includes a "goat milk" setting that automatically factors in a 27 percent reduction compared

Intramammary Antibiotic Infusions

In the 1970s, when I was milking and showing my two Jersey cows, you could "earn" a beautiful set of glasses with a matching water pitcher adorned with blue ribbons and a color image of the ideal dairy cow of your choice—Jersey, Guernsey, and so on. You earned these appealing items by buying a specific number of mastitis treatment udder infusions and then sending the box tops off to the company. In those days antibiotics were still miracle drugs—effective, cheap, and useful. I eagerly made my purchases and sent off for my set. Sadly, the glasses have all since broken (they were really pretty). But I still have the pitcher!

Most literature and, of course, manufacturers of dry and milking cow antibiotic treatments (or "therapies") still promote the regular use of these products, especially at dry-off time, as the only responsible way to manage a dairy herd. Their use is particularly widespread in the dairy cow industry. But do your goats need them? Probably not.

First, antibiotic resistance continues to increase in agriculture, as it does in the realm of human health. Second, which antibiotic to use to fight a particular pathogen can only be determined by a culture. Third, the cannula through which these medications are delivered causes its own damage to the teat orifice and is known to be associated with an increase in somatic cell counts after use. Think about organic producers, who seem to get along fine, albeit with a bit more thought and work, without using antibiotics.

The ready availability of these medications gives their use a false sense of harmlessness. But don't be fooled. You may be wasting your money, as well as increasing the resistance of bacteria on your farm—reducing the effectiveness of medications that would otherwise be available to one day help save a life.

---

with the cow's-milk setting. Somatic cell count can be monitored on the farm using a simple test called the California Mastitis Test or CMT. There are other tests that do the same thing as well. I included the CMT on the list of supplies for the advanced remedy kit in chapter 6.

Subclinical Mastitis

Chronic subclinical mastitis is common, and often undetected, in larger herds. It's usually caused by a low-grade bacterial infection. When bulk tank samples (meaning the pooled milk of the entire herd) are used as the only indicator of udder health for the entire herd, this type of infection goes undetected, which can lead to a false sense of security. The count may be within acceptable limits as a whole, but in reality one or more goats may have a very high somatic cell count and yet no visible symptoms. Subclinical mastitis leads to loss of milk production and increased odds of developing acute mastitis.

SIGNS AND SYMPTOMS

Slight loss of production on one side, or shrinkage of udder size on one side (often by the time this is noticed, the infection is gone). Normal milk production and udder appearance.

TESTING

Somatic cell count is elevated in one half compared with the other. Counts may only be slightly elevated or normal, having spiked for only a brief time and then gone back to normal. If a California Mastitis Test is performed, it is critical to read it based not just on the gel *thickness,* but on the *difference* between the two udder halves. If one shows more gelling than the other, that indicates a problem. If

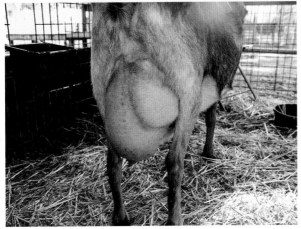

Figure 15.2. The blackened area on this udder (*top*) is due to gangrenous mastitis. Treatment of the infection was begun about a week before this photo was taken. Some months later, the doe (Wanduh) freshened again, and the damaged side of the udder still shows some milk production tissue that freshened with milk (*bottom*). This side later dried off, and she milks well from the remaining side (in fact, we now call her "Wanduh the one-teat wonder").

they are identical, then the goat may just have a naturally higher SCC.

TREATMENT

Many organic producers use garlic—given orally as a tincture, in cloves, or in capsules—as a treatment for mastitis. I have had good luck with it as well. I use commercially available garlic capsules manufactured for human use. Most goats will eat them out of your hand, and for those that don't, you can use a balling gun.

The most effective treatment—difficult to perform at large farms—is increasing the frequency of milking for an infected doe to keep her udder empty of bacteria. If kids are still nursing, that is one option. If they haven't nursed in a while, though, returning them to nurse may be too rough on the doe's teats, with damage to the skin and teat canal resulting.

When mastitis is widespread, consider nutritional deficiency along with milking equipment problems, sanitation, and water quality.

A homeopathic combination product called Mastoblast is available. I haven't had any luck with it, but I have tried it on only a couple of animals. Use other homeopathic remedies based on symptoms.

PREVENTION

Monitor the SCC of individual animals regularly, and then compare the difference in count between udder halves. Strict milking hygiene is important to prevent spreading bacteria between does. Inspect teat orifices and skin regularly for damage. Pay attention to overall herd nutrition, immune support, and environment hygiene. Consider regular feeding of probiotics and kelp meal, and address possible mineral deficiencies. Copper and vitamin E and selenium deficiencies are linked to elevated SCCs and to higher incidences of mastitis, but in reality any imbalance will make the goat more prone to all illnesses. Goats with chronic high SCC should be considered for culling if none of these strategies help.

CAE Mastitis or Hard Udder

This type of acute or clinical mastitis is associated with the CAE virus. Mastitis isn't as common as the arthritis form of the infection.

SIGNS AND SYMPTOMS

Firm, hard udder at time of delivery, but with loose skin (compare with symptoms of edema; see page

The Somatic Cell Count Paradox

I used to be under the impression that the lower an animal's SCC, the better. But then, during a presentation I was giving, an audience member related that he had heard the opposite—that if a cow or goat's SCC dropped too low, she could get mastitis very easily. Turns out, he might be right. The white blood cells are the bodyguards in the udder. If there are too few bodyguards, invading bacteria can grow rapidly. When reinforcement white blood cells appear on the scene, they are effective at destroying the bacteria, but as the bacteria are destroyed, they release toxins. Thus, if a large population of bacteria has developed before the white blood cells attack them, the resulting toxin levels are also greater than if the bacteria had been dealt with earlier on.

In cows, whose ideal SCC is believed to be about 200,000, some studies show that as the number drops below 100,000, a cow is much more likely to succumb to coliform mastitis.[1] I believe we witnessed this in our own herd with a LaMancha doe named Towanduh, or Wanduh for short. Her somatic cell count on the DHI milk test on a Tuesday was 79,000 (quite low for a heavy-producing goat); one week later, literally overnight, one half of her udder had turned blue with bloody serum expressed from the teat. She lost half of her udder. *Staphylococcus aureus* was not found in the sample, but of course by then it could have all been destroyed, with only toxins remaining. It also could have been another bacterium, such as a coliform or clostridium. Either way, I am now not as excited to see those counts on the goats at 100,000 or less!

304). No or little milk produced; gradually increasing over time. Swollen lymph nodes at top of escutcheon.

TREATMENT

If CAE is present in the herd, then the virus is the likely cause of the mastitis and the doe should be culled. The doe can be tested for CAE but may have a false positive. There is no effective treatment for this type of mastitis.

PREVENTION

Embark on a CAE prevention program (as described in chapter 9 and in the CAE Arthritis entry on page 283.)

Acute Bacterial Mastitis

In acute and clinical cases of udder infection, the symptoms can be seen by the farmer. The udder temperature, texture, and sensitivity might be affected; the milk quality might be visibly changed; and the goat's somatic cell count is likely to increase.

Table 15.2 covers several causes of mastitis, but it is by no means exhaustive. It includes the most common causes and a few identifying signs and actions to take that might help you troubleshoot the problem. A milk culture is the only definitive way to narrow down the cause of a specific case of mastitis, though. Keep in mind, too, that although it's uncommon, more than one bacterium at a time can be the cause of the animal's problem.

MILKSTONES

Occasionally a bit of minerals solidified from milk, called a milkstone or lactolith, will develop in the udder. The hard flake or formation will float in the teat cistern and work its way to the orifice, blocking the flow of milk. While the stone is harmless, getting it out of the teat can cause damage. If possible it should be worked to the bottom and then forced out by squeezing milk from above it; sometimes massaging the orifice at the same time is necessary. It isn't known why these "stones" form.

Table 15.2. Common Bacterial Causes of Mastitis

Bacteria	Source	Common Clinical or Acute Signs	Identifying Lab Characteristics	Actions
Coliforms (includes *E. coli*)	Normal in the environment—bedding, soil, manure	Occurs soon after kidding. Increased SCC. Udder warm, swollen, painful. May have fever, loss of appetite.	Gram negative	Improve bedding cleanliness. Dry teats well before milking. Improve teat skin condition. Dipping teats in sanitizer after milking doesn't reduce coliform counts.
Mycoplasma	Carrier animal, can be spread to others during milking procedures	Usually soon after kidding. Loss of appetite, looks sick, isolates from herd. Watery milk, later becoming clumpy or thick, total loss of production on affected side or both halves.	Special culture takes several weeks	Cull affected does or isolate asymptomatic carriers at end of milking lineup. Intramammary antibiotics will have no effect. For more on mycoplasma see page 290.
Staphylococcus aureus G2H—on skin or toxins produced in raw milk or other raw products	Carrier animal or hands of carrier human. Can be spread during milking procedures. The bacterium resides in micro-abcesses high in the udder and can't be eradicated.	Hot, painful udder half, fever, loss of appetite, abscess formation. Most common cause of gangrenous mastitis.	Coagulase positive [Most other *Staphylococcus* are coagulase negative, but two, *S. intermedius* and *S. hyicus*, are coagulase positive. They are occasionally found in goat milk and are harmless.]	Cull carrier animals, or, if kept, milk them last and do not feed the raw milk to kids.
Other staph	Normal on teat skin, human hands, and the environment	Usually mild—decreased milk production with no visible changes in milk and elevated somatic cell count. Resolves spontaneously.	Coagulase negative	Usually none needed
Streptococcus	Normal in the environment	Mild—flakes or clumps in milk, elevated somatic cell counts	Catalase negative	Improve environmental conditions—clean, dry bedding

PRECOCIOUS UDDER

Also known as inappropriate lactation syndrome, precocious udder sometimes occurs in an unbred younger doe. One side or both will enlarge and fill with milk. Sometimes there is only a bit of udder development; in other cases, the doe may look ready to milk. Often producers refer to such does as maiden milkers. Farmer wisdom suggests that only does with high production capability and genetics are likely to have a precocious udder, and this has been my experience as well. In most cases, it is best to ignore the mammary development and proceed with breeding, but in some cases hormonal therapy may be needed if the doe doesn't cycle with normal heat cycles. She might also be experiencing a false pregnancy. There are plenty of accounts of producers rotating such does into the milking string with a normal lactation resulting.

WARTS

CONTAGIOUS FROM GOAT TO GOAT

Warts on the udders of goats are most common on pink-skinned udders of animals raised in very sunny regions. Udder warts must be distinguished from other skin conditions. Udder warts are a problem because of interference with milking and milking hygiene, as well as the potential to develop into skin cancer after a number of years. There is no conventional treatment other than burning off the wart with liquid nitrogen, as is done in humans. There are homeopathic remedies, but they must be matched with the symptoms. Warts may subside on their own and never recur, or they may be seasonal. The best prevention is not introducing any animal that has or has had warts to your herd. Choosing darker-skinned goats for sunny climates has been shown to reduce the occurrence.

WEEPING TEAT AND TEAT WALL CYSTS

Occasionally when the unborn goat is forming inside the mom, milk-producing tissue will form

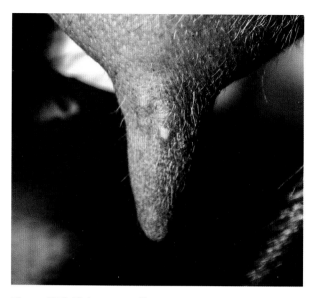

Figure 15.3. This teat wall cyst on a LaMancha-cross goat never wept and later disappeared without treatment.

in the mother's body in places other than it should. This is called ectopic (meaning "abnormal position or place"). Ectopic mammary tissue can occur on the vulva as well. When it occurs near the base of the teat, the bumps can interfere with milking and milking hygiene, because they often leak milk. The leakage can make the milker's hands wet or cause the milking equipment to slip. The leakage over milking equipment also compromises sanitation. Sometimes, however, the bumps are simply benign cysts that don't contain milk.

If a weeping teat is a problem, your vet can supply you with or treat the problem with a silver nitrate stick, which is applied after every milking to destroy the tissue.

WITCH'S MILK

The occasional goat kid is born with enough mammary development to produce a bit of milk even as a newborn. If present, this mammary development should be ignored and *never* milked out, because milking will remove the protective plug in the teat, which could lead to development of mastitis.

CHAPTER SIXTEEN

The Reproductive System

The goat's reproductive system is similar to that of other mammals. There are some unique differences, however, that are important to understand so that you will be able to ensure healthy new generations of goats and excellent survival rates of mother goats. Chapter 8 covers much of the hands-on information you will need to know to take care of your goats' reproductive health, including factors that influence fertility, and you may want to refer back to it as you read through the descriptions here of major reproductive disorders.

ABORTION

Several bacterial infections can lead to late-term abortions, those occurring in the last month of a goat's pregnancy (pregnancy in goats lasts almost five months). Depending on the bacteria, the abortion can be a direct result of damage to the fetus or it can be due to stress on the doe. Here, I focus on the most common causes, many of which are reportable diseases in the United States, including brucellosis, campylobacteriosis (a common bacterium, but rarely a cause of abortion for goats in North America), chlamydiosis (a sexually transmitted disease that is a fairly common cause of abortions), listeriosis, Q fever (becoming more common), salmonellosis (uncommon), and toxoplasmosis. Leptospirosis is another bacterial infection that can lead to abortion, but it's rare in goats. Caprine herpes virus is also known to cause abortions. There are a multitude of other causes, even more uncommon, of abortion as well. Any late-term abortion should be assessed for infectious causes.

In all cases of late-term abortion, wear gloves and possibly a mask when handling the fetus or doe until infections that can be passed to humans are ruled out. When the cause is unknown, the placenta should be sent for testing. If possible the fetus and any birthing fluids collected should also be tested. If an infectious cause is suspected or known, the fetuses, placenta, and soiled bedding should be burned or buried.

Because many of the causes of late-term abortion are bacterial, some producers give tetracycline, or another antibiotic in that family, to all the remaining pregnant does as a preventive in the hope of protecting the pregnancy. But some bacteria are showing resistance, which could make preventive dosing ineffective. Certified organic producers might choose to forgo this treatment and accept the potential loss of kids, knowing that the health of the mothers can still be addressed without the loss of organic status.

Retained placenta is a common side effect of many late-term abortions.

Brucellosis

CONTAGIOUS FROM GOAT TO GOAT
CONTAGIOUS FROM GOATS TO HUMANS

Two types of *Brucella* bacteria can infect goats and be transmitted to humans, usually through raw milk, raw milk products, or undercooked meat. Both are reportable in the United States. One is more common in goats and the other in cows, but the second type can be transmitted from infected cows to goats. The bacteria are shed not only in milk but also in urine, feces, fetuses, placentas, and vaginal discharges for several months after birth. In goats the only symptom is abortion. In humans flu-like symptoms of fever, headache, sweating, loss of appetite, fatigue, and aching muscles appear first. They can progress to arthritis, chronic fatigue, heart problems, continued fever, and more. Some nations have almost entirely eradicated brucellosis, also known as undulant fever, from domestic animals through testing and vaccination. In others, it is still widespread—an important thing to know when traveling. Regions considered by the United States Centers for Disease Control as high risk include the Mediterranean Basin (Portugal, Spain, southern France, Italy, Greece, Turkey, North Africa), as well as Mexico and all other countries in Central and South America, Eastern Europe, Asia, Africa, the Caribbean, and the Middle East.

SIGNS AND SYMPTOMS

Abortion in the last two months of pregnancy. Abortion storm when herd is first infected.

TREATMENT

There is no treatment, because it is not possible to know that a doe is infected.

PREVENTION

In countries where brucellosis is present, work with your vet to develop a strict vaccination program. Avoid contact with animals and milk products from high-risk countries when traveling.

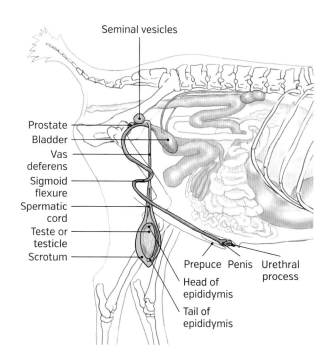

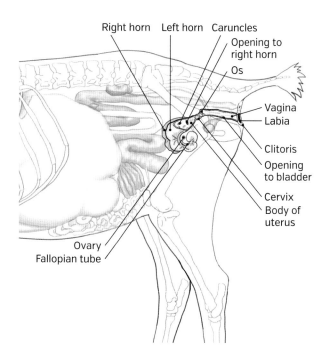

Figure 16.1. The male (*top*) and female (*bottom*) goat reproductive systems.

Table 16.1. Common Reproductive System Signs and Symptoms

Sign or Symptom	Possible Causes*
Abortion, early	Malnutrition
	Toxoplasmosis
Abortion, late	Any illness that causes high fever
	Brucellosis
	Campylobacteriosis
	Chlamydiosis
	Drugs
	Leptospirosis
	Listeriosis
	Malnutrition
	Plant toxicity
	Q fever
	Salmonellosis
	Selenium deficiency
	Stress
	Toxoplasmosis
	Trauma
	Vitamin A deficiency
Infertility, doe	Freemartin
	Intersex
	Malnutrition
	Nutritional imbalance
	Uterine infection
Infertility, buck	Age
	Malnutrition
	Nutritional imbalance
	Poor libido
	Pizzle rot

* Causes listed include the more common possibilities, but the lists are not exhaustive or intended to replace the diagnostic services of a licensed veterinarian.

Campylobacteriosis

CONTAGIOUS FROM GOATS TO HUMANS

Campylobacter bacteria are common in the intestines and feces of warm-blooded animals but are a source of foodborne illness in humans when fecal matter is ingested in contaminated food or by another route.

In goats, an infection can cause abortion, but it's an uncommon cause of miscarriage in North America.

SIGNS AND SYMPTOMS

Late-term abortion. On necropsy, kids show liver damage. Placenta swollen, with decaying cotyledons.

TREATMENT

There is no treatment for this problem.

PREVENTION

Avoid manure contamination of feeders and waterers.

Chlamydiosis

CONTAGIOUS FROM GOAT TO GOAT
CONTAGIOUS FROM GOATS TO HUMANS

Chlamydia are a type of bacteria that can reproduce only inside a host's cells. They inhabit the intestinal and genital tract as well as the conjunctiva of the eyes. Chlamydia can infect humans as well as other animals and are not species specific. In addition to causing abortions, chlamydial infections can also lead to conjunctivitis (pink eye), arthritis, and respiratory problems. In cases of abortion, the inflammation caused by the replication of the bacteria in the placenta stops the flow of nutrients to the fetus, and it dies. If the infection is new to the herd, an abortion storm is likely. Where the infection has been present before, only first fresheners are usually affected.

Infection is spread when an animal contacts the infected placenta or birthing fluids. The disease is known to be present in and spread through semen in sheep. The same is likely true for goats, but research is needed to verify this. Does that have been exposed, either bred or unbred, will develop immunity.

SIGNS AND SYMPTOMS

Abortion, usually during the last two months of pregnancy. Aborted kids look fresh, not mummified.

TREATMENT

Gloves should be worn when handling any abortion. If chlamydia is suspected, the fetus and placenta should be collected and sent for testing. If chlamydia is confirmed, consult with your vet to determine whether all other pregnant does should be treated (usually with tetracycline). Aborted fetuses and placentas not sent for testing should be destroyed. Does that abort should be isolated until vaginal discharge stops.

PREVENTION

Vaccination is effective, but in the United States it isn't regularly available. The vaccine, labeled for sheep, can be administered with a vet's prescription. New animals should not be introduced unless they're from a chlamydia-free herd. To prevent possible infection in humans, pregnant women should not work with aborting animals. If goats (and people) who are not pregnant contact infected tissues, however, immunity without disease can be the result.

Listeriosis

CONTAGIOUS FROM GOATS TO HUMANS

Although there are many types of harmless bacteria in the *Listeria* family, *L. monocytogenes* can be deadly to goats and humans alike. In goats an infection by these bacteria more typically causes central nervous problems (as described in the Listeriosis Encephalitis entry on page 299) but can also cause abortion in pregnant does. The two different types of the disease don't usually occur at the same time. For example, I once had a doe that developed the encephalitis form during her fourth month of pregnancy, but she didn't abort. It's important to remember that *Listeria* are generally present in the farm environment but don't always cause illness. Animals that are under stress, such as the rigors of pregnancy, are more prone to infection. Humans can become infected directly from the environment or from milk that has been contaminated either from feces or from *Listeria* that have entered the udder.

SIGNS AND SYMPTOMS

Abortion, usually after a system-wide (but not central nervous system) listeria infection, which

includes symptoms such as fever, loss of appetite, and decreased milk production.

TREATMENT

Follow the same measures as for listeriosis encephalitis (page 299).

PREVENTION

Follow the same measures as for listeriosis encephalitis (page 299).

Q Fever

CONTAGIOUS FROM GOAT TO GOAT
CONTAGIOUS FROM GOATS TO HUMANS

Short for Queensland or Query fever, Q fever is a bacterial disease (caused by *Coxiella burnetii*) that causes abortion in goats and humans. In humans it also causes long-term flu-like symptoms, pneumonia, liver disease, and even inflammation of the heart. This reportable disease is becoming more common in goats and other animals across the world. Humans who work with infected animals might not be sickened but will show antibodies for the disease. The bacteria are passed by ingestion, inhalation, or contact with a wound. The bacteria can be shed in milk and may be present in the placenta, birthing fluids, urine, and feces. They can also be passed through a tick bite. The bacteria remain viable in the environment for a long time and are resistant to many disinfectants.

SIGNS AND SYMPTOMS

Late-term abortion by otherwise healthy animals.

TESTING

If Q fever is suspected, send the placenta to a lab for testing. Cotyledons appear gray in an infected placenta.

TREATMENT

Aborting animals should be isolated, and your vet might suggest treating all other pregnant does with tetracycline to prevent abortion. Burn infected placentas and aborted fetuses. Wear gloves and masks when handling placentas, fetuses, and manure from infected animals.

PREVENTION

In herds that test positive for Q fever, ensuring high-quality nutrition and low-stress conditions for the pregnant does can help prevent abortion. Testing for Q fever is controversial because in many states where testing is required, there is no clearly defined and required action if an animal tests positive. Test all new animals before purchase. Milk and products made from positive herds must be pasteurized for use by humans or other animals. A Q fever vaccine for humans is available in France and Australia.

Toxoplasmosis
CONTAGIOUS FROM GOATS TO HUMANS

An infection by the protozoan parasite *Toxoplasma gondii* can cause abortion in a doe that hasn't previously been exposed to the organism. *Toxoplasma* is carried in the feces of cats that serve as the primary host for the organism. Cats aren't naturally infected but pick up the parasite by eating raw meat, the placenta of an infected animal, or rodents. Does exposed to *Toxoplasma* when they are young or not pregnant are likely to become immune, but when does are exposed during pregnancy, abortion or delivery of full-term dead or weak kids can result. The recent arrival of a new cat, followed by abortions with the following signs and symptoms, makes toxoplasmosis a prime suspect.

SIGNS AND SYMPTOMS

Late-term abortion, mummified kids, delivery of dead or weak kids.

TESTING

Unlike other infectious causes of abortion, it is not possible to reliably test an aborted fetus for *Toxoplasma*. However, if the blood of the doe is tested at the time of the abortion, it will show an absence of antibodies if toxoplasmosis wasn't the cause.

TREATMENT

There is no effective treatment for toxoplasmosis.

PREVENTION

Avoid introducing a new cat to the farm when does are pregnant. Keep stored feeds protected from cat feces and keep feeders clean if cats are present in the barn.

CRYPTORCHIDISM

In some species, the male's testicles don't descend into the scrotum until after the animal is born. Goats, however, are born with their parts in their final position. Occasionally a male goat will be born with only one testicle in the right place—or, rarely, with no descended testicles. The "missing" testicle is up in the abdominal cavity; it may be just inside the abdominal wall or closer to the kidney. Such testicles will usually not descend into proper position unless they are positioned just inside the inguinal canal—the opening into the abdomen.

Cryptorchidism is more common in Angora goats than in other breeds. Research has shown it to be a recessive trait in Angoras, but controlled by more than one pair of genes. (This simply means that it's a combination of factors that causes it to be expressed, unlike a simple recessive such as blue eyes in humans.) In other breeds it is also believed to be genetic, but the exact combination of circumstances that allows for it to occur isn't known.

Bucks with an undescended testicle are usually still fertile, but the semen volume and quality are lower because the internal testicle is too hot. A cryptorchid that is castrated by removing just the one testicle may or may not be fertile but will still act and smell like a buck. If the animal will be kept as a pet, it is advisable to have the other testicle surgically removed. A vet can use ultrasound to locate the missing organ and then perform the surgery.

Cryptorchid animals should never knowingly be used as breeding stock, both because of the genetic potential for increasing the occurrence of the problem and because of their reduced fertility.

Some books advise the culling of not only the cryptorchid, but also the parents. Early in our breeding program, we had one or more cryptorchid kids per year. One doe had quad boys, of which three were in this state. I either euthanized them at birth or found homes for them with owners who would have them surgically castrated. I never culled the parents. Even though I have not brought in many new genetics, it has been several years since we have had another cryptorchid kid. So while it is likely genetic, the history on my farm seems to indicate that other factors may also contribute to its occurrence.

INTERSEX

This term is used to describe goats that have external characteristics and internal organs that are a mixture of both genders. It includes true hermaphrodites—whose cells include chromosomes for both female, XX, and male, XY, genders—and pseudohermaphrodites, which are animals with mixed-gender appearance but only female chromosomes.

In the most common presentation, intersex kids will look like does at birth (pseudohermaphrodites). When puberty approaches, the clitoris protrudes and has a small, rounded protrusion. Behavior will be more buck-like, and a hair growth pattern that is long on the neck, back, and forehead will occur.

Pseudohermaphrodites are also called polled intersex (also polled intersex syndrome or PIS) because the condition is strictly associated with the polled (hornless) goat. Breeds that originated in Europe have a well-known connection to the intersex condition and being polled,[1] but Nubian and Angora goats don't. It's believed that the gene that determines whether a goat has horns (or a closely linked gene) also determines the goat's fertility. When two polled goats are bred to each other, the odds are good that some of the offspring will be intersex. For this reason, it's a recommended practice to cross a polled goat, whether doe or buck, *only* with a *horned* goat. Alternatively, the breeder can simply decide that any intersex offspring will be culled. Keep in mind that a polled goat that is an intersex could be a true hermaphrodite, or freemartin, instead of a pseudohermaphrodite resulting from polled genetics.

True hermaphroditism in cattle is fairly common. It occurs when two fetuses share a blood supply and therefore hormones and cells. When one is male and one is female, the female gains male (XY) cells and therefore male characteristics. The male is rarely affected by the exchange. The condition is much more common in cattle because it is normal for cattle twins to share the placental blood supply early in their development. In goats, it's thought that sharing doesn't occur or occurs only later in development. In

Table 16.2. Polled and Horned Goat Genetics

Parent 1	Parent 2	Polled Kids with Homozygous Dominant Polled Genes (PP)	Polled Kids with Heterozygous Genes (Ph)	Horned Kids with Homozygous Recessive Genes (hh)
PP	PP	100%	0%	0%
hh	hh	0%	0%	100%
Ph	Ph	25%	50%	25%
PP	hh	0%	100%	0%
Ph	hh	0%	50%	50%
Ph	PP	50%	50%	0%

genetic tests of horned (or disbudded or dehorned) intersex goats, they all appear to be genetically XX, XY. Another term for this mix of genes that occurs in a freemartin is blood chimerism; the resulting animal can also be called a chimera.

RETAINED PLACENTA

If the placenta isn't delivered within 12 hours (usually it is fully expelled within a few hours) the danger of its being retained must be addressed. You must determine whether it is truly retained or has just been eaten by the doe (which is a common occurrence) before you treat the animal. Placentas are most often retained when the birth is early, a cesarean has been performed, or a miscarriage has occurred.

SIGNS AND SYMPTOMS

Usually, a portion of the placenta hanging from the vulva 12 or more hours after delivery.

TREATMENT

Follow the steps in the Manually Dilating the Cervix section on page 208, then verify that the cervix is still open and try to verify that there is no kid remaining in the udder—dead or alive. If the cervix is still dilated and there are no kids present, then oxytocin should be administered to cause the uterus to contract and expel the remaining parts of the placenta. Infuse the uterus with calendula tincture and warm water. Give arnica, pustilla. Check the doe's tetanus vaccination status or consider giving a dose of antitoxin to help prevent the risks associated with introducing tetanus bacteria while internally examining the doe.

UTERINE PROLAPSE

Fortunately this situation is rare in goats. When it does occur, the entire uterus or one horn will protrude from the vaginal canal after delivery of the kids. If the condition is caught early, call your vet, and while waiting, wash it off with body-temperature water, sprinkle it with sugar to gently draw out

Figure 16.2. When the entire placenta hasn't been expelled, a portion or shred might be visible (here seen as blood-colored tissue) dangling from the doe's vagina.

excess fluid (making it easier for the vet to replace), and cover it with a plastic bag. The procedure brings risks, including tearing the uterus, so don't attempt it without professional help. If the prolapse is older than several days, your vet may advise its amputation.

UTERINE TORSION
(Twisted Uterus)

The uterus is suspended in the body by several bands of tissue, but during the end of pregnancy its weight and mass make it prone to shifting position. If a large kid is present on one side of the uterus and there is no kid or a small kid on the other side, the uterus is vulnerable to a change in position. If the doe moves suddenly, rolls over, is pushed, or even has an overfull or underfull rumen, the heavier side of the uterus may rotate up and over the lighter side. This condition usually occurs between the first stages of labor and active labor (see chapter 8 for description of stages of labor). The body and one horn of the uterus, or simply one horn, becomes positioned over the top of the other horn. This causes a twist (torsion) in the body of the uterus, cervix, and sometimes the vaginal canal. When torsion occurs, the doe will fail to dilate and cannot deliver. (This was described in the Ohmnom sidebar on page 209.)

SIGNS AND SYMPTOMS

Obvious first stages of labor and even attempts at pushing. On palpation, non-dilated cervix that cannot be manually dilated. Possibly, cervix far to the right or left and directed downward. Vaginal canal may feel shorter than expected. Possible palpable or visible (when viewed using a speculum) twist in vaginal canal.

TESTING

A vet can use portable ultrasound equipment to determine if torsion has occurred. Manual palpation or visualizing the vaginal canal with a speculum and AI light might also reveal torsion.

TREATMENT

Torsions may or may not be correctable without performing a cesarean section. Some that are ignored may resolve if the doe rolls herself over in the throes of labor, but it's risky to pin your hopes on that occurring. The plank-on-flank method, in which the uterus is held in place while the goat is rotated to untwist the uterus, can be successful. If not, and sometimes even after successful untwisting, a cesarean section should be performed. You must be certain of the direction of the twist before attempting this! In the plank-on-flank method the doe is laid on the side to which the uterus has twisted (for example, if the left

horn has shifted over and to the right side, she is laid on her right side) and a length of wooden board or other rigid material is laid across her flank so that it extends on either side of her body. A person should stand at each end of the plank and use the plank to apply firm downward pressure onto the doe's flank. Two other people then slowly roll the doe over onto her other side. She should then be returned to the starting side (without using the plank) and the roll repeated. If successful this will untwist the uterus. After two rolls you should manually check the cervix, or verify with ultrasound, to see whether it has been repositioned.

VAGINAL PROLAPSE

Various factors can cause all or most of the vagina to protrude outward during the late stages of pregnancy and especially during delivery. The causes are not well understood but are likely to include genetics, increased pressure inside the doe from a large litter of kids, and lack of exercise. I've seen a couple of does experience a partial vaginal prolapse. One doe

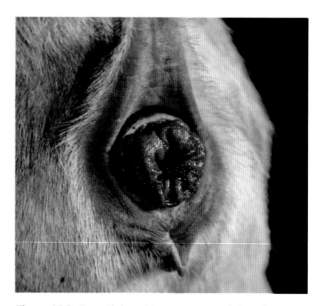

Figure 16.3. A partial and temporary rectal prolapse might occur when a pregnant or overweight doe coughs. Rectal prolapses are usually mild and don't require veterinary care.

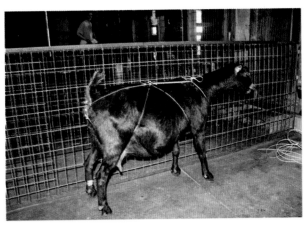

Figure 16.4. A very effective truss can be rigged up quickly to hold a vaginal prolapse in place.

prolapsed only once, the other twice (I retired her after the second time). The truss shown in figure 16.4 proved quite effective at keeping the prolapse from recurring during the rest of the pregnancy.

SIGNS AND SYMPTOMS

Portion of vaginal canal protruding through vulva. Often appears only when doe is lying down. Usually begins during final two months of pregnancy.

TREATMENT

The prolapse must be managed or it will make delivery difficult or impossible. A vet can suture the opening closed to keep it from happening—leaving enough room for urine to pass—but the sutures must be removed during delivery. A coated clothesline can be used to rig up a truss (see figure 16.4) that is quite effective and can be left in place even during kidding. Here's how to truss a pregnant goat:

Step 1. Place a long length of plastic-coated clothesline (it won't rub) over the doe's neck, and run both ends between the doe's front legs, crossing them as you do.

Step 2. Bring the ends up over her withers and cross again.

Step 3. Run the ends of the lines down in front of her hind legs and up around her udder.

Step 4. Bring the ends up over her back and tie them loosely to the neck loop. Leave some extra length for adjustment.

Step 5. Cut three additional short sections of clothesline.

Step 6. Tie one at the point where the lines cross over her back, and connect the two lines that come up from her tail to this cross section.

Step 7. Tie another just below her tail to connect the two lines on either side of the tail.

Step 8. Use the last as a connector between the two lines, just below her vulva.

Step 9. Adjust the length where the line is tied to the neck loop to allow the doe freedom of movement.

PREVENTION

Be sure that pregnant does get plenty of exercise. If a doe repeatedly suffers prolapses, she should be culled.

CHAPTER SEVENTEEN

The Urinary System

Urinary tract problems are pretty rare in working goats. Pet goats, especially castrated males, are much more prone to specific problems, as are goats that are overfed or fed without thought to the mineral balance in their diet. Problems with the urinary system often originate from nutritional imbalances. Herds of working goats that are healthy and in nutritional homeostasis rarely experience urinary system issues.

CAPRINE HERPES VIRUS

This virus, also called caprine herpesvirus 1 (CpHV-1), is a growing problem for goats throughout the world. It first appeared in the United States in the 1970s and then in New Zealand and Australia. It can now be found in many parts of Europe and the Mediterranean. The virus, which is believed to be passed primarily from infected bucks to does during breeding, causes a sequence of problems that can result in abortion, infertility, death of young kids, and possibly pneumonia. Symptoms in bucks are similar to those of posthitis (see the Posthitis entry on page 323).

SIGNS AND SYMPTOMS

Does: Onset of symptoms just after beginning of breeding season. Initial swelling and redness of vulva and vagina (vulvovaginitis) followed by yellow to gray vaginal discharge and many shallow sores. Possible abortion.

Table 17.1. Common Urinary System Signs and Symptoms

Sign or Symptom	Possible Causes*
Lower abdominal distension (water belly)	Ruptured bladder
Difficulty urinating	Bladder infection (cystitis) Pizzle rot (posthitis) Urinary stones (urolithiasis) Vaginal infection (vaginitis)
Abnormal urine color	Poisoning Urinary tract infection Vaginal infection (vaginitis) from caprine herpes virus or other cause White muscle disease (brown urine)
Swollen vulva	Caprine herpes Due to kid In heat (estrus)
Swelling, redness, sores around tip of penis	Pizzle rot (posthitis)

* Causes listed include the more common possibilities, but the lists are not exhaustive or intended to replace the diagnostic services of a licensed veterinarian.

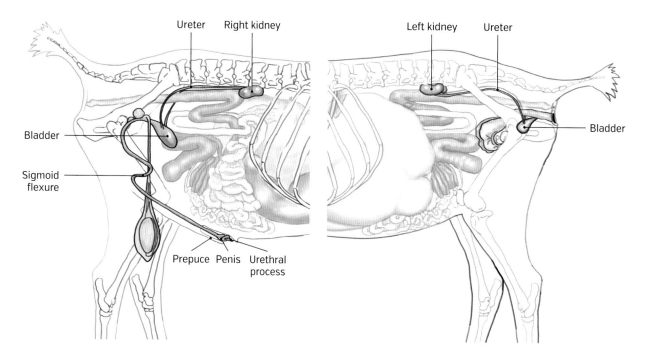

Figure 17.1. The male [*left*] and female [*right*] goat urinary systems.

Bucks: Infected animals may not show symptoms. When present, small sores on head of penis (balanitis); swelling and redness of prepuce (skin encasing penis).

Kids: Death by age one to two weeks.

TREATMENT

There is no effective treatment.

PREVENTION

It is currently believed that the virus is only passed sexually, but in experiments the virus was able to infect new goats when administered nasally. Ask your vet about current research if the virus has been detected in your herd. At the time of writing, the recommendation is to test all new animals, especially bucks. In infected herds seeking elimination of the virus, kids should be separated before breeding age, and buck kids should test negative (before four months they may test positive if their dams were positive) before being used for breeding. Adults should be tested and positive animals culled.

CYSTITIS
(Bladder Infection)

Bladder infections occasionally occur in does. Some authors suggest that does that have had assisted deliveries are more prone to cystitis, but this hasn't been my experience. I have only had one doe with a bladder infection. The first time it was treated successfully with antibiotics. A recurrence a year later was successfully eliminated with cranberry extract tablets (made for humans); it hasn't come back.

SIGNS AND SYMPTOMS

Frequent squatting to urinate. Small amounts of blood on labia.

TREATMENT

Provide access to plenty of water. You can try offering cranberry extract or feeding ammonium chloride at the same rate as for urinary stones (see the Urolithiasis entry on page 323).

PREVENTION

Provide access to clean, palatable water at all times. There is a possible correlation between a goat's anatomy and bladder infection. If the angle of the vulva is such that it helps prevent feces from collecting in the labia, chances of infection are lessened. This underscores the importance of looking for breeding stock with ideal conformation.

POSTHITIS (Pizzle Rot)

Pizzle rot is an infection of the prepuce, the skin surrounding the penis (also called a sheath or foreskin), caused by bacteria that break down excess nitrogen (in the form of urea) in the urine and convert it into ammonia. The technical term *posthitis* refers to any infection or inflammation of the prepuce. The problem is similar to urine scald, but with more severe symptoms.

SIGNS AND SYMPTOMS

Swelling of prepuce; when severe, inability to extend penis. If left untreated, urine flow is blocked, which leads to death of animal.

TREATMENT

Decrease the protein content of the diet. Trim excess hair from the surrounding areas to prevent urine soaking the hair and skin.

PREVENTION

Preventive measures are the same as those for treating an infected animal.

URINE SCALD

In urine scald, the skin is burned by overexposure to urine. It's typically seen on the front legs and faces of mature bucks, and sometimes around the vulva of fiber goat does. The buck's habit of spraying his legs and face with urine makes this a common sight in the buck pen. In fiber does, urine might be caught in fiber or splatter up onto the tender skin of the labia and surrounding perineum. In bucks, washing the

Figure 17.2. Urine scald is common on the front legs and faces of bucks, especially during breeding season.

areas, drying them, and applying a water-resistant ointment can help. In fiber does, clipping the excess hair around the perineum will help.

UROLITHIASIS (Urinary Stones)

Urinary stones, calculi, are a common problem for wethers and bucks. Stones form in the bladder of does as well but are passed easily through a doe's short, straight urethra. Bucks have a urethra that includes some sharp turns and ends with a small opening. Wethers, because they are castrated, don't develop as large a penis or urethra as an uncastrated goat, which leaves them more prone to blockage by mineral stones. Young wethers being grown for meat are particularly at risk because they are often fed to promote fast growth.

Calculi are made up of minerals, including phosphorus, calcium, magnesium, and silica. The most common culprit is phosphorus, followed by calcium. Typically these stones form from an excess of one of these minerals in the goat's diet. Grain is high in phosphorus, legumes are high in calcium; a diet too high in magnesium, even when the calcium-to-phosphorus ratio is correct, can cause the formation of calcium stones. Some grazing forages are high in silica, which can lead to stones composed of this mineral.

There are several other important issues to consider when you're dealing with urinary calculi or seeking to prevent them. Vitamin A deficiency has been linked to obstruction by stones. Bladder infections can raise the pH of the urine, which is already alkaline, making the formation of mineral stones more likely. A low-fiber diet can limit rumination and consequently increase the amount of phosphorus excreted by the kidneys (when a goat ruminates, phosphorus is used in the saliva and then processed). Goats that are fed one or two large meals a day experience the sudden need for more moisture in their digestive system, making the urine more concentrated.

SIGNS AND SYMPTOMS

Straining to urinate. Little or no urine. Standing with the back hollowed and the rear legs extended back. Grinding of the teeth and other signs of pain. If the bladder has ruptured—as will happen if the blockage isn't cleared—pain will diminish for a while, but then the goat will begin to show signs of uremia—the buildup of waste products in the bloodstream. Visible swelling in the belly (where urine is collecting) is possible. Death will follow, but euthanasia should be done to prevent extended suffering.

TREATMENT

If you catch them early, you can encourage some stones—those lodged low in the system—to pass via manipulating the urethra. More stones may be higher in the urethra, though, or at the opening of the bladder to the urethra. Warm hot towels massaged over the prepuce and backward toward the scrotal area might help. Pain medication and tranquilization with acepromazine or diazepam from your vet can help with the process by both relaxing the goat and having the side effect of relaxing the muscles of the urethra. A urinary catheter can be inserted, which may allow the stone to pass. If the stone is lodged in the urethral process—the little protrusion at the end of the penis—your vet might decide to snip it off. If

Figure 17.3. This wether is straining to urinate because of a blockage. This sort of posture and behavior should always be closely observed to rule out a urinary tract issue. Unfortunately this fellow, named Click, didn't survive.

the stone is dislodged and the goat able to urinate, follow the treatment with a urine acidifier such as ammonium chloride at 200 to 300 mg/kg per day or, if that's not available (it usually has to be ordered), ammonium sulfate at 0.6 to 0.7 percent of the feed ration. (This can be found in the fertilizer section of nurseries and farm stores.)

PREVENTION

The diet must be assessed for proper mineral balance. Ideally the calcium-to-phosphorus balance should be 2:1 or 2.5:1. Magnesium should not exceed 0.3 percent of the dry matter intake. Goats should have access to fresh, palatable water and be encouraged to drink. In herds with a history of calculi, the addition of extra salt to the feed is sometimes suggested. You can give goats a urine deacidifier, but some experts feel that, over time, deacidifiers lose their effectiveness. It's unknown whether intermittent use will prove effective and preserve effectiveness of the treatment. Vitamin A supplementation, especially at times when the level of the vitamin might be low in feed (see chapter 4 for more on vitamin A), a high-fiber diet, and access to grazing or browse throughout the day are other factors that can help prevent urinary calculi.

ACKNOWLEDGMENTS

You would think that after writing a few books, an author would be prepared for the amount of work these projects involve—especially after the manuscript is "finished." With this book, my fifth, I feel especially grateful for the incredible contributions that a top-notch publishing company and editor make toward the creation of a work that will, I hope, never end up in the discount bin near the checkout stand. I never thought I'd find any reason to compare myself to a supermodel, but it struck me one day that, like a model, an author's work is really just a solid base on which a makeup artist, stylist, and designer work their magic. For this book, my biggest undertaking to date, I am so grateful to the Chelsea Green staff for the transformative effect that their work had on the book's outcome. Many thanks to my editor, Fern Marshall Bradley, for her insightful questions and suggestions, and to the production and design team of Pati Stone, Melissa Jacobson, and Alex Bullett for bringing together all the details of text, photos, and art in the book's wonderful layout. Yes, it's still my work, but whipped into shape with a significant layer of concealer, coat of mascara, and gloss of lipstick, all wrapped in a beautiful designer frock. Thanks also to Margo Baldwin for believing in this book!

I always rely on the objective and honest input of a handful of people who, naively perhaps, volunteer for the monumental task of reading the unfinished manuscript and adding their expert comments. For this book I was fortunate to have the input of some super readers: Dr. Susan Beal, DVM; Dr. Mark Wustenberg, DVM; Dr. Kristen Mason, DVM; Scott Bice, herd manager at Redwood Hill Farm; and Amanda Nuñez, goat farmer and former intern at Pholia Farm. Early readers, before the book went through one of its three incarnations, include Roy Chapin, Jeanne Sexton, and Amanda Nuñez (yes, she was gullible—I mean kind—enough to read it twice).

My husband, Vern, got off "easy" this time—I spared him the task of being my first reader. He still contributed greatly, often by dropping me off to write at my favorite coffee shop where the music is loud, the lattes are perfect, and the seats are comfy, while he ran errands.

And lastly, I thank the goats, who won my heart, forced me to move outside my comfort zone, and always keep me humble. Goats, you are the most goofy, giving, challenging, and worthy beasts that humans have had the good fortune to share their lives with.

APPENDIX A

Sample Pedigrees

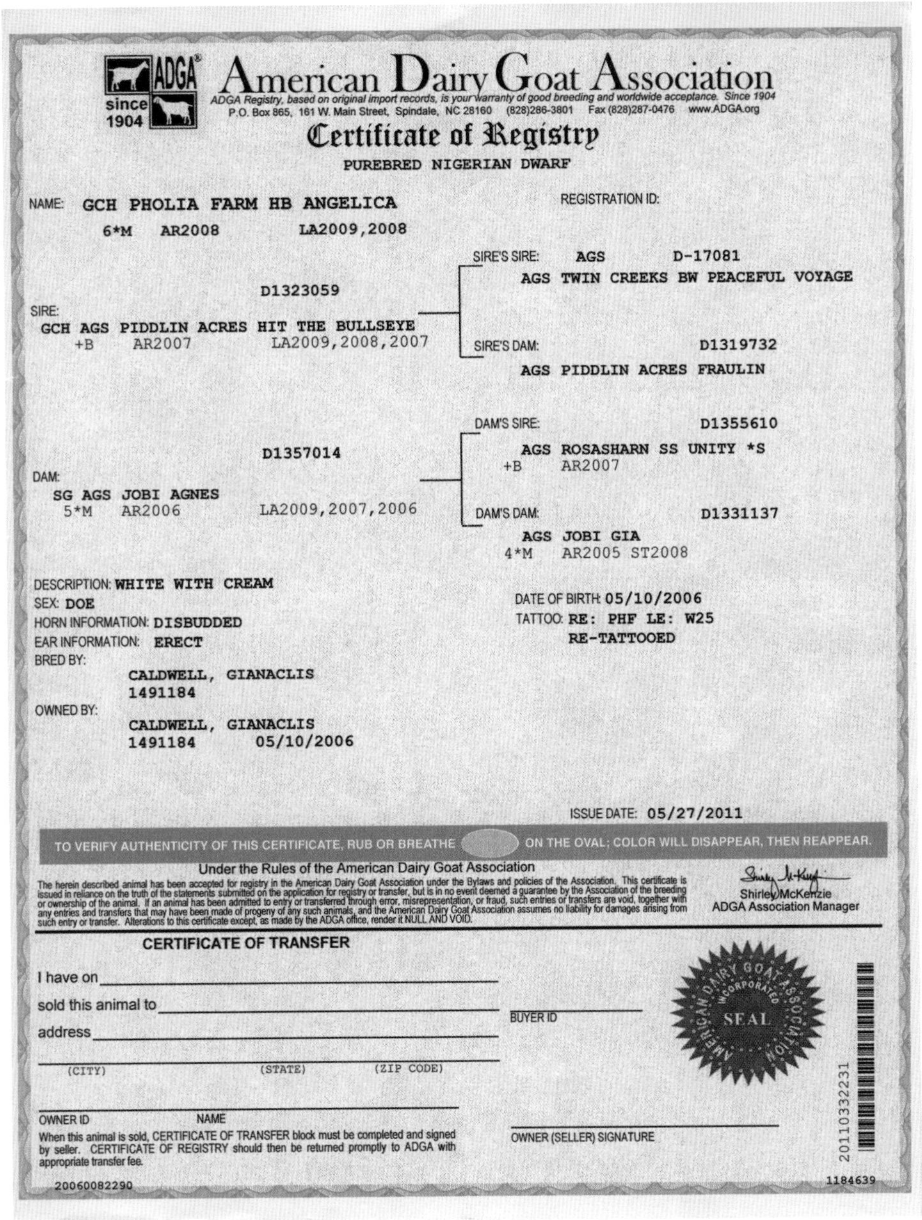

American Dairy Goat Association

ADGA since 1904

ADGA Registry, based on original import records, is your warranty of good breeding and worldwide acceptance. Since 1904
P.O. Box 865, 161 W. Main Street, Spindale, NC 28160 (828)286-3801 Fax (828)287-0476 www.ADGA.org

Certificate of Registry

PUREBRED NIGERIAN DWARF

NAME: GCH PHOLIA FARM HB ANGELICA
6*M AR2008 LA2009,2008

REGISTRATION ID:

SIRE: D1323059
GCH AGS PIDDLIN ACRES HIT THE BULLSEYE
+B AR2007 LA2009,2008,2007

SIRE'S SIRE: AGS D-17081
AGS TWIN CREEKS BW PEACEFUL VOYAGE

SIRE'S DAM: D1319732
AGS PIDDLIN ACRES FRAULIN

DAM: D1357014
SG AGS JOBI AGNES
5*M AR2006 LA2009,2007,2006

DAM'S SIRE: D1355610
AGS ROSASHARN SS UNITY *S
+B AR2007

DAM'S DAM: D1331137
AGS JOBI GIA
4*M AR2005 ST2008

DESCRIPTION: WHITE WITH CREAM
SEX: DOE
HORN INFORMATION: DISBUDDED
EAR INFORMATION: ERECT
BRED BY:
CALDWELL, GIANACLIS
1491184
OWNED BY:
CALDWELL, GIANACLIS
1491184 05/10/2006

DATE OF BIRTH: 05/10/2006
TATTOO: RE: PHF LE: W25
RE-TATTOOED

ISSUE DATE: 05/27/2011

TO VERIFY AUTHENTICITY OF THIS CERTIFICATE, RUB OR BREATHE ON THE OVAL; COLOR WILL DISAPPEAR, THEN REAPPEAR.

Under the Rules of the American Dairy Goat Association

The herein described animal has been accepted for registry in the American Dairy Goat Association under the Bylaws and policies of the Association. This certificate is issued in reliance on the truth of the statements submitted on the application for registry or transfer, but is in no event deemed a guarantee by the Association of the breeding or ownership of the animal. If an animal has been admitted to entry or transferred through error, misrepresentation, or fraud, such entries or transfers are void, together with any entries and transfers that may have been made of progeny of any such animals, and the American Dairy Goat Association assumes no liability for damages arising from such entry or transfer. Alterations to this certificate except, as made by the ADGA office, render it NULL AND VOID.

Shirley McKenzie
ADGA Association Manager

CERTIFICATE OF TRANSFER

I have on ____________
sold this animal to ____________ BUYER ID ________
address ____________
(CITY) (STATE) (ZIP CODE)

OWNER ID NAME

When this animal is sold, CERTIFICATE OF TRANSFER block must be completed and signed by seller. CERTIFICATE OF REGISTRY should then be returned promptly to ADGA with appropriate transfer fee.

OWNER (SELLER) SIGNATURE

AMERICAN DAIRY GOAT ASSOCIATION INCORPORATED SEAL

20060082290 20110332231 1184639

Figure A.1. An example of the registration papers of a purebred dairy goat registered with the American Dairy Goat Association. This doe, Angelica, was one of Pholia Farm's all-time best animals.

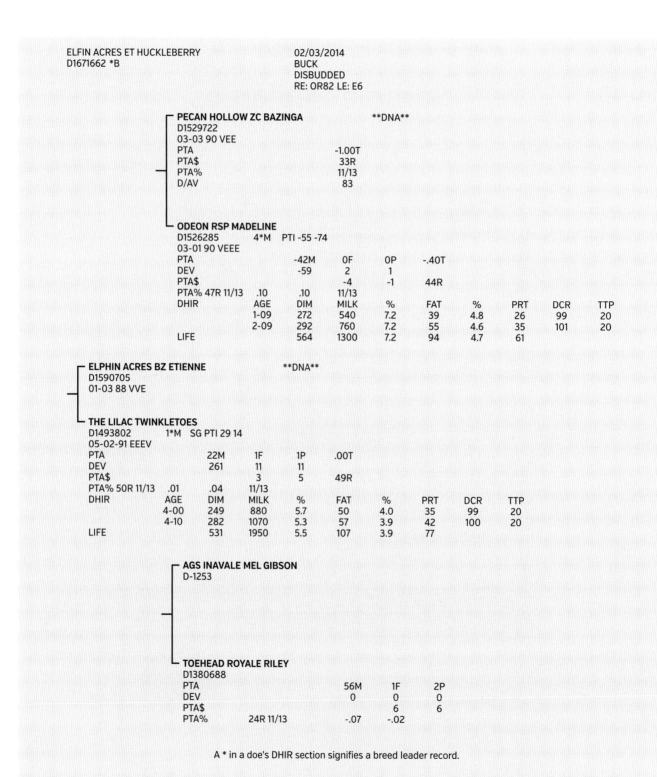

ELFIN ACRES ET HUCKLEBERRY
D1671662 *B

02/03/2014
BUCK
DISBUDDED
RE: OR82 LE: E6

PECAN HOLLOW ZC BAZINGA   **DNA**
D1529722
03-03 90 VEE
PTA   -1.00T
PTA$ 33R
PTA%   11/13
D/AV   83

ODEON RSP MADELINE
D1526285   4*M   PTI -55 -74
03-01 90 VEEE

	PTA			
PTA	-42M	0F	0P	-.40T
DEV	-59	2	1	
PTA$		-4	-1	44R
PTA% 47R 11/13	.10	.10	11/13	

DHIR	AGE	DIM	MILK	%	FAT	%	PRT	DCR	TTP
	1-09	272	540	7.2	39	4.8	26	99	20
	2-09	292	760	7.2	55	4.6	35	101	20
LIFE		564	1300	7.2	94	4.7	61		

ELPHIN ACRES BZ ETIENNE   **DNA**
D1590705
01-03 88 VVE

THE LILAC TWINKLETOES
D1493802   1*M   SG PTI 29 14
05-02-91 EEEV

PTA	22M	1F	1P	.00T
DEV	261	11	11	
PTA$		3	5	49R
PTA% 50R 11/13	.01	.04	11/13	

DHIR	AGE	DIM	MILK	%	FAT	%	PRT	DCR	TTP
	4-00	249	880	5.7	50	4.0	35	99	20
	4-10	282	1070	5.3	57	3.9	42	100	20
LIFE		531	1950	5.5	107	3.9	77		

AGS INAVALE MEL GIBSON
D-1253

TOEHEAD ROYALE RILEY
D1380688

PTA	56M	1F	2P
DEV	0	0	0
PTA$		6	6
PTA% 24R 11/13	-.07	-.02	

A * in a doe's DHIR section signifies a breed leader record.

Figure A.2. An example of a buck's production pedigree produced by the American Dairy Goat Association showing production records and genetic data for his ancestors.

PECAN HOLLOW PS ZUZAK
D1397512

AGS DILL PICKLE LB CONSTANTINE

		CH		
D1386731P				
PTA		-43M	-2F	-1P
DEV		0	0	0
PTA$			-6	-6
PTA%	16R 11/13	.00	.01	

ROSASHARN SP TRIUMPH

		+*B	PTI -145 -153		
D1499644					
PTA		-90M	-3F	-2P	-.70T
PTA$			-11	-9	34R
PTA%	29R 11/13	.06	.08	11/13	
D/AV		718	48	33	84

ODEON ATS MICHAEL

		3*M	PTI -43 -102				
D1477056							
03-05 84 ++VV							
PTA		-20M	1F	1P	1.70T		
DEV		-34	3	3			
PTA$			0	3	37R		
PTA%	56R 11/13	.14	.10	11/13			
LIFE		567	1300	6.4	83	4.5	58

AGS TWIN CREEKS BRAVE HEART
D-6212

AGS ENCHANTED HILL DOUBLE LILY
D-7073

AGS ORCHARD VIEW GT CROWN ROYALE
DNA
D1332880
05-02 83 ++E

PTA			-.90T
PTA$		10R	
PTA%			11/11
D/AV			86

AGS TOEHEAD MADISON

		PTI -7 -73			
D1332891					
PTA		53M	1F	1P	-.60T
DEV		0	0	0	
PTA$			6	5	33R
PTA%	11R 11/13	-.09	-.03	11/13	

150-Day Goat Gestation Chart

Table B.1. 150-Day Goat Gestation Chart

Breeding	Kidding	Breeding	Kidding	Breeding	Kidding	Breeding	Kidding	Breeding	Kidding	Breeding	Kidding
1-Jan	30-May	27-Jan	25-Jun	22-Feb	21-Jul	20-Mar	16-Aug	15-Apr	11-Sep	11-May	7-Oct
2-Jan	31-May	28-Jan	26-Jun	23-Feb	22-Jul	21-Mar	17-Aug	16-Apr	12-Sep	12-May	8-Oct
3-Jan	1-Jun	29-Jan	27-Jun	24-Feb	23-Jul	22-Mar	18-Aug	17-Apr	13-Sep	13-May	9-Oct
4-Jan	2-Jun	30-Jan	28-Jun	25-Feb	24-Jul	23-Mar	19-Aug	18-Apr	14-Sep	14-May	10-Oct
5-Jan	3-Jun	31-Jan	29-Jun	26-Feb	25-Jul	24-Mar	20-Aug	19-Apr	15-Sep	15-May	11-Oct
6-Jan	4-Jun	1-Feb	30-Jun	27-Feb	26-Jul	25-Mar	21-Aug	20-Apr	16-Sep	16-May	12-Oct
7-Jan	5-Jun	2-Feb	1-Jul	28-Feb	27-Jul	26-Mar	22-Aug	21-Apr	17-Sep	17-May	13-Oct
8-Jan	6-Jun	3-Feb	2-Jul	1-Mar	28-Jul	27-Mar	23-Aug	22-Apr	18-Sep	18-May	14-Oct
9-Jan	7-Jun	4-Feb	3-Jul	2-Mar	29-Jul	28-Mar	24-Aug	23-Apr	19-Sep	19-May	15-Oct
10-Jan	8-Jun	5-Feb	4-Jul	3-Mar	30-Jul	29-Mar	25-Aug	24-Apr	20-Sep	20-May	16-Oct
11-Jan	9-Jun	6-Feb	5-Jul	4-Mar	31-Jul	30-Mar	26-Aug	25-Apr	21-Sep	21-May	17-Oct
12-Jan	10-Jun	7-Feb	6-Jul	5-Mar	1-Aug	31-Mar	27-Aug	26-Apr	22-Sep	22-May	18-Oct
13-Jan	11-Jun	8-Feb	7-Jul	6-Mar	2-Aug	1-Apr	28-Aug	27-Apr	23-Sep	23-May	19-Oct
14-Jan	12-Jun	9-Feb	8-Jul	7-Mar	3-Aug	2-Apr	29-Aug	28-Apr	24-Sep	24-May	20-Oct
15-Jan	13-Jun	10-Feb	9-Jul	8-Mar	4-Aug	3-Apr	30-Aug	29-Apr	25-Sep	25-May	21-Oct
16-Jan	14-Jun	11-Feb	10-Jul	9-Mar	5-Aug	4-Apr	31-Aug	30-Apr	26-Sep	26-May	22-Oct
17-Jan	15-Jun	12-Feb	11-Jul	10-Mar	6-Aug	5-Apr	1-Sep	1-May	27-Sep	27-May	23-Oct
18-Jan	16-Jun	13-Feb	12-Jul	11-Mar	7-Aug	6-Apr	2-Sep	2-May	28-Sep	28-May	24-Oct
19-Jan	17-Jun	14-Feb	13-Jul	12-Mar	8-Aug	7-Apr	3-Sep	3-May	29-Sep	29-May	25-Oct
20-Jan	18-Jun	15-Feb	14-Jul	13-Mar	9-Aug	8-Apr	4-Sep	4-May	30-Sep	30-May	26-Oct
21-Jan	19-Jun	16-Feb	15-Jul	14-Mar	10-Aug	9-Apr	5-Sep	5-May	1-Oct	31-May	27-Oct
22-Jan	20-Jun	17-Feb	16-Jul	15-Mar	11-Aug	10-Apr	6-Sep	6-May	2-Oct	1-Jun	28-Oct
23-Jan	21-Jun	18-Feb	17-Jul	16-Mar	12-Aug	11-Apr	7-Sep	7-May	3-Oct	2-Jun	29-Oct
24-Jan	22-Jun	19-Feb	18-Jul	17-Mar	13-Aug	12-Apr	8-Sep	8-May	4-Oct	3-Jun	30-Oct
25-Jan	23-Jun	20-Feb	19-Jul	18-Mar	14-Aug	13-Apr	9-Sep	9-May	5-Oct	4-Jun	31-Oct
26-Jan	24-Jun	21-Feb	20-Jul	19-Mar	15-Aug	14-Apr	10-Sep	10-May	6-Oct	5-Jun	1-Nov

Breeding	Kidding	Breeding	Kidding	Breeding	Kidding	Breeding	Kidding	Breeding	Kidding	Breeding	Kidding
6-Jun	2-Nov	11-Jul	7-Dec	15-Aug	11-Jan	19-Sep	15-Feb	24-Oct	22-Mar	28-Nov	26-Apr
7-Jun	3-Nov	12-Jul	8-Dec	16-Aug	12-Jan	20-Sep	16-Feb	25-Oct	23-Mar	29-Nov	27-Apr
8-Jun	4-Nov	13-Jul	9-Dec	17-Aug	13-Jan	21-Sep	17-Feb	26-Oct	24-Mar	30-Nov	28-Apr
9-Jun	5-Nov	14-Jul	10-Dec	18-Aug	14-Jan	22-Sep	18-Feb	27-Oct	25-Mar	1-Dec	29-Apr
10-Jun	6-Nov	15-Jul	11-Dec	19-Aug	15-Jan	23-Sep	19-Feb	28-Oct	26-Mar	2-Dec	30-Apr
11-Jun	7-Nov	16-Jul	12-Dec	20-Aug	16-Jan	24-Sep	20-Feb	29-Oct	27-Mar	3-Dec	1-May
12-Jun	8-Nov	17-Jul	13-Dec	21-Aug	17-Jan	25-Sep	21-Feb	30-Oct	28-Mar	4-Dec	2-May
13-Jun	9-Nov	18-Jul	14-Dec	22-Aug	18-Jan	26-Sep	22-Feb	31-Oct	29-Mar	5-Dec	3-May
14-Jun	10-Nov	19-Jul	15-Dec	23-Aug	19-Jan	27-Sep	23-Feb	1-Nov	30-Mar	6-Dec	4-May
15-Jun	11-Nov	20-Jul	16-Dec	24-Aug	20-Jan	28-Sep	24-Feb	2-Nov	31-Mar	7-Dec	5-May
16-Jun	12-Nov	21-Jul	17-Dec	25-Aug	21-Jan	29-Sep	25-Feb	3-Nov	1-Apr	8-Dec	6-May
17-Jun	13-Nov	22-Jul	18-Dec	26-Aug	22-Jan	30-Sep	26-Feb	4-Nov	2-Apr	9-Dec	7-May
18-Jun	14-Nov	23-Jul	19-Dec	27-Aug	23-Jan	1-Oct	27-Feb	5-Nov	3-Apr	10-Dec	8-May
19-Jun	15-Nov	24-Jul	20-Dec	28-Aug	24-Jan	2-Oct	28-Feb	6-Nov	4-Apr	11-Dec	9-May
20-Jun	16-Nov	25-Jul	21-Dec	29-Aug	25-Jan	3-Oct	1-Mar	7-Nov	5-Apr	12-Dec	10-May
21-Jun	17-Nov	26-Jul	22-Dec	30-Aug	26-Jan	4-Oct	2-Mar	8-Nov	6-Apr	13-Dec	11-May
22-Jun	18-Nov	27-Jul	23-Dec	31-Aug	27-Jan	5-Oct	3-Mar	9-Nov	7-Apr	14-Dec	12-May
23-Jun	19-Nov	28-Jul	24-Dec	1-Sep	28-Jan	6-Oct	4-Mar	10-Nov	8-Apr	15-Dec	13-May
24-Jun	20-Nov	29-Jul	25-Dec	2-Sep	29-Jan	7-Oct	5-Mar	11-Nov	9-Apr	16-Dec	14-May
25-Jun	21-Nov	30-Jul	26-Dec	3-Sep	30-Jan	8-Oct	6-Mar	12-Nov	10-Apr	17-Dec	15-May
26-Jun	22-Nov	31-Jul	27-Dec	4-Sep	31-Jan	9-Oct	7-Mar	13-Nov	11-Apr	18-Dec	16-May
27-Jun	23-Nov	1-Aug	28-Dec	5-Sep	1-Feb	10-Oct	8-Mar	14-Nov	12-Apr	19-Dec	17-May
28-Jun	24-Nov	2-Aug	29-Dec	6-Sep	2-Feb	11-Oct	9-Mar	15-Nov	13-Apr	20-Dec	18-May
29-Jun	25-Nov	3-Aug	30-Dec	7-Sep	3-Feb	12-Oct	10-Mar	16-Nov	14-Apr	21-Dec	19-May
30-Jun	26-Nov	4-Aug	31-Dec	8-Sep	4-Feb	13-Oct	11-Mar	17-Nov	15-Apr	22-Dec	20-May
1-Jul	27-Nov	5-Aug	1-Jan	9-Sep	5-Feb	14-Oct	12-Mar	18-Nov	16-Apr	23-Dec	21-May
2-Jul	28-Nov	6-Aug	2-Jan	10-Sep	6-Feb	15-Oct	13-Mar	19-Nov	17-Apr	24-Dec	22-May
3-Jul	29-Nov	7-Aug	3-Jan	11-Sep	7-Feb	16-Oct	14-Mar	20-Nov	18-Apr	25-Dec	23-May
4-Jul	30-Nov	8-Aug	4-Jan	12-Sep	8-Feb	17-Oct	15-Mar	21-Nov	19-Apr	26-Dec	24-May
5-Jul	1-Dec	9-Aug	5-Jan	13-Sep	9-Feb	18-Oct	16-Mar	22-Nov	20-Apr	27-Dec	25-May
6-Jul	2-Dec	10-Aug	6-Jan	14-Sep	10-Feb	19-Oct	17-Mar	23-Nov	21-Apr	28-Dec	26-May
7-Jul	3-Dec	11-Aug	7-Jan	15-Sep	11-Feb	20-Oct	18-Mar	24-Nov	22-Apr	29-Dec	27-May
8-Jul	4-Dec	12-Aug	8-Jan	16-Sep	12-Feb	21-Oct	19-Mar	25-Nov	23-Apr	30-Dec	28-May
9-Jul	5-Dec	13-Aug	9-Jan	17-Sep	13-Feb	22-Oct	20-Mar	26-Nov	24-Apr	31-Dec	29-May
10-Jul	6-Dec	14-Aug	10-Jan	18-Sep	14-Feb	23-Oct	21-Mar	27-Nov	25-Apr	1-Jan	30-May

Kid Tracking Charts

Table C.1. Pholia's Kid Tracking Chart

Parents	Name/ Gender	DOB	Ease of Delivery	Color	Tattoo	Registration	Tetanus or Other Vaccination	Destination

Table C.2. Extensive Management Kidding Chart

DOB	Dam	Sire	Kids in Shelter?	Kid's Condition: Wet, Damp, Dry?	Kids Nursed?	Gender	ID/ Ear Tag	Navel Dipped/ Weight/Other	Dam's Udder Quality

Resources

There are a plethora of resources available for the proactive goat farmer. In these listings, I've included those that I've used myself or have had recommended to me as a resource. I've grouped information into several categories, including general goat and health books and websites; herd management and pedigree software; grazing and livestock management books and websites; livestock and pasture management supplies, equipment, and services; and goat health supplies and services.

General Goat and Health Books and Websites

A Holistic Vet's Prescription for a Healthy Herd by Richard Holliday, DVM, and Jim Helfter (Acres USA, 2015)

This is the book that inspired me to try the free-choice buffet-style approach to mineral feeding that I am now sold on.

Natural Goat Care by Pat Coleby (Acres USA, 2001)

A wonderful, holistic approach that helped guide me to my current beliefs.

The Complete Herbal Handbook for Farm and Stable by Juliette de Baïracli Levy (Faber and Faber, 1991)

This book is a wonderful classic and worth owning but does include some suggestions and ideas that, in light of current knowledge, might be considered questionable.

Ethnoveterinary Botanica Medicine edited by David R. Katerere and Dibungi Luseba (CRC Press, 2010)

A fascinating and daunting look at traditional herbal medicines for animals from cultures around the world. No doubt the way of the future as conventional medicines continue to lose strength and the interest in organics continues to increase.

Herbs for Pets by Gregory by L. Tilford and Mary L. Wulff (i-5 Publishing, 2009)

My favorite herbal guide, easy to read and enjoyable.

Alternative Treatments for Ruminant Animals by Paul Dettloff, DVM (Acres USA, 2009)

This is a great guide for homeopathy and alternative treatments. Dettloff mainly writes about cattle, but his intent is to offer information relevant for all ruminants.

Treating Dairy Cows Naturally by Hubert J. Karreman, VMD (Acres USA, 2004)

Yes, this book is about cows, but it is full of very helpful information for caring for all ruminants.

Goat Medicine, 2nd edition, by Mary C. Smith and David M. Sherman (Wiley-Blackwell, 2009)

Hands down the best goat medicine book around, an academic book intended primarily for vet students but worthwhile for all serious producers. It was my primary resource while writing my own book. It's not cheap, so put it on your Christmas list. (Note: The first edition can be purchased used for a great price and is still pertinent. It also has a better index and better binding.)

***Diseases of the Goat* by John Matthews (Wiley-Blackwell, 2009)**

A great resource that takes a different approach from that of *Goat Medicine*. England-focused, but still worth owning.

***The Goatkeeper's Veterinary Book* by Peter Dunn (Old Pond Publishing, 2007)**

Another great book but with an English focus, including the fun spelling differences. Great photos and illustrations.

***Livestock Protection Dogs: Selection, Care, and Training* by Orysia Dawydiak and David E. Sims (Alpine Publications, 2004)**

This is the livestock guardian book that I learned from.

***Nutrient Requirements of Small Ruminants*, National Research Council of the National Academies (2007)**

Not for the faint of heart, but this volume has a great appendix that lists nutrient contents for nearly every type of feed possible.

***Goats: Homoeopathic Remedies* by George Macleod (Saffron Walden, 1991)**

This book is hard to find (I had to order mine from England). It's a great resource, but a little hard to navigate.

***Homeopathic Care for Dogs and Cats* by Don Hamilton, DVM (North Atlantic Books, 1999)**

I recommend this book as the place to start if you're serious about learning more about homeopathy. Unlike Macleod's book, it's easy to follow and quite complete. You can fairly easily apply the symptoms described for cats and dogs to goats.

***The Pet Lover's Guide to Natural Healing for Cats and Dogs* by Barbara Fougère, BVSc (Elsevier Saunders, 2006)**

Don't be fooled by the title: You can apply most of the information in this book to goats. What I love about it is that for each health issue described, you'll find nutritional, herbal, acupuncture, and homeopathic remedy suggestions.

***Goat Husbandry* by David Mackenzie (Faber and Faber, 1993)**

This book has been around for a long time and should stay around. I love it for its diagrams of goat gardens and more.

***Guide to Regional Ruminant Anatomy Based on the Dissection of the Goat* by Gheorghe M. Constantinescu (Iowa State University Press, 2001)**

This is a gem of a book that those of us who are fascinated by the inner workings of the body are sure to enjoy. My used copy includes notes by a vet student. Wonderful!

***The Mysterious Goat: Images and Impressions* by Dr. C. Naaktgeboren (B&B Press, 2006)**

This book is a bit hard to find but absolutely a *must*-own for goat lovers. You can find it online in English or Dutch. I can't say enough about the wonder of this book.

***The Meat Goat Production Handbook*, 2nd edition, edited by R. C. Merkel, T. A. Gipson, and T. Sahlu (Langston University Press, 2015)**

You have to order this book directly from Langston University (www2.luresext.edu/goats), but it's a must-have for anyone serious about meat goats. It's a good deal for what you get.

***The Dairy Goat Production Handbook*, edited by R. C. Merkel, T. A. Gipson, and T. Sahlu (Langston University Press, 2016)**

Another wonderful compilation assembled by Langston University. (I wrote the cheesemaking chapter.) You have to order directly from them, but it's worth owning.

***The Small-Scale Dairy* by Gianaclis Caldwell (Chelsea Green Publishing, 2014)**

Not about goats in particular, but all the practices and information you will need for running any small dairy.

***Angora Goats the Northern Way* by Susan Black Drummond (Stoney Lonesome Farm, 2005)**

This book is focused on the challenges of successfully raising Angora goats in the northern latitudes, but I recommend it for anyone raising fiber goats.

Butchering by Adam Danforth (Storey Publishing, 2014)

This is a comprehensive, well-photographed guide to butchering poultry, rabbits, lamb, goat, and pork. A must-have if you are going to be doing any home or custom butchering.

Building Small Barns, Sheds and Shelters by Monte Burch (Garden Way Publishing, 1983)

This was one of the first books I bought, many years ago, when I was teaching myself how to do construction.

The Biology of the Goat

www.goatbiology.com

Fantastic animations and information unlike any other site I've found.

Goat Tracks magazine

www.goattracksmagazine.org

Langston University's Goat Feed Calculator

www2.luresext.edu/goats/research/nutrition module1.htm

This helpful tool can give you some idea of the nutrient content of your feed ration. It's probably more helpful for more intensively managed farms but can be a decent estimate for other management approaches as well.

Livestock Guardian Dogs

www.lgd.org

Information on breeds and more.

Barn and Shelter Plans

www.goatworld.com/articles/shelters/movable shelter1.pdf

Movable field shelter.

American Goat Federation

http://www.americangoatfederation.org

An organization worth joining. I'm new to it but have the impression that it will serve the greater goat world well.

Oklahoma State University, Breeds of Goats

www.ansi.okstate.edu/breeds/goats

Wonderful information on and images of goat breeds around the world.

The North American Packgoat Association

www.napga.org

Northwest Pack Goats and Supplies

www.northwestpackgoats.com

USDA Natural Resources Conservation Services

"Prescribed Grazing Information Sheet," www.nrcs .usda.gov/wps/PA_NRCSConsumption/download ?cid=stelprdb1261918&ext=pdf

"Brush Management with Goats," www.nrcs.usda.gov /Internet/FSE_DOCUMENTS/stelprdb1117286.pdf

Two very helpful web pages if you're managing brush goats.

"Nutritional Characteristics and Feeding Strategies for Fibre Producing Goats" by P. Morand-Fehr

www.ladocumentationcaprine.net/plan/alimentation /art/artali12.pdf

This online PDF is a very helpful document for those focusing on high-quality fiber production.

USDA Institutional Meat Purchase Specifications

www.ams.usda.gov/grades-standards/imps

A very helpful site if you are thinking of selling meat to institutions and retailers.

Sheep and Goat Marketing

http://sheepgoatmarketing.info

This website from Cornell University is designed to help fiber and meat producers find good market options. I would say it is helpful for some regions and will likely expand to others over time.

The Livestock Conservancy (USA)

www.livestockconservancy.org

Information and contacts for rare and threatened breeds, including the San Clemente and Spanish goat.

Rare Breeds Survival Trust (UK)

www.rbst.org.uk

Dedicated to the preservation of rare livestock breeds in the United Kingdom.

Kinne's Minis

www.kinne.net

This is Maxine Kinne's information webpage. It's a wonderful collection of articles that cover many topics for goat owners.

Online Herd Health Resources

Scrapie Identification Requirements by State
www.eradicatescrapie.org/State%20ID%20
Requirements.html
**USDA List of Reportable Diseases
and Other Regulations**
www.aphis.usda.gov/aphis/ourfocus/animalhealth
Biosecurity Information
University of Guelph, Ontario, "Managing Zoonotic
Diseases in Goats," http://www.uoguelph.ca
/~pmenzies/PDF/Goat_Zoonoses_PMenzies_2009
_GreyBruce.pdf
University of Pennsylvania, "Major Zoonoses of
Ruminants," www.upenn.edu

Herd Management Pedigree and Records Software and Resources

Basic Goat Manager, by Centrics Software, Inc.
www.centricsoftwareinc.com/basicgoatmanager.htm
 I haven't used this, but it looks quite thorough
 and well thought out. Reasonably priced.
The Boer and Meat Goat Information Center
www.boergoats.com/clean/articles/recordkeeping
/software.php
 This is a very helpful overview of other software
 options at Boergoats.com, with reviews and
 prices. Worth spending some time looking over.
**Langston University's Goat
Genetics Training module**
www2.luresext.edu/goats/training/genetics.html
 You can also find this information in the
 appendix of Langston's *Meat Goat Production
 Handbook*, 2nd edition.
National Sheep Improvement Program
www.nsip.org/nsip-search-breeding-stock/goat
-breeding-stock
 Much like the Dairy Herd Improvement program,
 which was designed for cattle but now includes
 goats, this meat sheep program for genetic
 improvement offers its services to the meat goat
 industry.

Grazing and Pasture Management Books and Websites

The Art and Science of Grazing by Sarah Flack
(Chelsea Green Publishing, 2016)
 This book covers a great deal to do with understanding animals' needs, the plants' needs, and
 how to manage both in a pasture setting.
Fences for Pasture and Garden by Gail Damerow
(Storey Publishing, 1992)
 I've had this book for years and still find it helpful
 and easy to read.
*Restoration Agriculture: Real-World Permaculture
for Farmers* by Mark Shepard (Acres USA, 2013)
 This is my favorite permaculture book to
 date—even though it mentions that goats are
 just about the most difficult animal to integrate
 into a permaculture plan! It still helps you figure
 it out.
Fertility Pastures by Newman Turner
(Acres USA, 2009)
 This book was originally published in 1955 (at
 that time the author went by F. Newman Turner).
 I love it for its insight into the nutritional aspects
 of "weeds" and their role in soil fertility. It also
 has a nice herbal seed mixture recommendation
 for goats in the chapter "Herbal Ley Mixtures."
USDA Natural Resources Conservation Service
www.nrcs.usda.gov/wps/portal/nrcs/main
/national/landuse
 This service is, in my opinion, underutilized. As
 described to me by one of its employees, who is
 also a farmer, "It's the carrot branch of the USDA—
 not the stick—and the two don't talk to each other."
 If you're looking for input on how to manage your
 land for the good of your farm and the planet, talk
 to these guys.
**Manure Management, University of
Connecticut**
www.animalscience.uconn.edu/extension
/publications/manuremanagement.htm
 Terrific information to help you understand the
 considerations of manure management.

Mortality and Waste Management, Cornell University

http://cwmi.css.cornell.edu

> This booklet is very helpful specifically when you're designing a mortality management program.

Forage Identification, Purdue University

www.agry.purdue.edu/ext/forages/ForageID/forageid.htm

> Helpful when trying to ascertain what is growing in your pastures.

Poisonous Plants

www.poisonousplants.ansci.cornell.edu/goatlist.html

> A very thorough list of poisonous plants and how they harm animals.

Livestock and Pasture Management—Equipment, Supplies, and Services

Caprine Supply

800-646-7736

www.caprinesupply.com

> Goat milking and home dairy supplies.

Dairy One

800-344-2697

www.dairyone.com

> My preferred lab in the United States for having feed and forage tested.

Farm and Ranch Depot

928-951-8332

www.farmandranchdepot.com

> Many barn and dairy supplies.

Farm Tek

800-327-6835

www.farmtek.com

> High tunnel housing, fodder-growing systems, housing solutions, and more.

Hamby Dairy Supply

800-306-8937

www.hambydairysupply.com

> Supplies for dairies of all sizes. Very knowledgeable and helpful.

Hoegger Farmyard Supply Company

800-221-4628

www.hoeggerfarmyard.com

> Mostly carries goat supplies but also has a nice selection of home dairy supplies, such as butter churns and cream separators.

Kencove Farm Fence, Inc.

800-536-2683

www.kencove.com

> Electric fencing options for pasture rotation.

Premier1

800-282-6631

www.premier1supplies.com

> Ear tags, electric fencing for pasture rotation, footbath supplies and more.

Sydell

800-842-1369

www.sydell.com

> Manufacturer of sheep and goat handling equipment.

USDA Sheep and Goat Identification

www.aphis.usda.gov/aphis/ourfocus/animalhealth/animal-disease-information/sheep-and-goat-health

> Click on the link for Sheep and Goat Identification. This is also where you register for a premises ID number (the first step in the scrapie eradication program). The ear tags are free.

Goat Health Supplies and Services

Testing for Disease and Pregnancy

BioPRYN

208-882-9736

www.biopryn.com

> Blood-based pregnancy test. There are many labs in the United States offering this service; you can locate one on this site.

University of Minnesota Laboratory for Udder Health

800-605-8787

www.vdl.umn.edu/services-fees/udder-health-mastitis

Offers a wide spectrum of milk testing that's easy to choose from. It's also relatively simple to send in samples.

Washington Animal Disease Diagnostic Lab

www.waddl.vetmed.wsu.edu

Provides testing services for producers and veterinarians for diseases in all species, including options of test types such as ELISA and PCR.

Herbal and Other Remedies

Fedco Seeds, Clinton, ME

207-426-9900

www.fedcoseeds.com

Regano oil, CEG Remedy, more.

Fresh Start Growers, Louisville, KY

502-442-7883

www.freshstartgrowers.com/store

Diatomaceous earth, Fertrell Herbal Capsules and garlic tincture, more.

Molly's Herbals

www.fiascofarm.com/herbs

A great source for herbal and botanical blends and remedies.

Ralco Animal Health

www.ralcoanimalhealth.com

Producer of Regano and other natural livestock products. Not a dealer, but they can help you locate one.

Minerals and Supplements

Advanced Biological Concepts

800-373-5971

www.abcplus.biz

Individual free-choice mineral mixes and other supplements.

Free Choice Enterprise, Ltd.

608-723-7977

www.freechoiceminerals.com

Individual free-choice mineral mixes.

Rich Earth

757-635-0933

www.richearth.net/livestock/index.html

This humate products manufacturer doesn't sell directly but can help you find a retailer.

Parasite Monitoring

FAMACHA Training and Scorecard

http://www.wormx.info/workshops

A list of training opportunities, including online.

Fecal Egg Count Tutorial and Egg Identification

www.youtube.com/watch?v=ZZQymZKe_hs&feature=youtu.be

www.merck-animal-health-usa.com/binaries/18475_Worm_Parasite_Atlas_tcm96-86736.pdf

Patterson Veterinary

www.pattersonvet.com

McMaster slides, fecalyzers, and other monitoring supplies.

Artificial Insemination

BIO-Genics, Ltd.

www.biogenicsltd.com/index.html

AI semen collection, classes, and supplies.

Capra Gia Semen Company

www.capragiasemen.weebly.com

AI semen collection and semen sales.

Veterinary Associations

American Association of Small Ruminant Practitioners

www.aasrp.org

Veterinarians who join this organization treat one or more of the following: sheep, goats, camelids (such as llamas), and cervids (such as commercially farmed elk). If you attend an American Dairy Goat Association convention, you can sit in on the AASRP continuing education program and get up to date on what's new in goat medicine.

American Holistic Veterinary Medical Association

www.ahvma.org

Use this site to help locate a veterinarian who practices a holistic approach.

Academy of Veterinary Homeopathy

www.theavh.org

Use this site to help locate a veterinarian who also practices homeopathy.

Veterinary Botanical Medicine Association

www.vbma.org

Use this site to help locate a veterinarian who utilizes herbs and botanicals in their practice.

NOTES

CHAPTER 1

1. Saeid Naderi et al. "The Goat Domestication Process Inferred from Large-Scale Mitochondrial DNA Analysis of Wild and Domestic Individuals," *Proceedings of the National Academy of Sciences of the United States of America* 105, no. 46 (2008): 17659–64.
2. Oklahoma State University, "Breeds of Livestock—Goat Breeds," http://www.ansi.okstate.edu/breeds/goats.
3. Tatiana Stanton, "An Overview of the Goat Meat Market 2012," *Sheep and Goat Marketing*, Cornell University, http://www.sheepgoatmarketing.info/education/meat goatmarket.php.
4. Roger Merkel and Terry Gipson, *The Meat Goat Production Handbook*, 2nd ed. (Langston, OK: Langston University, 2015), p. 9.
5. Gary Cutrer, "Boer Goats for Beginners," *Ranch & Rural Living Magazine* 77, no. 2 (November 1995).

CHAPTER 2

1. Roger Merkel and Terry Gipson, *The Meat Goat Production Handbook*, 2nd ed. (Langston, OK: Langston University, 2015), p. 199.

CHAPTER 3

1. Gail Damerow, *Fences for Pasture and Garden* (North Adams, MA: Storey Publishing, 1992), p. 97.
2. https://ahdc.vet.cornell.edu/docs/GoatNAHMSDisease.pdf.
3. Roger Merkel and Terry Gipson, *The Meat Goat Production Handbook*, 2nd ed. (Langston, OK: Langston University, 2015), p. 427.

CHAPTER 4

1. Sandra Solaiman, *Goat Science and Production,* ed. (Ames, IA: Wiley-Blackwell, 2010), 159.
2. Ibid.
3. "Nutritional Requirements of Goats," *Merck Veterinary Manual Online*, http://www.merckvetmanual.com/mvm /management_and_nutrition/nutrition_goats/nutritional _requirements_of_goats.html, accessed October 8, 2016.
4. National Research Council, *Nutrient Requirements of Small Ruminants* (Washington, DC: National Academies Press, 2007), p. 129.
5. Ibid.

6. Mary C. Smith and David M. Sherman, *Goat Medicine*, 2nd ed. (Ames, IA: Wiley-Blackwell, 2009), p. 748.
7. David Mackenzie, *Goat Husbandry* (London: Faber and Faber, 1993), p. 101.
8. Smith and Sherman, *Goat Medicine*, p. 266.
9. USDA Natural Resources Conservation Services, "Rotational Grazing," http://www.nrcs.usda.gov/wps /portal/nrcs/detail/ky/newsroom/factsheets/?cid =stelprdb1101721, accessed October 1, 2016.

CHAPTER 5

1. Mary C. Smith and David M. Sherman, *Goat Medicine*, 2nd ed. (Ames, IA: Wiley-Blackwell, 2009), p. 306.
2. "Terminology for Grazing Lands and Grazing Animals," *Journal of Production Agriculture* (1992), www.agron. iastate.edu/Courses/agron515/Terminology%20for%20 Grazing.pdf, accessed October 9, 2016.
3. "Mold and Mycotosin Issues in Dairy Cattle: Effect, Prevention and Treatment," http://articles.extension.org /pages/11768/mold-and-mycotoxin-issues-in-dairy -cattle:-effects-prevention-and-treatment, p. 11, accessed October 9, 2016.
4. Ibid., p. 15.
5. Y. H. Hui, Wai-Kit Nip, and Robert Roger, eds., *Meat Science and Applications* (New York: Marcel Dekker, 2001), p. 129.
6. Code of Federal Regulations, Title 21, Volume 6, revised April 1, 2016, http://www.accessdata.fda.gov/scripts/cdrh /cfdocs/cfcfr/CFRSearch.cfm?fr=558.355.
7. T. R. Callaway et. al., "What Are We Doing About *Escherichia coli* 0157:H7 in Cattle?," *Journal of Animal Science* 82, E. Suppl. (2004), p. 95.
8. Ibid.
9. M. Kruger et al., "Field Investigations of Glyphosate in Urine in Danish Dairy Cows," *Environmental and Analytical Toxicology* 3, no. 186 (2013), doi: 10.4172/2161-0525.1000186.
10. Henning Gerlach et al., "Oral Application of Charcoal and Humic Acids Influence Selected Gastrointestinal Microbiota, Enzymes, Electrolyes, and Substrates in the Blood of Dairy Cows Challenged with Glyphosate in GMO Feeds," *Environmental and Analytical Toxicology* 4, no. 256 (2014), doi: 10.4172/2161-0525.1000256.

11. Callaway et al., "What Are We Doing About *Escherichia coli* 0157:H7 in Cattle?"

12. L. Q. Vieira et. al., "Probiotics Protect Mice Against Experimental Infections," *Journal of Clinical Gastroenterology* 42, Suppl 3 Part 2 (September 2008), S168–69, doi: 10.1097/MCG.0b013e31818063d4.

CHAPTER 6

1. Rajeev Singh, et al., "Total Phenolic, Flavonoids and Tannin Contents in Different Extracts of *Artemisia absinthium*," *Journal of Intercultural Ethnopharmacology* 1 (2012), pp. 101–04. doi: 10.5455/jice.20120525014326.

2. Mary C. Smith and David M. Sherman, *Goat Medicine*, 2nd ed. (Ames, IA: Wiley-Blackwell, 2009), p. 412.

3. SARE, Sustainable Agriculture Research and Education, "Oregano Oil for Internal Parasite Control in Sheep, Goats, and Beef Cattle," http://mysare.sare.org/sare_project/one 08-088.

4. G. C. Waghorn et. al., "Condensed Tannins and Herbivore Nutrition," www.internationalgrasslands.org/files/igc /publications/1997/iii-153.pdf, accessed October 10, 2016.

5. Byeng Ryel Min, "Pine Bark and Other Natural Dewormers for Small Ruminants," American Consortium for Small Ruminant Parasite Control, www.wormx.info/pinebark, accessed October 12, 2016.

CHAPTER 8

1. Tatiana Stanton, "Out of Season Breeding" fact sheet, Cornell University, goatdocs.ansci.cornell.edu /Resources/GoatArticles/Factsheets/OutOfSeason FactSheet.pdf.

2. Mary C. Smith and David M. Sherman, *Goat Medicine*, 2nd ed. (Ames, IA: Wiley-Blackwell, 2009), p. 796.

3. John Matthews, *Diseases of the Goat*, 3rd ed. (Oxford: Wiley-Blackwell, 2009), p. 10.

4. Smith and Sherman, *Goat Medicine*, p. 599.

CHAPTER 10

1. John Matthews, *Diseases of the Goat*, 3rd ed. (Oxford: Wiley-Blackwell, 2009), p. 263.

2. Mary C. Smith and David M. Sherman, *Goat Medicine*, 2nd ed. (Ames, IA: Wiley-Blackwell, 2009), p. 483.

3. Ibid., p. 477.

4. Ibid., p. 412.

5. Ibid., p. 768.

6. Roger Merkel and Terry Gipson, *The Meat Goat Production Handbook*, 2nd ed. (Langston, OK: Langston University, 2015), p. 143.

7. Smith and Sherman, *Goat Medicine*, p. 762.

8. Ibid., p. 521.

9. F. Paolicchi et al., "Relationship Between Paratuberculosis and the Microelements Copper, Zinc, Iron, Selenium and Molybdenum in Beef Cattle," Brazilian Journal of Microbiology 44, no. 1 (2013): 153–60, www.scielo.br/pdf /bjm/v44n1/540322.pdf, accessed January 15, 2016.

10. Matthews, *Diseases of the Goat*, p. 136

CHAPTER 11

1. Mary C. Smith and David M. Sherman, *Goat Medicine*, 2nd ed. (Ames, IA: Wiley-Blackwell, 2009), p. 356.

CHAPTER 12

1. "Caseous Lymphadenitis of Sheep and Goats," *Merk Veterinary Manual Online*, http://www.merckvetmanual .com/mvm/circulatory_system/lymphadenitis_and _lymphangitis/caseous_lymphadenitis_of_sheep_and _goats.html, accessed January 28, 2016

2. Colorado Serum Company, http://www.colorado-serum .com/vets/vol_2/vol2_10.htm, accessed September 28, 2016.

CHAPTER 13

1. Roger Merkel and Terry Gipson, *The Meat Goat Production Handbook*, 2nd ed. (Langston, OK: Langston University, 2015), p. 118.

2. Mary C. Smith and David M. Sherman, *Goat Medicine*, 2nd ed. (Ames, IA: Wiley-Blackwell, 2009), p. 30.

3. Ibid., p. 106.

4. Ibid., p. 140.

CHAPTER 14

1. Mary C. Smith and David M. Sherman, *Goat Medicine*, 2nd ed. (Ames, IA: Wiley-Blackwell, 2009), p. 228.

2. Ibid., p. 189.

3. Ibid., p. 208.

CHAPTER 15

1. The SCC, Ontario Ministry of Agriculture, Food, and Rural Affairs, http://www.omafra.gov.on.ca/english /livestock/dairy/facts/scc.htm, accessed October 26, 2016.

CHAPTER 16

1. Mary C. Smith and David M. Sherman, *Goat Medicine*, 2nd ed. (Ames, IA: Wiley-Blackwell, 2009), p. 609.

GLOSSARY

Abomasum: The fourth chamber of a ruminant's upper digestive system, also known as the true stomach.

Acidosis: A condition in which the rumen and/or blood and cell fluids are too acidic.

Acute: Having a sudden onset and severe symptoms.

Agroforestry: The growing of fodder trees for the purpose of feeding livestock.

American: Referring to the pedigree status of a goat registered with the American Dairy Goat Association that conforms to its breed's standards, but whose parentage includes one or more goats from a different breed.

Amnion, amniotic sac, amniotic fluid: The membrane that surrounds the fetus and is filled with fluid produced by the mother and, later, also urine from the fetus.

Anthelmintic: A remedy or medication that limits the growth of or kills internal parasites.

Ataxia: Discoordination or stumbling. The term also refers to the disorder that can occur in kids as a result of copper deficiency.

Balling gun: A tool used to administer solid oral medications, such as pills and capsules, to the goat.

Bezoar ibex/goat: The wild goat to which most if not all domestic goats trace their roots. Native to the mountainous parts of the Middle East, Asia, and Eastern Europe.

Biologics: A term used in medicine to describe medications and treatments derived from biological substances. Biologics include vaccinations, antitoxins, serum, toxoids, plasma, and so on.

Biosecurity: Procedures and practices designed to shield animals, including humans, and facilities from exposure to contamination by disease pathogens.

Bloat: A condition arising from the buildup of gas in the goat's rumen.

Bolus: A dose of medication or treatment delivered orally via a capsule or pill. The term also refers to a wad of food (cud) eructated during rumination.

Botanicals: Plants used as medicine. In the United States, the term *herbs* is used synonymously.

Buffers: Substances able to reduce or neutralize acid, or simply prevent pH changes. They're important in the goat as a part of balancing the rumen pH and preventing acidosis.

Cabrito: The Spanish word for "goat meat," usually applied to the meat of a young goat, or kid.

Capra: The genus to which goats belong. It includes the domestic goat *Capra aegagrus hircus* and several wild goat species, including the one from which all domestic goats originated, the Bezoar, *Capra aegagrus*.

Carotene: Also beta-carotene. The provitamin, or precursor, of vitamin A. Associated with orange and yellow feeds. Its absence in goat's milk gives this milk a whiter appearance than cow's milk.

Cashmere: The fine fiber from the undercoat of a goat.

Chevon: The French term for "goat meat."

Chèvre: Literally French for "goat" but used specifically for goat's-milk cheese and in the United States even more specifically for soft, fresh goat's-milk cheese.

Chronic: Existing or developing over an extended period of time, as opposed to peracute or acute.

CIDR: Controlled internal drug release. A vaginal device used to induce ovulation.

Clinical: Easily able to be diagnosed by symptoms and signs, as opposed to subclinical.

Closed herd: A management approach under which no new animals are brought onto the farm. New genetics are obtained via artificial insemination or embryo transplant. Some herds are modified closed herds, where the only new animals allowed are bucks, which are housed separately.

Cloudburst: A term used by some goat producers to describe the natural end of a false pregnancy (hydrometra) and the expulsion of fluid that has developed in the uterus.

Colostrum: The first "milk" produced in the udder, colostrum contains essential antibodies for the protection of the newborn, as well as dense nutrients.

Concentrate: A feed, usually grain, that is high in energy.

Conjunctivitis: Inflammation of the mucous membranes (conjunctiva) on the inside of the eyelids and surrounding the eyeball. Also called pink eye.

Copper oxide wire particle boluses, COWP: Gelatin capsules filled with tiny, copper oxide wire particles given to a goat orally. COWP embed in the rumen lining and are absorbed slowly by the animal. Used primarily to control gastrointestinal parasites.

Cryptorchid: A goat with one or no external testicles. The condition is called cryptorchidism.

Cud: Regurgitated rumen contents that are rechewed and swallowed as a part of rumination.

Diatomaceous earth: Finely ground fossilized remains of tiny creatures called diatoms.

Drenching: Administering a liquid medication to a goat orally.

Drenching syringe: A specially designed syringe, usually with a bent tip, for giving liquid oral medications.

Dry off: To stop milking a goat or stop allowing nursing in order to cause the mammary gland to stop producing milk, usually in preparation for giving birth.

Dysphagia: Difficulty swallowing.

Dystocia: Difficult birthing.

Emasculator: A tool for the sterilization of bucks by crushing the blood supply to the testicles without breaking the skin. A popular brand is Burdizzo.

Enzootic: Occurring regularly in a particular area or herd (synonymous with the term *endemic*, which applies to humans).

Epithelial: Relating to cells that line the surface of glands, organs (including skin), and blood vessels.

Epizootic: Affecting animals in a widespread population (synonymous with the term *epidemic*, which applies to humans).

Esophageal groove: A feature in the goat's upper digestive system that closes off the reticulorumen so that milk flows directly to the abomasum, the true stomach, and swallowed cud is deposited into the omasum.

Estrus: The time when the female animal prepares for and then ovulates. In the goat, estrus occurs about every 18 to 21 days, usually during the fall months. Also called heat, heat cycle, or being in heat.

Extensive farming: Management practices that seek to replicate the goat's natural environment, including grazing and browsing over large tracts of land, as well as minimal intervention for preventive medicine, such as parasite control.

FAMACHA: A method for assessing anemia by inspecting inner eyelids for color and comparing with a FAMACHA scorecard.

Fecal float: A procedure in which a manure sample is combined with a dense solution in which parasite eggs will float over time and be collected on the surface of a glass slide or coverslip.

Feed mat, feed raft: A collection of fiber and forage matter that sits or floats on top of the liquid contents of the rumen.

Feral: Reverted to a wild state, as a previously domesticated animal or species of animal.

Flagging: Term used to describe the quick movement of a doe's tail in response to the presence of a buck, or his odor, when the doe is in heat.

Flehmen response, flehmening: The behavior of an animal lifting its upper lip and pulling air through its mouth so that pheromones can be detected by the vomeronasal organ.

Flight zone: The area around an animal that when entered by something interpreted as a threat, such as a human or more aggressive animal, will cause the animal to attempt to flee.

Flushing: The management practice of feeding extra food, usually grain or legume hay, to does for a few weeks before breeding to attempt to increase the number of eggs produced, as well as improve energy stores for the initial stage of pregnancy.

Fodder: Forage and roughage-type feeds harvested and fed to animals. The term *fresh fodder* often refers to grains that are sprouted and allowed to grow, without soil, for a short period of time, and then fed to livestock.

Forage: Plants that animals consume where they grow. Also, the act of seeking out these plants.

Forb: A broad-leafed plant, sometimes considered a weed.

Free-choice: Referring to any feed or supplement left out in amounts that allow the animal to eat as much as desired.

Freshen: To give birth and begin to produce milk.

Gestation: The period of time from fertilization of the egg to birth—for goats, approximately 147 to 150 days.

Heat: See *estrus*.

Hemolytic crisis: A rapid, life-threatening body-wide destruction of red blood cells.

Hydrometra: A false pregnancy that includes the development of a large amount of fluid in the uterus.

Intramuscular, IM: Occurring or administered within a muscle.

Intravenous, IV: Occurring in or penetrating a vein.

Kid box: A box designed to restrain a young kid with only its head protruding while procedures such as tattooing are performed.

Lactation: The act of giving milk.

Laminae: Internal structures in the hoof that attach the exterior hoof to the internal living tissue of the foot.

Landrace: An organism that has developed specific characteristics due to regional isolation or influences; often a species of livestock or cultivated plants.

Legume: Any plant that has the capacity to associate with soil-dwelling bacteria that convert nitrogen from the air into a form that remains in the soil and can be absorbed by plant roots.

Linebreeding: The purposeful mating of animals with similar genetics.

Macro minerals: Minerals essential to metabolism and growth that are used in greater quantity than micro or trace minerals and generally available in feed in greater quantity.

Mastitis: An infection of the mammary gland.

Meconium: The first feces of the newborn, formed in the intestines from cells, bile, mucus, amniotic fluid swallowed by the baby, et cetera. Dark green in color.

Metabolism: The processes in the body that convert food to energy and other nutrients. Metabolic disorders affect the entire function of the body.

Micro minerals: Minerals essential to metabolism and growth but needed in much smaller quantities than macro minerals; typically they're available in feed in very small amounts.

Necropsy: The inspection of a deceased animal's body for signs of the cause of death or illness.

NSAID: Non-steroidal anti-inflammatory drugs—examples include aspirin and flunixin (Banamine)—used for treating pain and inflammation.

Omasum: The third compartment of a ruminant's upper digestive system or stomachs.

Parity: The number of times a doe has given birth (or carried kids to full term).

Parturition: The act of giving birth.

Peracute: Literally "very acute," this term is used to describe disorders; the peracute form of a disorder arises suddenly and often ends in death.

Peristalsis: Movement of the intestinal tract.

Pica: Describes both the condition and the behavior of an animal eating unusual things, such as dirt, to satisfy a craving.

Placenta, placentome: The tissue connecting the mother's uterus to the blood supply to the fetus. Goats have multiple placentomes, instead of a single placenta, as in humans.

Prepuce: The skin covering the penis, also often called the foreskin.

Prey animals: Animals that serve as a natural food source for other animals—predators—and are hunted by those animals.

Reportable disease: A contagious disease that is deemed of high risk to human or animal health and when detected must be reported to regulatory officials at state and federal levels.

Resilience: In goats, the ability to carry a significant internal parasite load with no ill effects to health.

Resistance: In goats, the ability to remain healthy in the presence of pathogens or to limit the number of internal parasites to very few.

Reticulorumen: The first two chambers of the goat's upper digestive system, or stomachs, consisting of the reticulum and the rumen.

Reticulum: The first chamber of the goat's upper digestive system, or stomachs.

Rhinitis: An upper respiratory infection, runny nose.

Rumen: The second and largest chamber of the goat's upper digestive system, or stomachs. Serves as a large fermentation vessel for processing feed.

Scours: A term used to describe diarrhea, usually in young animals.

Scur: The horny growth that sometimes occurs after a goat has been incompletely disbudded or dehorned.

Senior kid: A young goat over six months of age and under one year.

Short cycling: Estrus cycles occurring approximately 6 days apart.

Somatic cells: Literally "body cells"; in livestock, the term is often used to refer to white blood cells and milk duct lining cells found in milk. A somatic cell count (SCC) is a lab test that counts these cells.

Subclinical: Having no observable signs or symptoms.

Subcutaneous, sub-Q: Under the skin.

Subtherapeutic: Below a level that will treat illness (be therapeutic). Used to describe measures taken to prevent illness or for another desired side effect.

Tannins, condensed tannins: Natural substances found in many plants. Condensed tannins are linked to better gastrointestinal health and limiting internal parasites.

Teaser buck: A buck that has been made infertile—usually by vasectomy, but he could also be naturally infertile—and is used to help induce estrus or identify does in heat.

Thiaminase: An enzyme that blocks the production of thiamine.

Total mixed ration, TMR: A feed formulation fed exclusively that contains all the recommended nutrients for the animal.

Transfaunation: The transfer of microbes (fauna) from one animal to another. One example is the collection of rumen contents (cud) from the mouth of a healthy goat and administration of the material orally to a sick goat.

Trocar: An instrument used in emergency situations to relieve bloat. It consists of a sharp, puncturing center surrounded by a hollow tube, or cannula.

Urolith: Urinary tract stones, or calculi.

Wattle: A fleshy appendage often found on the necks of goats (and occasionally in other locations).

Wether: A castrated male goat.

Yearling: A goat between the ages of one and two years.

Zoonoses: Diseases that can be passed from animals to humans.

INDEX

ABOUT THE AUTHOR

Gianaclis Caldwell and her husband operate Pholia Farm, an off-grid goat dairy in Oregon. She is the author of the award-winning *Mastering Artisan Cheesemaking*, as well as other books on cheesemaking and small dairy topics. For a decade Pholia Farm commercially produced aged, raw-milk cheeses that were applauded by some of the world's foremost authorities on cheese. In 2016 the couple shifted gears and changed the farm's focus to one of education and homestead-style sustainability.

Applying a holistic, organic management approach to the health of her goats comes naturally for Gianaclis. Her parents organically managed the farm on which she was raised (where Pholia Farm is now located) and were chiropractors and skeptics of conventional medicine, believing instead in the power of nutrition and healthy living. Gianaclis's own career as a nurse reinforced these values.

Gianaclis began her animal husbandry education in 1976 with the purchase of her first dairy cow. This eventually led to leading a local dairy 4-H club and a study of dairy cattle science. She often describes her transition to a passion for goats as "seeing the light" and appreciates the amazing sustainability aspects of goats, as well as their personalities and intelligence.

Gianaclis travels nationwide and occasionally to other countries to present workshops on cheesemaking, small-dairy management, and goat husbandry. Pholia Farm also offers workshops and courses, including a week-long Goat Academy and a three-day Farmstead Cheesemaker Short Course. You can learn more about Gianaclis and Pholia Farm at www.gianaclis caldwell.com and www.pholiafarm.com.

the politics and practice of sustainable living

CHELSEA GREEN PUBLISHING

Chelsea Green Publishing sees books as tools for effecting cultural change and seeks to empower citizens to participate in reclaiming our global commons and become its impassioned stewards. If you enjoyed *Holistic Goat Care*, please consider these other great books related to agriculture and homesteading.

THE ART AND SCIENCE OF GRAZING
How Grass Farmers Can Create Sustainable Systems
for Healthy Animals and Farm Ecosystems
SARAH FLACK
9781603586115
Paperback • $39.95

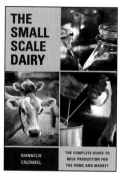

THE SMALL-SCALE DAIRY
The Complete Guide to Milk Production
for the Home and Market
GIANACLIS CALDWELL
9781603585002
Paperback • $34.95

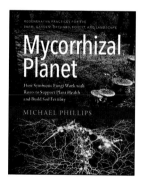

MYCORRHIZAL PLANET
How Symbiotic Fungi Work with Roots to
Support Plant Health and Build Soil Fertility
MICHAEL PHILLIPS
9781603586580
Hardcover • $40.00

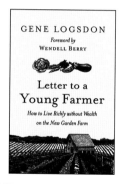

LETTER TO A YOUNG FARMER
How to Live Richly without Wealth
on the New Garden Farm
GENE LOGSDON
9781603587259
Hardcover • $22.50

CHELSEA GREEN PUBLISHING
the politics and practice of sustainable living

For more information or to request a catalog,
visit **www.chelseagreen.com** or
call toll-free **(800) 639-4099**.